Arithmetic, Geometry, Cryptography, and Coding Theory 2021

CONTEMPORARY MATHEMATICS

779

Arithmetic, Geometry, Cryptography, and Coding Theory 2021

18th International Conference
Arithmetic, Geometry, Cryptography, and Coding Theory
May 31–June 4, 2021
Centre International de Rencontres Mathématiques,
Marseille, France

Samuele Anni
Valentijn Karemaker
Elisa Lorenzo García
Editors

AMS AMERICAN MATHEMATICAL SOCIETY

2020 *Mathematics Subject Classification.* Primary 11G20, 11G30, 11G32, 11G40, 11T71, 14G10, 14H40, 14Q05, 20C20, 20G41.

For additional information and updates on this book, visit
www.ams.org/bookpages/conm-779

Library of Congress Cataloging-in-Publication Data

Names: International Conference on Arithmetic, Geometry, Cryptography and Coding Theory (18th: 2021: Marseille, France), author. | Anni, Samuele, 1985- editor. | Karemaker, Valentijn, 1990- editor. | Lorenzo Garcí, Elisa, 1987- editor. | Centre national de rencontres mathématiques (France), host institution.

Title: Arithmetic, geometry, cryptography and coding theory 2021 : 18th International Conference on Arithmetic, Geometry, Cryptography, and Coding Theory, May 31-June 4, 2021, Centre International de Rencontres Mathématiques, Marseille, France / Samuele Anni, Valentijn Karemaker, Elisa Lorenzo García, editors.

Description: Providence, Rhode Island : American Mathematical Society, [2022] | Series: Contemporary mathematics, 0271-4132 ; volume 779 | Includes bibliographical references.

Identifiers: LCCN 2022008520 | ISBN 9781470467944 (paperback) | 9781470470890 (ebook)

Subjects: LCSH: Coding theory–Congresses. | Geometry, Algebraic–Congresses. | Cryptography–Congresses. | Number theory–Congresses. | AMS: Number theory – Arithmetic algebraic geometry (Diophantine geometry) – Curves over finite and local fields. | Number theory – Arithmetic algebraic geometry (Diophantine geometry) – Curves of arbitrary genus or genus $\neq 1$ over global fields. | Number theory – Arithmetic algebraic geometry (Diophantine geometry) – Arithmetic aspects of dessins d'enfants, Belyĭ theory. | Number theory – Arithmetic algebraic geometry (Diophantine geometry) – L-functions of varieties over global fields; Birch-Swinnerton-Dyer conjecture. | Number theory – Finite fields and commutative rings (number-theoretic aspects) – Algebraic coding theory; cryptography (number-theoretic aspects). | Algebraic geometry – Zeta functions and related questions in algebraic geometry (e.g., Birch-Swinnerton-Dyer conjecture). | Algebraic geometry – Curves in algebraic geometry – Jacobians, Prym varieties. | Group theory and generalizations – Representation theory of groups – Modular representations and characters. | Group theory and generalizations – Linear algebraic groups and related topics – Exceptional groups.

Classification: LCC QA268 .I57 2021 | DDC 512.7/4–dc23/eng20220528

LC record available at https://lccn.loc.gov/2022008520

∞ The paper used in this book is acid-free and falls within the guidelines
established to ensure permanence and durability.
Visit the AMS home page at `https://www.ams.org/`

10 9 8 7 6 5 4 3 2 1   27 26 25 24 23 22

Contents

Preface

The 18th edition of the AGC^{2}T conference (Arithmetic, Geometry, Cryptography, and Coding Theory), planned to take place at CIRM (Centre International de Rencontres Mathématiques) in Marseille, France, as did the previous editions, finally went ahead online between May 31 and June 4, 2021.

This workshop is part of a series dating back to 1987. Since then, these workshops have become a major event in the area of arithmetic geometry and its applications to cryptography and coding theory.

The online setting allowed us to invite more participants than we would have otherwise been able to host at CIRM, and we thank all of the 130 participants for their active contributions during the week and afterwards, on the various online platforms that were used. Despite the unusual circumstances, we were very happy to still experience the stimulating and welcoming atmosphere that is typical for the AGC^{2}T community.

We would like to especially thank the speakers — Jeroen Sijsling, Davide Lombardo, Angela Ortega, Türkü Çelik, Nirvana Coppola, Leonardo Colò, Luca Notarnicola, Sorina Ionia, Jean Kieffer, Tomoyoshi Ibukiyama, Stefano Marseglia, Ernst-Ulrich Gekeler, Chia-Fu Yu, Sergey Rybakov, Kate Stange, Monika Trimoska, Richard Griffon, Fabien Narbonne, Annamaria Iezzi, Joachim Rosenthal, Beth Malmskog, Pietro Speziali, Maria Chara, Luciane Quoos, Cecília Salgado, Peter Beelen, Gunther Cornelissen, Kaloyan Slavov, Nathan Kaplan, Wei Ho, Gabor Wiese, Marco Streng, Tony Ezome, Francesco Campagna, Elisa Gorla, Sudhir Ghorpade, Marc Perret, Stefano Lia, and Elena Berardini — for their lectures. The conference centred around interactions between pure mathematics (in particular arithmetic and algebraic geometry) and information theory (especially cryptography and coding theory). The topics of the talks ranged from the study of relations between curves and their Jacobians to the study of endomorphism rings and isogeny graphs of supersingular elliptic curves and their applications to cryptography; and from classifying abelian varieties over finite fields to classifying different properties of convolutional, linear or algebraic codes.

The editors are indebted to the staff of CIRM, and of the Institut de Mathématiques de Marseille for their patience in dealing with constantly changing circumstances and their consistent help.

We gratefully acknowledge the financial support of the local sponsors of the event (Aix-Marseille Université, the Ville de Marseille, the FRUMAM, the Institut de Mathématiques de Marseille, the Institut Archimède, the CNRS through the GDR JC2A, and the ANR project MELODIA ANR-20-CE40-0013), as well as the international funding bodies (the Foundation Compositio Matematica and the NWO).

We would also like to thank Christine Thivierge at the American Mathematical Society for guiding us through the Contemporary Mathematics production process. And last but certainly not least, we are very grateful to the authors of the articles contained in this volume for their mathematical creativity and their kind cooperation in the editorial process.

Contemporary Mathematics
Volume **779**, 2022
https://doi.org/10.1090/conm/779/15667

Numerical reconstruction of curves from their Jacobians

Daniele Agostini, Türkü Özlüm Çelik, and Demir Eken

ABSTRACT. We approach the Torelli problem of reconstructing a curve from its Jacobian from a computational point of view. Following Dubrovin, we design machinery to solve this problem effectively, which builds on methods in numerical algebraic geometry. We verify this method via numerical experiments with curves up to genus 7.

1. Introduction

The Torelli theorem is a classical and foundational result in algebraic geometry, stating that a Riemann surface, or smooth algebraic curve, C is uniquely determined by its Jacobian variety $J(C)$. More concretely, the theorem says that a Riemann surface of genus g can be recovered from one Riemann matrix τ that represents its Jacobian. The key object is the Riemann theta function:

$$\theta\colon \mathbb{C}^g \times \mathbb{H}_g \longrightarrow \mathbb{C}, \qquad \theta(\mathbf{z},\tau) := \sum_{n\in\mathbb{Z}^g} \exp\left(\pi i n^t \tau n + 2\pi i n^t \mathbf{z}\right) \tag{1.1}$$

where $\mathbb{H}_g$ is the Siegel upper-half space of $g \times g$ symmetric complex matrices with positive definite imaginary part. There are various proofs of Torelli's theorem, which can be even made concrete in computational terms. Most proofs rely on the geometry of the theta divisor. This is the locus inside the Jacobian variety $J(C) = \mathbb{C}^g/(\mathbb{Z}^g + \tau\mathbb{Z}^g)$ which is cut out by the theta function:

$$\Theta = \{\mathbf{z} \in J(C) \,|\, \theta(\mathbf{z},\tau) = 0\}.$$

For example, suppose that the Riemann surface C is not hyperelliptic, so that we can identify C with a canonical model $C \subseteq \mathbb{P}^{g-1}$. Then for any singular point $\mathbf{z} \in \Theta_{\text{sing}}$ of the theta divisor, the corresponding Hessian matrix $(\frac{\partial^2\theta}{\partial z_i \partial z_j}(\mathbf{z},\tau))$ defines a quadric in the projective space space $\mathbb{P}^{g-1}$. By a result of Green **[11]**, such quadrics span the space of quadrics in the ideal of the curve. Hence, if the curve is not trigonal, or a smooth plane quintic, these quadrics generate the whole canonical ideal. This result has been extended by Kempf and Schreyer, which gave a way to recover the curve from a single singular point **[13]**. In particular, this

---

2020 *Mathematics Subject Classification.* Primary 14Q05; Secondary 14H42, 14H70.
Key words and phrases. Algebraic curves, theta functions, Torelli theorem, Jacobian variety.
The second author was supported by Turkish Scientific and Technological Research Council (TÜBİTAK) – TÜBİTAK 2236, project number 1119B362000396.

gives a powerful effective reconstruction of the curve, provided that we are able to solve the system

$$\theta(\mathbf{z},\tau) = \frac{\partial\theta}{\partial z_1}(\mathbf{z},\tau) = \cdots = \frac{\partial\theta}{\partial z_g}(\mathbf{z},\tau) = 0. \tag{1.2}$$

This has been implemented numerically for curves of genus 4 in [**7**], but it is a rather hard task in general since the theta function is inherently transcendental. Moreover, this problem is also quite sensitive to the precision of the data: for example, if we move τ a bit, the corresponding theta divisor will not have singular points. There are various other proofs of the Torelli theorem, but many involve solving system of equations such as (1.2).

Hence, we look for different, more algebraic methods. Such a strategy was proposed by Dubrovin [**8**], building on Krichever's work [**14**] on algebraic curves and the Kadomtsev-Petviashvili (KP) equation:

$$\frac{\partial}{\partial x}\left(4u_t - 6uu_x - u_{xxx}\right) = 3u_{yy}. \tag{1.3}$$

More precisely, for each Riemann surface C of genus g there exists a threefold $\mathcal{D}_C$ in a weighted projective space $\mathbb{WP}^{3g-1}$ parametrizing triples (U,V,W) such that the function

$$u(x,y,t) = 2\frac{\partial^2}{\partial x^2}\log\tau(x,y,t)+c, \qquad \tau(x,y,z) := \theta(Ux+Vy+Wz+D,\tau) \tag{1.4}$$

is a solution to the KP equation (1.3) for any $D\in\mathbb{C}^g$ and some $c\in\mathbb{C}$. This threefold was called the *Dubrovin threefold* in [**3**] and it was studied there from a computational point of view. The important properties of this object for our point of view are two: first, $\mathcal{D}_C$ is cut out by some explicit equations whose coefficient are derivatives of theta functions (with characteristic) evaluated at zero. These can be computed explicitly with software for the evaluation of the theta functions, such as `Theta.jl` in Julia [**1**]. Second, the projection of $\mathcal{D}_C$ onto the projective space $\mathbb{P}^{g-1}_U$ of the coordinates $u_1,\ldots,u_g$ consists exactly of the canonical model for the curve $C\subseteq\mathbb{P}^{g-1}_U$. Hence, equations for the canonical model of C can be obtained by eliminating the variables V,W from the equations of the Dubrovin threefold $\mathcal{D}_C$, a purely algebraic process. In conclusion, this allows recovering the curve from the Riemann matrix τ without having to solve a transcendental system such as (1.2).

In this note, we explain how to implement this strategy effectively, using the methods of numerical algebraic geometry. In Section 2, we explain the background behind the Dubrovin threefold and we state the key Lemma 2.1, which explains how to recover equations for the curve. In particular, this allows us to recover quartic equations, but we discuss also the case of quadrics and cubics. Furthermore, we also comment on applications of these methods to the classical Schottky problem. In Section 3 we state the algorithm and analyze its complexity. Moreover, even if our focus is on methods that avoid finding singular points of the theta divisor, the latter can be very useful when we are able to solve the transcendental system (1.2), and we comment on this in Section 3.1. We conclude by presenting numerical experiments with curves from genera from 3 to 7, which we carried out with the packages `RiemannSurfaces` in Sage and `Theta.jl` and `Homotopycontinuation.jl` in Julia.

2. The Dubrovin threefold

We start by recalling some background on the Dubrovin threefold, following [**3**, **8**]. Let C be a smooth projective algebraic curve, or compact Riemann surface, of genus g. We fix a symplectic basis $a_1, b_1, \ldots, a_g, b_g$ for the first homology group $H_1(C, \mathbb{Z})$ and we choose a *normalized basis* $\omega_1, \omega_2, \ldots, \omega_g$ of holomorphic differentials, meaning that

$$\int_{a_i} \omega_j = \delta_{ij}. \tag{2.1}$$

The corresponding Riemann matrix $\tau \in \mathbb{H}_g$ is defined as

$$\tau = \left(\int_{b_i} \omega_j \right)_{1 \le i,j \le g}. \tag{2.2}$$

One can see that the function (1.4) is a solution to the KP equation (1.3) for all values of $D \in \mathbb{C}^g$ and a certain value of $c \in \mathbb{C}$ if and only if there exists $d \in \mathbb{C}$ such that the following quartic PDE, known as the *Hirota bilinear equation*, is satisfied:

(2.3)

$$(\tau_{xxxx}\tau - 4\tau_{xxx}\tau_x + 3\tau_{xx}^2) + 4(\tau_x\tau_t - \tau\tau_{xt}) + 6c\,(\tau_{xx}\tau - \tau_x^2) + 3(\tau\tau_{yy} - \tau_y^2) + 8d\tau^2 = 0.$$

We now introduce a weighted projective space $\mathbb{WP}^{3g+1}$ with variables (U, V, W, c, d) where the $U = (u_1, \ldots, u_g)$ have degree 1, the $V = (v_1, \ldots, v_g)$ have degree 2, the $W = (w_1, \ldots, w_g)$ have degree 3 and finally c, d have degree 2 and 4 respectively. The *big Dubrovin threefold* $\mathcal{D}_C^{\text{big}}$ parametrizes all elements (U, V, W, c, d), with $U \neq \mathbf{0}$, such that $\tau(x, y, z)$ in (1.4) is a solution to the Hirota bilinear equation (2.3) for all $D \in \mathbb{C}^g$. The projection of this variety to the space $\mathbb{WP}^{3g-1}$ of the (U, V, W) is called simply the *Dubrovin threefold* $\mathcal{D}_C$.

Equations for $\mathcal{D}_C^{\text{big}}$ can be obtained directly from (2.3) as follows. Given any $\mathbf{z}$ in $\mathbb{C}^g$, we write the Riemann theta function as $\theta(\mathbf{z}) = \theta(\mathbf{z}, \tau)$. Then we consider the differential operator $\partial_U := u_1 \frac{\partial}{\partial z_1} + \cdots + u_g \frac{\partial}{\partial z_g}$, and the analogous operators ∂_V, ∂_W. For any fixed vector $\mathbf{z} \in \mathbb{C}^g$, the *Hirota quartic* $H_{\mathbf{z}}$ is defined as:

$$\begin{aligned} &(\partial_U^4\theta(\mathbf{z}) \cdot \theta(\mathbf{z}) - 4\partial_U^3\theta(\mathbf{z}) \cdot \partial_U\theta(\mathbf{z}) + 3\{\partial_U^2\theta(\mathbf{z})\}^2) \\ &+ 4 \cdot (\partial_U\theta(\mathbf{z}) \cdot \partial_W\theta(\mathbf{z}) - \theta(\mathbf{z}) \cdot \partial_U\partial_W\theta(\mathbf{z})) + 6c \cdot (\partial_U^2\theta(\mathbf{z}) \cdot \theta(\mathbf{z}) - \{\partial_U\theta(\mathbf{z})\}^2) \\ &\qquad + 3 \cdot (\theta(\mathbf{z}) \cdot \partial_V^2\theta(\mathbf{z}) - \{\partial_V\theta(\mathbf{z})\}^2) + 8d \cdot \theta(\mathbf{z})^2. \end{aligned} \tag{2.4}$$

This is exactly the expression obtained by combining (2.3) and (1.3), hence the big Dubrovin threefold $\mathcal{D}_C^{\text{big}}$ is cut out by the Hirota quartics $H_{\mathbf{z}}$, as $\mathbf{z}$ runs over all vectors in $\mathbb{C}^g$, see [**3**, Proposition 4.2]. The coefficients $H_{\mathbf{z}}(U, V, W, c, d)$ are the values of the theta function θ and its partial derivatives of certain order at $\mathbf{z}$ and they can be computed using numerical software for evaluating theta functions and their derivatives. We use the `Julia` package that is introduced in [**1**].

This yields an infinite number of equations that vanish on the big Dubrovin threefold. A finite set of equations can be derived for the threefold via the addition formula [**8**, §VI.1] theta functions with characteristics $\varepsilon, \delta \in \{0,1\}^g$:

$$\theta\begin{bmatrix}\varepsilon\\ \delta\end{bmatrix}(\mathbf{z} \,|\, \tau) = \sum_{n \in \mathbb{Z}^g} \exp\left(\pi i \left(n + \frac{\varepsilon}{2}\right)^T \tau \left(n + \frac{\varepsilon}{2}\right) + \left(n + \frac{\varepsilon}{2}\right)^T \left(z + \frac{\delta}{2}\right) \right). \tag{2.5}$$

This function in (2.5) coincides with the Riemann theta function (1.1) for $\varepsilon = \delta = 0$ and in general it differs from it by an exponential factor. We consider the following function

$$\hat{\theta}[\varepsilon](\mathbf{z}) := \theta \begin{bmatrix} \varepsilon \\ 0 \end{bmatrix} (\mathbf{z} \,|\, 2\tau). \tag{2.6}$$

For fixed τ, these complex numbers $\hat{\theta}[\varepsilon](0)$ at $\mathbf{z} = \mathbf{0}$ are called *theta constants.* We use the term theta constant also for evaluations at $\mathbf{z} = \mathbf{0}$ of derivatives of (2.6). With these conventions, we define the *Dubrovin quartic* in (U, V, W, c, d) associated to the half-characteristic ε as:

(2.7)

$$F[\varepsilon](U,V,W,c,d) := \partial_U^4\hat{\theta}[\varepsilon](\mathbf{0}) - \partial_U\partial_W\hat{\theta}[\varepsilon](\mathbf{0}) + \frac{3}{2}c\partial_U^2\hat{\theta}[\varepsilon](\mathbf{0}) + \frac{3}{4}\partial_V^2\hat{\theta}[\varepsilon](\mathbf{0}) + d\hat{\theta}[\varepsilon](\mathbf{0}).$$

The Dubrovin and the Hirota quartics span the same vector subspace of the complex vector space of homogeneous polynomials of degree 4 as shown in [**3**, Proposition 4.3], hence they also provide defining equations for $\mathcal{D}_C^{\mathrm{big}}$. We note that the proof of [**3**, Proposition 4.3] relies on Riemann's Addition Formula, and that's where the argument 2τ in (2.6) comes from.

We come to the crucial point: the projection of the big Dubrovin threefold onto the projective space $\mathbb{P}^{g-1} = \mathbb{P}_U^{g-1}$ coincides exactly with the canonical model of the curve C induced by the basis of holomorphic differentials of (2.1):

$$C \longrightarrow \mathbb{P}_U^{g-1}, \qquad p \mapsto [\omega_1(p), \omega_2(p), \ldots, \omega_g(p)]\,. \tag{2.8}$$

In particular, if the curve C is not hyperelliptic, the canonical model is isomorphic to the curve C itself. In algebraic terms, this means that the canonical model of (2.8) can be recovered by eliminating the variables V, W, c, d from the equations of the big Dubrovin threefold. This is reduced to a problem of linear algebra as follows: for any half-characteristic $\varepsilon \in \{0,1\}^g$ write $Q[\varepsilon]$ for the Hessian matrix of the function $\hat{\theta}[\varepsilon](\mathbf{z})$ at $\mathbf{z} = \mathbf{0}$, then combining [**3**, Lemma 4.6] and [**3**, Proposition 4.7] we have:

LEMMA 2.1. *Suppose that C is a curve given by way of its Riemann matrix τ. Let us denote by $V_\tau \subseteq \mathbb{C}[u_1, \ldots, u_g]$ the vector space of linear combinations*

$$\sum_{\varepsilon \in \{0,1\}^g} \lambda_\varepsilon \cdot \partial_U^4 \hat{\theta}[\varepsilon], \tag{2.9}$$

where the 2^g complex scalars λ_ε satisfy the linear equations

$$\sum_\varepsilon \lambda_\varepsilon \cdot Q[\varepsilon] = 0 \qquad \textit{and} \qquad \sum_\varepsilon \lambda_\varepsilon \cdot \hat{\theta}[\varepsilon] = 0. \tag{2.10}$$

Then a linear combination of the Dubrovin quartics is independent of c, d if and only if it belongs to V_τ. Furthermore, V_τ has dimension $2^g - \frac{g(g+1)}{2} - 1$ and the corresponding quartics (2.9) *cut out the canonical model* (2.8) *of the curve C.*

Hence, if the curve C is not hyperelliptic, this lemma gives a way to recover the curve from the Riemann matrix τ, which depends only on the evaluation of the theta function and its derivatives.

2.1. Recovering quadrics and cubics. Lemma 2.1 allows us to recover a linear space of quartics that cut out a canonical model of C. However, it is also possible to recover quadric equations. We start with the following basic observation: if in the space V_τ we can find a quartic of the form $Q(U)^2$, then Q is a quadric containing the curve C. We can actually find such special quartics inside V_τ: indeed, suppose that $\mathbf{z}_0 \in \mathbb{C}^g$ is a singular point of the theta divisor Θ. Then the Hirota quartic $H_{\mathbf{z_0}}$ becomes:

$$H_{\mathbf{z_0}} = 3\left(\partial_U^2\theta(\mathbf{z_0})\right)^2 \tag{2.11}$$

and since this is independent of c, d, Lemma 2.1 tells us that $(\partial_U^2\theta(\mathbf{z_0}))^2 \in V_\tau$. Furthermore, we know by Green's result mentioned in the introduction, that the quadrics $\partial_U^2\theta(\mathbf{z_0})$ appearing in (2.11) span the whole vector space of quadrics in the ideal of the canonical curve. Hence, if C is not hyperelliptic, trigonal, or a smooth plane quintic, such quadrics generate the canonical ideal of the curve. We again point out that, at least in principle, such quadrics can be computed by algebraic and not transcendental methods. Indeed, this corresponds to intersecting the space V_τ with the subvariety in $\mathbb{C}[U]_4$ given by quartics of the form Q^2, so it amounts to solving a polynomial system of equations in the space V_τ.

We discuss briefly also the case of cubics, which can appear if the curve is trigonal or a smooth plane quintic. In general, if $\mathbf{z}_0 \in \Theta$ is a singular point in the theta divisor, the cubic equation $\partial_U^3\theta(\mathbf{z}_0)$ belongs to the canonical ideal of the curve [**13**]. If we apply the operator ∂_U to the Hirota quartic $H_{\mathbf{z}}$ and we evaluate it at $\mathbf{z} = \mathbf{z}_0$, we obtain the quintic equation

$$\partial_U H_{\mathbf{z}|\mathbf{z}=\mathbf{z_0}} = 2(\partial_U^2\theta(\mathbf{z_0}))(\partial_U^3\theta(\mathbf{z_0})). \tag{2.12}$$

The quintic (2.12) is a linear combination of the quintics $u_i \cdot F[\varepsilon]$, for $i = 1, \ldots, g$ and $\varepsilon \in \{0,1\}^g$, so in principle we could try to proceed as for quadrics, and look for reducible quintics of the form $Q(U) \cdot T(U)$, where $\deg Q(U) = 2$ and $\deg T(U) = 3$.

2.2. Applications to the Schottky problem. Up to now we have discussed the Torelli problem of reconstructing a smooth curve C from a Riemann matrix τ of its Jacobian $J(C)$. Another fundamental question in this area is the Schottky problem [**12**], which asks, given a matrix $\tau \in \mathbb{H}_g$, whether this represents the Jacobian of a curve. This can be formulated in different ways with different possible solutions: see for example [**9**] for a very recent one. In particular, one of these was given by Krichever [**14**] and Shiota [**15**] via the KP equation. This solution can be formulated in terms of the Dubrovin threefold [**8**, Section IV.4] by saying that $\tau \in \mathbb{H}_g$ represents a Jacobian if and only if the Dubrovin quartics (2.7) cut out a threefold.

In particular, we can check that a matrix $\tau \in \mathbb{H}_g$ does not represent a Jacobian, by computing the quartics of Lemma 2.1 and then checking that they do not define a curve in $\mathbb{P}^{g-1}$. We verified this experimentally in Example 3.6.

3. Numerical recovery

We can sum up the discussion of the previous section in the following algorithm. We have implemented it in Julia, which can be found at `https://turkuozlum.wixsite.com/tocj`.

Algorithm 1: Recovery through the Dubrovin threefold

Input: A matrix $\tau \in \mathbb{H}_g$ representing the Jacobian of a non-hyperelliptic curve.

Output: Quartics that cut out the canonical model of the algebraic curve C whose Riemann matrix is τ.

Step 1: Set up the linear system in (2.10) by computing the theta constants via the Julia package `Theta.jl`.

Step 2: Solve the linear system in (2.10).

Step 3: Write the quartics (2.9) and return them.

The algorithm is straightforward and we can easily analyze its complexity in terms of the genus g. In Step 1, we need to evaluate $2^g \cdot (g(g+1)/2+1)$ theta constants, coming from the matrices $Q[\varepsilon]$ and the scalars $\hat{\theta}[\varepsilon]$. Then, in Step 2, we need to solve a $(g(g+1)/2+1) \times 2^g$ linear system of maximal rank. Finally, in Step 3, we need to compute the quartics (2.9), which involves the evaluation of $2^g \cdot (g+3)(g+2)(g+1)g/24$ theta constants. In our experiments, we considered examples, taken from the literature, up to genus 7, so that the linear system of Step 2 is of relatively small size and can be solved very quickly in Julia. What takes most of the time is the evaluation of the theta constants: the following table presents the approximated times to compute the theta constants in the examples below, with 12 digits of precision. In the table, ∂^i indicates the order of the partial derivative of θ that we compute. The last column denotes the time needed to run the entire algorithm.

genus	∂^0	∂^2	∂^4	total
3	0.0009 sec	0.001 sec	0.002 sec	5 sec
4	0.008 sec	0.015 sec	0.02 sec	11 sec
5	0.07 sec	0.15 sec	0.23 sec	9 min
6	2.1 sec	4.2 sec	6.9 sec	12 h
7	6 sec	8 sec	10 sec	60 h

3.1. Computing the singular points. As we explained before, one of the advantages of the Dubrovin threefold is that it allows us to recover the curve without computing a singular point of the theta divisor. However, this is also a very useful method, if we manage to solve the transcendental system (1.2).

A Sage code that computes a singular point of the theta divisor in genus 4 is presented in the article [**7**]. The idea, that can be extended to any genus, is to solve system (1.2) by numerical optimization, starting from a random input $z = a + \tau b$, where a, b are real vectors with entries between 0 and 1. In our implementation, we use the function `optimize.root` from the `SciPy` package. We call this function with the method `lm`, based on the Levenberg-Marquardt algorithm, which speeds up the computation substantially in comparison with the `hybr` method. The function `optimize.root` evaluates the partial derivatives (1.2) of the given function via estimating the limits of the function. Instead, we used the partial derivatives that is implemented in the Sage package `abelfunctions` [**6**], which gave more accurate results. In our experiments, it took about 30 minutes for one singular point to be computed in the case of genus 4 and about 1.5 hours in the case of genus 5.

REMARK 3.1. Before presenting our experiments, we observe that it is often convenient to work with an arbitrary basis of differentials instead of a normalized one as in (2.1). For such an arbitrary basis $\widetilde{\omega}_1, \ldots, \widetilde{\omega}_g$, we consider the corresponding $g \times g$ period matrices

$$\Pi_a = \left(\int_{a_j} \widetilde{\omega}_i \right)_{ij} \quad \text{and} \quad \Pi_b = \left(\int_{b_j} \widetilde{\omega}_i \right)_{ij}. \tag{3.1}$$

Then we obtain a normalized basis of differentials as in (2.1) and the corresponding Riemann matrix by taking

$$(\omega_1, \omega_2, \ldots, \omega_g)^T = \Pi_a^{-1} (\widetilde{\omega}_1, \widetilde{\omega}_2, \ldots, \widetilde{\omega}_g)^T. \tag{3.2}$$

$$\tau = \Pi_a^{-1} \Pi_b. \tag{3.3}$$

3.2. Numerical experiments. Finally, we present some examples illustrating our algorithm.

In our experiments, we start with an explicit plane affine model for a non-hyperelliptic curve and possibly also its canonical model $C \subseteq \mathbb{P}^{g-1}$, and then we use the package `RiemannSurface` of Sage [**5**] to compute a Riemann matrix τ on which we run the Algorithm 1. We then verify that the resulting quartics cut out the canonical curve we started with. We can do this explicitly in genus 3, when the curve itself is a smooth plane quartic. In higher genera, we first verify that the quartics belong to the ideal of the curve by running the polynomial division algorithm, which returns a remainder of zero, up to a certain numerical approximation. Furthermore, to verify that the quartics cut out the curve set-theoretically, we compute the intersection with a hyperplane in $\mathbb{P}^{g-1}$ by adding a random linear form and solving the resulting polynomial system via homotopy continuation. This is the primary computational method in numerical algebraic geometry, and we used the Julia implementation of `HomotopyContinuation.jl` [**4**]. This computation returns $2g-2$ solutions, confirming that the quartics cut out a curve of degree $2g-2$.

We also tried to recover the quadrics vanishing on the curve using the method of Section 2.1. We set up the problem of finding elements of the form $Q(U)^2$ in the space of quartics returned by Algorithm 1, and we solved it again via `HomotopyContinuation.jl`. We could do this in genus 4.

In genera 4 and 5, we could also compute singular points of the theta divisor, using the methods of Section 3.1. With these singular points, we could compute quadric and cubic equations for the curve, as described in Section 2.1.

EXAMPLE 3.2 (Genus three). The Trott curve is a smooth plane quartic with affine model $C = \{f(x,y) = 0\}$, where

$$f(x,y) = 12^2(x^4 + y^4) - 15^2(x^2 + y^2) + 350x^2y^2 + 81.$$

In particular, this is already the canonical model, and the curve is of genus 3 and not hyperelliptic. We compute a Riemann matrix using `RiemannSurface` in Sage [**5**]: in particular, the package uses the basis of differentials:

$$\widetilde{\omega}_1 = \frac{1}{f_y} dx, \quad \widetilde{\omega}_2 = \frac{x}{f_y} dx, \quad \widetilde{\omega}_3 = \frac{y}{f_y} dx,$$

where f_y denotes the derivative $\frac{\partial f}{\partial y}$, and then it computes the period matrices Π_a and Π_b as in (3.1). The entries of the corresponding symmetric normalized Riemann matrix $\tau := (\tau_{ij})$ are as follows:

$$\tau_{11} = 1.06848368471179 + 0.723452867814272i, \tau_{12} = -0.305886633614305 + 0.123618182281837i,$$
$$\tau_{13} - 0.160517941389541 - 0.206682546926085i, \tau_{22} = 0.776859918461210 + 1.25292663517205i,$$
$$\tau_{23} = -0.626922516393387 - 0.289746911570334i, \tau_{33} = 0.376235735801471 + 0.484440302728207i.$$

With this matrix, we set the linear system (2.10) by computing the theta constants appearing in the expressions of the system [**1**]. As expected, this has a unique solution, up to scalar multiplication, and we compute the corresponding quartic polynomial (2.9):

$$
\begin{gathered}
(0.440553382315733 27 - 0.117125218955325 13i)u_1^4 + (2.094882287195226 + 7.879664904010854i)u_1^3u_2 \\
- (5.316458517368645 - 1.4134300016965646i)u_1^3u_3 + (61.49338091003442 - 16.348587918073555i)u_1^2u_2^2 \\
+ (27.505923029039046 + 105.6412469122926i)u_1^2u_2u_3 - (43.67750279381081 - 12.658628276584892i)u_1^2u_3^2 \\
- (0.20611709900405373 + 0.7752863638524854i)u_1u_2^3 + (142.137577271911 + 22.777083502115772i)u_1u_2^2u_3 \\
+ (101.16905240593528 + 146.6228999954985i)u_1u_2u_3^2 - (28.214458865117336 - 92.58798535078905i)u_1u_3^3 \\
-(0.06519271764094459 - 0.017332091034810038i)u_2^4 - (0.016856400506870983 + 0.8256030828721883i)u_2^3u_3 \\
+ (64.66553470742735 + 38.49587006148285i)u_2^2u_3^2 + (94.88897578016996 + 81.18194430047456i)u_2u_3^3 \\
+ (33.080420780163195 + 41.521570514217885i)u_3^4.
\end{gathered}
$$

At a first glance, this might not look like the Trott curve. However, this equation is for the canonical model of C with respect to a basis of normalized differentials $\omega_1, \omega_2, \omega_3$. If we go back to the differentials $\widetilde{\omega}_1, \widetilde{\omega}_2, \widetilde{\omega}_3$ via the change of coordinates in (3.2), we obtain the following quartic, after scaling the coefficients.

$$
\begin{gathered}
81u_1^4 + (1.2223597321441586 \cdot 10^{-13} - 9.454838005323456 \cdot 10^{-14}i)u_1^3u_2 \\
+(2.9124976279639876 \cdot 10^{-13} + 1.1282283371974781 \cdot 10^{-13}i)u_1^3u_3 \\
-(225.00000000000017 - 4.607401108070593 \cdot 10^{-13}i)u_1^2u_2^2 \\
+(3.669767553538813 \cdot 10^{-13} - 3.017230893609506 \cdot 10^{-13}i)u_1^2u_2u_3 \\
-(224.99999999999986 + 5.357443148919295 \cdot 10^{-13}i)u_1^2u_3^2 \\
fj - (4.1371303331328177 \cdot 10^{-13} - 3.5463573271644895 \cdot 10^{-13}i)u_1u_2^3 \\
-(8.382029113614384 \cdot 10^{-13} + 3.97078497125283 \cdot 10^{-13}i)u_1u_2^2u_3 \\
+(7.725484571929981 \cdot 10^{-13} + 2.34428275709395 \cdot 10^{-13}i)u_1u_2u_3^2 \\
-(8.239810406206657 \cdot 10^{-13} + 2.6625152861265 \cdot 10^{-13}i)u_1u_3^3 \\
+(143.99999999999918 - 7.607569271465399 \cdot 10^{-13}i)u_2^4 \\
-(9.177234341211958 \cdot 10^{-13} - 8.304604428951876 \cdot 10^{-13}i)u_2^3u_3 \\
+(350.0000000000026 + 1.2750714694427922 \cdot 10^{-12}i)u_2^2u_3^2 \\
-(1.3400119435300388 \cdot 10^{-13} - 5.996042934502803 \cdot 10^{-13}i)u_2u_3^3 \\
+(143.99999999999895 - 5.357443148919295 \cdot 10^{-14}i)u_3^4.
\end{gathered}
$$

This is nothing but the quartic defining the Trott curve, up to an error of 10^{-12}. In particular, we can recover the exact equation if we round up the coefficients to the nearest integer. We emphasize that this example is treated slightly different than as it has been in [**3**, Example 4.8]. Indeed, in [**3**, Example 4.8] we obtained an numerical equation of the curve via the method presented in this paper, and then we checked that the Dixmier-Ohno invariants of the new curve agreed with the invariants of the Trott curve up to numerical round-off. Here, instead, we could recover an approximate equation of the original Trott curve.

Inspired by this example, we repeated the same experiments with 20 plane quartics with integer coefficients. The coefficients were bounded in absolute value by 100. We computed the period and the Riemann matrix with 53 bits of precision, we computed the theta constants with 12 digits of precision, and at the end we could recover the exact equation of the curve by rounding up the coefficients to the closest integer. Each experiment took approximately 4 seconds.

EXAMPLE 3.3 (Genus four). Moving on to the case of genus 4, we consider the canonical curve

$$
C = \left\{ u_1u_4 - u_2u_3 = 0 \,,\; u_1^3 - u_2^3 - u_3^3 - u_4^3 = 0 \right\}. \tag{3.4}
$$

This has an affine plane model given by $\{f(x,y)=0\}$, where $f(x,y)=1-x^3-y^3-x^3y^3$. We can recover the previous canonical model via the basis of differentials

$$\widetilde{\omega}_1=-\frac{1}{f_y}dx,\quad \widetilde{\omega}_2=-\frac{x}{f_y}dx,\quad \widetilde{\omega}_3=-\frac{y}{f_y}dx,\quad \widetilde{\omega}_4=-\frac{xy}{f_y}dx.$$

We can compute the 4×4 Riemann matrix τ via the plane model of the curve with the Sage package [**5**]. This takes approximately 677 milliseconds for 53 bits, or about 16 digits, of precision.

To reconstruct the canonical model of the curve back from τ, we compute the 5 quartics in (2.9) by solving the linear system in (2.10).

By Lemma 2.1, these 5 quartics cut out the canonical curve (3.4) after the basis change (3.2). We can first verify that the transformed quartics belong to the ideal of C by the polynomial division algorithm. We did it in Sage, working over the complex field with 200 digits of precision. The coefficients of the remainder of the division algorithm were all of size 10^{-15}.

Then, to verify that these quartic equations cut out the curve, we use the Julia package `HomotopyContinuation.jl` [**4**]. We add a random linear form to the polynomial system of our 5 quartics and then `HomotopyContinuation.jl` returns 6 solutions, which is what we expect from a curve of degree 6 in $\mathbb{P}^3$. Moreover, again via homotopy continuation methods, we can find a quadratic polynomial $Q(U)$ such that $Q(U)^2$ is in the linear space generated by the 5 quartics, as in Section 2.1. After applying the change of basis in (2.1) and rescaling we get the following expression for the quadric:

$$\begin{aligned}u_1u_4 &-(1.4829350744889013\cdot 10^{-15}-1.6682847904065378\cdot 10^{-15}i)u_1u_2\\ &-(3.5425309660018567\cdot 10^{-15}+6.403641669846521\cdot 10^{-16}i)\\ &+u_1u_3(2.3679052278901118\cdot 10^{-15}+1.6728691462607347\cdot 10^{-15}i)u_1^2\\ &+(1.8423363133604865\cdot 10^{-15}-3.1265312370929112\cdot 10^{-15}i)u_2^2\\ &-(1.0000000000000007-2.3672426468663847\cdot 10^{-15}i)u_2u_3\\ &-(1.739413822171687\cdot 10^{-15}-2.775912191360744\cdot 10^{-15}i)u_2u_4\\ &+(3.3764916290020825\cdot 10^{-15}-6.001818720458345\cdot 10^{-15}i)u_3^2\\ &-(3.777550457759975\cdot 10^{-16}-1.4453231221486755\cdot 10^{-15}i)u_3u_4\\ &-(1.1268542006403671\cdot 10^{-15}+1.7990461567600933\cdot 10^{-15}i)u_4^2.\end{aligned}$$

In genus four, we can also compute numerically a singular point of the theta divisor. We do it in Sage, as described in Section 3.1 and we find the point:

$$\mathbf{z}_0=(0.75+0.54819629i,0.75-0.54819629i,0.5+0.33618324i,0.75+0.2120130i).$$

The theta function and its derivatives vanish at this point up to 13 digits. With this, we can compute the quadric $\partial_U^2\theta(\mathbf{z}_0)$ and the cubic $\partial_U^3\theta(\mathbf{z}_0)$. After the usual change of coordinates (3.2), the quadric becomes

$$\begin{aligned}u_1u_4 &+(9.977112210552615\cdot 10^{-8}+6.939529950175681\cdot 10^{-8}i)u_1^2\\ &+(6.74409346713264\cdot 10^{-8}-2.3021247380555947\cdot 10^{-15}i)u_1u_2\\ &-(2.3274471968848503\cdot 10^{-8}+5.037739720772648\cdot 10^{-8}i)u_1u_3\\ &+(9.977111730319195\cdot 10^{-8}-6.93953067335553\cdot 10^{-8}i)u_2^2\\ &-(0.9999999999999997-5.892639748496844\cdot 10^{-8}i)u_2u_3\\ &+(2.3274465793326793\cdot 10^{-8}-5.037739901039536\cdot 10^{-8}i)u_2u_4\\ &+(3.9887820808350887\cdot 10^{-8}+2.7942580923377634\cdot 10^{-8}i)u_3^2\\ &+(1.3142269019133975\cdot 10^{-7}-6.865574771844027\cdot 10^{-15}i)u_3u_4\\ &+(3.988781716962214\cdot 10^{-8}-2.7942586271411155\cdot 10^{-8}i)u_4^2,\end{aligned}$$

which coincides with the quadric $u_1u_4 - u_2u_3$ of (3.4), up to about 10 digits of precision. The cubic equation that we obtain is:

$$
\begin{gathered}
u_1^3 - (1.0000000001244782 + 4.8426070737375934 \cdot 10^{-11} i)u_2^3 \\
-(1.0000000001244924 + 4.8417923378918607 \cdot 10^{-11} i)u_3^3 \\
-(1.0000000002161082 + 3.561043256396786 \cdot 10^{-11} i)u_4^3 \\
+(4.5851667064228973 \cdot 10^{-11} + 3.883459414267029 \cdot 10^{-11} i)u_1^2u_2 \\
-(5.655054824682744 \cdot 10^{-11} - 2.0264603303086512 \cdot 10^{-11} i)u_1^2u_3 \\
-(4.562873576948296 \cdot 10^{-11} + 1.0723355355290677 \cdot 10^{-10} i)u_1u_2^2 \\
+(4.416105308034146 \cdot 10^{-11} + 6.485802277816666 \cdot 10^{-11} i)u_2^2u_4 \\
-(2.4095153783271284 \cdot 10^{-11} - 3.821774264257014 \cdot 10^{-11} i)u_2u_4^2 \\
-(2.1061545135963428 \cdot 10^{-11} + 3.9985852337772294 \cdot 10^{-11} i)u_3u_4^2 \\
+(3.407561378730717 \cdot 10^{-11} - 7.066441338212726 \cdot 10^{-11} i)u_3^2u_4 \\
-(7.006134562951018 \cdot 10^{-11} - 9.313087455058934 \cdot 10^{-11} i)u_1u_3^2 \\
+(5.3419725864118215 - 2.3068027651869776 i)u_1^2u_4 - (5.341972586241411 - 2.3068027651197216 i)u_1u_2u_3 \\
-(0.861775203997493 - 1.4926384376919832 i)u_1u_2u_4 + (0.8617752039804856 - 1.492638437827363 i)u_2^2u_3 \\
-(0.8617752037076406 + 1.4926384378593538 i)u_1u_3u_4 + (0.8617752038333869 + 1.4926384379123137 i)u_2u_3^2 \\
-(2.396786904582313 - 2.79440847281551 i)u_1u_4^2 + (2.3967869046516648 - 2.794408472851858 i)u_2u_3u_4.
\end{gathered}
$$

And we see that, with an approximation of 9 digits, this is the cubic $u_1^3 - u_2^3 - u_3^3 - u_4^3$ of (3.4), plus a linear combination of $u_i(u_1u_4 - u_2u_3)$, for $i = 1, 2, 3, 4$.

EXAMPLE 3.4. Let C be the genus 5 curve with an affine plane equation given by the polynomial $f(x, y) = x^2y^4 + x^4 + x + 3$. The differentials $\frac{1}{f_y}dx$, $\frac{x}{f_y}dx$, $\frac{xy}{f_y}dx$, $\frac{xy^2}{f_y}dx$, $\frac{x^2}{f_y}dx$ form a basis of the space of holomorphic differentials. The corresponding canonical model is given by the complete intersection of the three quadrics:

$$
u_4^2 + u_5^2 + u_2u_1 + 3u_1^2, \quad u_3^2 - u_2u_4, \quad u_2^2 - u_5u_1. \tag{3.5}
$$

We compute the sixteen Dubrovin quartics (2.9) and we check that, after the change of coordinates (3.2), they belong to the ideal of C. We do this by polynomial division in Sage, over the complex field with 200 digits of precision as in Example 3.3. The coefficients of the remainder are of size 10^{-10}.

We also check whether the quartics define a curve by adding a random linear form and solving the corresponding system via `HomotopyContinuation.jl`. We obtain 8 solutions, which is nothing but the degree of our canonical genus 5 curve. Here, one needs to increase the precision to about 15 digits while computing the Riemann matrix, which is required for computing the Dubrovin quartics.

In this example, we could compute also singular points of the theta divisor as explained in Section 3.1. We computed three points $\mathbf{z}_1, \mathbf{z}_2, \mathbf{z}_3$ where the theta function and all its derivative vanish up to an error of 10^{-10}. Then we obtain three quadrics $\partial_U^2\theta(\mathbf{z}_1), \partial_U^2\theta(\mathbf{z}_2), \partial_U^2\theta(\mathbf{z}_3)$, which, after the usual change of variables (3.2), can be expressed as three independent linear combinations of the quadrics in (3.5), again up to an error of 10^{-10}.

In the following examples, we push experimenting our methods to higher genera.

EXAMPLE 3.5 (Genus 6 and 7). Here we choose the curves of genus 6 and 7 known as Wiman's sextic [**16**] and the butterfly curve [**10**]. Their respective plane affine equations are:

$$
x^6 + y^6 + 1 + (x^2 + y^2 + 1)(x^4 + y^4 + 1) = 12x^2y^2,
$$
$$
x^6 + y^6 = x^2.
$$

We first compute their Riemann matrices numerically in Sage. Then, we estimate the corresponding 42 and 99 quartics (2.9) in $\mathbb{P}^5$ and $\mathbb{P}^6$ respectively. Using the homotopy continuation method in Julia, we could verify that they define curves of degree 10 and 12 as expected. We point out that in these cases we needed to increase the precision in the Riemann matrix computation: 200 bits of precision in genus 6 and 500 bits in genus 7 were enough for the homotopy continuation computation to terminate.

EXAMPLE 3.6. Finally, we discuss some numerical experiments related to the Schottky problem, as in Section 2.2. We choose 100 random Riemann matrices in genus 4, we computed the corresponding quartics as in Lemma 2.1, we added a random linear form and we solved numerically the resulting system via `Homotopycontinuation.jl`. As expected, we found no solutions, confirming the fact that the quartics do not cut out a curve in $\mathbb{P}^3$. We expect that this circle of ideas would lead to an effective numerical solution to the Schottky problem, and we will investigate this in future work.

Acknowledgments

We would like to thank Nils Bruin, Bernard Deconinck, Bernd Sturmfels and André Uschmajew for their useful comments and their support. We would also like to thank the anonymous referees for their careful reading and for their remarks.

References

[1] Daniele Agostini and Lynn Chua, *Computing theta functions with Julia*, J. Softw. Algebra Geom. **11** (2021), no. 1, 41–51, DOI 10.2140/jsag.2021.11.41. MR4285763
[2] Daniele Agostini and Lynn Chua, *On the Schottky problem for genus-five Jacobians with a vanishing theta-null*, Ann. Sc. Norm. Super. Pisa Cl. Sci. (5) **22** (2021), no. 1, 333–350. MR4288659
[3] Daniele Agostini, Türkü Özlüm Çelik, and Bernd Sturmfels, *The Dubrovin threefold of an algebraic curve*, Nonlinearity **34** (2021), no. 6, 3783–3812, DOI 10.1088/1361-6544/abf08c. MR4281432
[4] P. Breiding, S. Timme, *HomotopyContinuation.jl: A Package for Homotopy Continuation in Julia*, Mathematical Software – ICMS 2018, Lecture Notes in Computer Science, Springer, Cham **10931** (2018), 458–465.
[5] Nils Bruin, Jeroen Sijsling, and Alexandre Zotine, *Numerical computation of endomorphism rings of Jacobians*, Proceedings of the Thirteenth Algorithmic Number Theory Symposium, Open Book Ser., vol. 2, Math. Sci. Publ., Berkeley, CA, 2019, pp. 155–171. MR3952010
[6] C. Swierczewski et al.:, Abelfunctions: A library for computing with Abelian functions, Riemann surfaces, and algebraic curves, github.com/abelfunctions/abelfunctions, 2016.
[7] Lynn Chua, Mario Kummer, and Bernd Sturmfels, *Schottky algorithms: classical meets tropical*, Math. Comp. **88** (2019), no. 319, 2541–2558, DOI 10.1090/mcom/3406. MR3957905
[8] B. A. Dubrovin, *Theta-functions and nonlinear equations* (Russian), Uspekhi Mat. Nauk **36** (1981), no. 2(218), 11–80. With an appendix by I. M. Krichever. MR616797
[9] Hershel M. Farkas, Samuel Grushevsky, and Riccardo Salvati Manni, *An explicit solution to the weak Schottky problem*, Algebr. Geom. **8** (2021), no. 3, 358–373, DOI 10.14231/ag-2021-009. MR4206440
[10] H. T. Fay, *The Butterfly Curve*, American Mathematical Monthly **96** 5 (1989), 442–443.
[11] M. L. Green, *Quadrics of rank four in the ideal of a canonical curve*, Invent. Math. **75** (1984), no. 1, 85–104, DOI 10.1007/BF01403092. MR728141
[12] Samuel Grushevsky, *The Schottky problem*, Current developments in algebraic geometry, Math. Sci. Res. Inst. Publ., vol. 59, Cambridge Univ. Press, Cambridge, 2012, pp. 129–164. MR2931868
[13] George R. Kempf and Frank-Olaf Schreyer, *A Torelli theorem for osculating cones to the theta divisor*, Compositio Math. **67** (1988), no. 3, 343–353. MR959216
[14] I. M. Krichever, *Methods of algebraic geometry in the theory of non-linear equations*, Russian Mathematical Surveys **32** (1977), 185–213.

[15] Takahiro Shiota, *Characterization of Jacobian varieties in terms of soliton equations*, Invent. Math. **83** (1986), no. 2, 333–382, DOI 10.1007/BF01388967. MR818357

[16] A. Wiman, *Zur Theorie der endlichen Gruppen von birationalen Transformationen in der Ebene* (German), Math. Ann. **48** (1896), no. 1-2, 195–240, DOI 10.1007/BF01446342. MR1510931

Max-Planck-Institut für Mathematik in den Naturwissenschaften, Inselstraße 22, 04103 Leipzig, Germany
Email address: daniele.agostini@mis.mpg.de

Department of Mathematics, Boğaziçi University, 34342 Bebek, İstanbul, Turkey
Email address: turkuozlum@gmail.com

Department of Mathematics, Bilkent University, 06800 Bilkent, Ankara, Turkey
Email address: demir.eken@ug.bilkent.edu.tr

Contemporary Mathematics
Volume **779**, 2022
https://doi.org/10.1090/conm/779/15668

A strategy to optimize the complexity of Chudnovsky-type algorithms over the projective line

Stéphane Ballet, Alexis Bonnecaze, and Bastien Pacifico

Abstract. Chudnovsky-type algorithms of multiplication in finite fields are well known for their good bilinear complexity. Recently, two advances have been obtained in the study of these algorithms: a strategy to optimize the scalar complexity of the original algorithm and the development of a generic recursive construction over the projective line. The construction of recursive Chudnodvsky-type algorithms over the projective line makes possible an efficient generic strategy to optimize their complexity (number of scalar and bilinear multiplications and additions in the base field). Then, several examples are given. In particular, considering Baum-Shokrollahi's experiment (1992), this constructive method provides a Chudnovsky-type algorithm of multiplication in $\mathbb{F}_{256}/\mathbb{F}_4$ with the best known complexity, while being much more efficient than existing optimization methods.

1. Introduction

The search for finite field multiplication algorithms with good algebraic complexity (cf. [**BCS97**]) is still a major issue in algorithmics and in cryptography. In this paper, we are interested in the number of arithmetic operations in the base field when multiplying in an extension of finite degree of a finite field. Several more general remarkable methods are known (for example [**Kar63**], [**Fü09**], and [**SS71**]) and can be used to address this problem. Recently, Harvey and van der Hoeven [**HvdH19**] have proven that such a multiplication can be computed with $\mathcal{O}(n \log n)$ operations (when q is fixed), assuming a widely-believed hypothesis. Similarly to the latter, many works focus on multiplication algorithms with efficient asymptotic complexities, giving estimations of the total number of operations in the base field relatively to the degree of the extension using the $\mathcal{O}$ notations. But these methods may not be optimal at finite distance (i.e. not from the point of view of asymptotic complexity), in particular for moderate-sized parameters (around a few thousands of bits). For example, Schönhage-Strassen's algorithm [**SS71**] has a better asymptotic complexity than Karatsuba's algorithm [**Kar63**], but outperforms the latter method only when the parameters become huge (around millions of bits), exceeding many usage sizes. As for the Fürer algorithm [**Fü09**], it is competitive for even larger numbers. Moreover, it is well known that different operations do not have the same cost in terms of bit operations. In particular, multiplication is more expensive than addition and bilinear multiplication is itself more expensive

Key words and phrases. Multiplicative complexity, finite fields, Chudnovsky-type algorithms.

than scalar multiplication ([**STV92**], [**BCP$^+$21**] Section 1). We therefore choose to consider the algebraic complexity model while taking into account the different costs of these operations. To do so, we consider the multiplication method of D.V. and G.V. Chudnovsky [**CC88**] which admits the best bilinear complexity, both in finite distance and in asymptotics. We propose a strategy for the construction of this method in order to optimize the number of scalar multiplications and the number of additions while keeping the same bilinear complexity. Our goal is to be able to obtain in practice, i.e. at finite distance, a total complexity competitive with the best algorithms. It should also be noted that determining the asymptotic complexity of our method remains an open problem.

Let q be a prime power, $\mathbb{F}_q$ the finite field with q elements and $\mathbb{F}_{q^n}$ the degree n extension of $\mathbb{F}_q$. Let $\mathcal{B} = \{e_1, \ldots, e_n\}$ be a basis of $\mathbb{F}_{q^n}$ over $\mathbb{F}_q$ then for $x = \sum_{i=1}^n x_i e_i$ and $y = \sum_{j=1}^n y_j e_j$, the direct calculation of the product is given by

$$z = xy = \sum_{h=1}^n z_h e_h = \sum_{h=1}^n \left(\sum_{i,j=1}^n t_{ijh} x_i y_j \right) e_h, \tag{1}$$

where $e_i e_j = \sum_{h=1}^n t_{ijh} e_h$, $t_{ijh} \in \mathbb{F}_q$ being some constants.

One can distinguish two types of multiplications in this product: the bilinear ones, that are depending of the two elements being multiplied (i.e. the $x_i y_j$); and the scalar ones that are multiplications by a constant in $\mathbb{F}_q$. At first glance, the latest computation requires n^2 bilinear multiplications, n^3 scalar multiplications and $n^3 - n$ additions.

DEFINITION 1.1. Let $\mathcal{U}_{q,n}$ be an algorithm for the multiplication in $\mathbb{F}_{q^n}$ over $\mathbb{F}_q$.

- The number of non-trivial scalar multiplications in $\mathbb{F}_q$ (i.e. multiplications by $\alpha \in \mathbb{F}_q$ with $\alpha \neq 0, 1$), used in $\mathcal{U}_{q,n}$ is called its scalar complexity, and is denoted by $\mu_s(\mathcal{U}_{q,n})$.
- The number of bilinear multiplications in $\mathbb{F}_q$ used in $\mathcal{U}_{q,n}$ is called its bilinear complexity, denoted by $\mu_b(\mathcal{U}_{q,n})$.

We also denote by $a(\mathcal{U}_{q,n})$ the number of additions in $\mathbb{F}_q$ in the algorithm. Consequently, the total complexity of $\mathcal{U}_{q,n}$, denoted by $\mu(\mathcal{U}_{q,n})$ is given by $\mu(\mathcal{U}_{q,n}) = \mu_b(\mathcal{U}_{q,n}) + \mu_s(\mathcal{U}_{q,n}) + a(\mathcal{U}_{q,n})$. Note that bilinear multiplications are known to be computationally heavier than the scalar ones. Algorithms with good bilinear complexity are interpolation algorithms. Among them, the method introduced by D.V. and G.V. Chudnovsky [**CC88**] makes it possible to obtain the best known bilinear complexity. The original Chudnovsky-Chudnovsky Multiplication Algorithm (CCMA) is an interpolation algorithm over rational places of a function field. This construction has been generalized in different ways, for instance with the use of places of arbitrary degrees or the use of derivative evaluations. A detailed review on the topic is given in [**BCP$^+$21**].

Nevertheless, the total complexity of these algorithms has not been deeply studied yet. The first step in this direction has been made by Ballet et al. [**BBD19**, **BBD21**], giving a strategy to optimize the scalar complexity of the original CCMA. This strategy can be summarized as follows. The algorithm involves two matrices. For each coefficient distinct from zero and one of a matrix, a scalar multiplication is performed. Therefore, in order to reduce the number of scalar multiplications, these matrices must have a maximum number of zeros and ones.

In this paper, we focus on the optimization of the recursive Chudnovsky-type algorithms over the projective line, proposed in [**BBP20**]. First, we rely on the work of [**BBD21**] to extend it to the use of places of arbitrary degrees. Then, a constructive method is developed to optimize the complexity of the algorithms on the projective line. The paper is organized as follows. In Section 2, we introduce Chudnovsky-type algorithms with evaluations at places of arbitrary degrees, and then the recursive construction over the projective line. In Section 3, we deal with the scalar complexity of Chudnovsky-type algorithms when non-rational places are used. Then, we propose a strategy to improve the scalar complexity of recursive Chudnovsky-type algorithms over the projective line. In Section 4, we give several examples. In particular, we illustrate this process on the extension of degree 4 of $\mathbb{F}_4$, and obtain an algorithm of same bilinear complexity and better total complexity than the Baum-Shokrollahi experiment [**BS91**], and its optimizations given in [**BBD21**].

2. Chudnovsky-type algorithms

Let $F/\mathbb{F}_q$ be a function field of genus g over $\mathbb{F}_q$. For $\mathcal{O}$ a valuation ring, the place P is defined to be $P = \mathcal{O} \setminus \mathcal{O}^\times$. We denote by $F_P = \mathcal{O}_P/P$ the residue class field at the place P, that is isomorphic to $\mathbb{F}_{q^d}$, d being the degree of the place. A rational place is a place of degree 1. A divisor $\mathcal{D}$ is a formal sum $\mathcal{D} = \sum_i n_i P_i$, where P_i are places and n_i are relative integers. The support *supp* $\mathcal{D}$ of $\mathcal{D}$ is the set of the places P_j for which $n_j \neq 0$, and $\mathcal{D}$ is effective if all the n_i are positive. The degree of $\mathcal{D}$ is defined by $\deg \mathcal{D} = \sum_i n_i$. The Riemann-Roch space associated to the divisor $\mathcal{D}$ is denoted by $\mathcal{L}(\mathcal{D})$. A divisor $\mathcal{D}$ is said to be non-special if $\dim \mathcal{L}(\mathcal{D}) = \deg(\mathcal{D}) + 1 - g$. Details about algebraic function fields can be found in [**Sti08**].

2.1. CCMA with evaluation at places of arbitrary degrees.

The latest generalization of CCMA is given in [**BCP$^+$21**]. Before introducing the algorithm, let us give a definition of the generalized Hadamard product.

DEFINITION 2.1. Let q be a prime power and $d_1, \ldots, d_N$ be positive integers. The generalized Hadamard product in $\mathbb{F}_{q^{d_1}} \times \cdots \times \mathbb{F}_{q^{d_N}}$, denoted by $\odot$, is given for all $(a_1, \ldots, a_N), (b_1, \ldots, b_N) \in \mathbb{F}_{q^{d_1}} \times \cdots \times \mathbb{F}_{q^{d_N}}$ by

$$(a_1, \ldots, a_N) \odot (b_1, \ldots, b_N) = (a_1 b_1, \ldots, a_N b_N).$$

With this notation, we recall the version of the Chudnovsky-Chudnovsky algorithm useful for our study, namely the one allowing evaluations at places of arbitrary degrees (see [**BCP$^+$21**, Corollary 5.4]).

THEOREM 2.2 (CCMA at places of arbitrary degrees). *Let*

- n *be a positive integer,*
- $F/\mathbb{F}_q$ *be an algebraic function field of genus* g*,*
- Q *be a degree* n *place of* $F/\mathbb{F}_q$*,*
- $\mathcal{D}$ *be a divisor of* $F/\mathbb{F}_q$*,*
- $\mathcal{P} = \{P_1, \ldots, P_N\}$ *be an ordered set of places of arbitrary degrees of* $F/\mathbb{F}_q$*.*

We suppose that supp $\mathcal{D} \cap \{Q, P_1, \ldots, P_N\} = \emptyset$ *and that*

(i) *the evaluation map*

$$\begin{array}{rccc} Ev_Q: & \mathcal{L}(\mathcal{D}) & \rightarrow & F_Q \\ & f & \mapsto & f(Q) \end{array}$$

is surjective,

(ii) *the evaluation map*

$$\begin{array}{rccc} Ev_{\mathcal{P}}: & \mathcal{L}(2\mathcal{D}) & \rightarrow & \mathbb{F}_{q^{\deg P_1}} \times \cdots \times \mathbb{F}_{q^{\deg P_N}} \\ & f & \mapsto & (f(P_1), \ldots, f(P_N)) \end{array}$$

is injective.

Then,

(1) *we have a multiplication algorithm $\mathcal{U}_{q,n}^{F,\mathcal{P}}(\mathcal{D}, Q)$ such that for any two elements x, y in $\mathbb{F}_{q^n}$:*

$$xy = E_Q \circ {Ev_{\mathcal{P}}|_{ImEv_{\mathcal{P}}}}^{-1} \left(E_{\mathcal{P}} \circ Ev_Q^{-1}(x) \underline{\odot} E_{\mathcal{P}} \circ Ev_Q^{-1}(y) \right), \tag{2}$$

where E_Q denotes the canonical projection from the valuation ring $\mathcal{O}_Q$ of the place Q in its residue class field F_Q, $E_{\mathcal{P}}$ the extension of $Ev_{\mathcal{P}}$ on the valuation ring $\mathcal{O}_Q$ of the place Q, ${Ev_{\mathcal{P}}|_{ImEv_{\mathcal{P}}}}^{-1}$ the restriction of the inverse map of $Ev_{\mathcal{P}}$ on its image, $\underline{\odot}$ the generalized Hadamard product and $\circ$ the standard composition map;

(2) *the algorithm $\mathcal{U}_{q,n}^{F,\mathcal{P}}(\mathcal{D}, Q)$ defined by (2) has bilinear complexity*

$$\mu_b(\mathcal{U}_{q,n}^{F,\mathcal{P}}(\mathcal{D}, Q)) = \sum_{i=1}^{N} \mu_b(\mathcal{U}_{q,\deg P_i}(P_i)),$$

where $\mathcal{U}_{q,\deg P_i}(P_i)$ is the algorithm used to multiply the evaluations at P_i, in $\mathbb{F}_{q^{\deg P_i}}$.

Sufficient application conditions are given in the following.

Proposition 2.3 (Criteria for CCMA at places of arbitrary degrees). *Let q be a prime power and let $n > 1$ be an integer. If there exists an algebraic function field $F/\mathbb{F}_q$ of genus g with a set of places $\mathcal{P} = \{P_1, \ldots, P_N\}$ and an effective divisor $\mathcal{D}$ of degree $n + g - 1$ such that*

1) *there exists a place Q of degree n (which is always the case if $2g + 1 \leq q^{\frac{n-1}{2}}(q^{\frac{1}{2}} - 1)$),*
2) *$Supp\ \mathcal{D} \cap (\mathcal{P} \cup Q) = \emptyset$, and $D - Q$ is non-special,*
3) *$\sum_{i=1}^{N} \deg P_i = 2n + g - 1$ and $2\mathcal{D} - \sum P_i$ is non-special,*

then,

(i) *the evaluation map*

$$\begin{array}{rccc} Ev_Q: & \mathcal{L}(\mathcal{D}) & \rightarrow & F_Q \\ & f & \mapsto & f(Q) \end{array}$$

is an isomorphism of vector spaces over $\mathbb{F}_q$,

(ii) *and the evaluation map*

$$\begin{array}{rccc} Ev_{\mathcal{P}}: & \mathcal{L}(2\mathcal{D}) & \rightarrow & \mathbb{F}_{q^{\deg P_1}} \times \cdots \times \mathbb{F}_{q^{\deg P_N}} \\ & f & \mapsto & (f(P_1), \ldots, f(P_N)) \end{array}$$

is an isomorphism of vector spaces of dimension $2n + g - 1$ over $\mathbb{F}_q$.

2.2. Recursive Chudnovsky-type algorithm over the projective line. In this paper, we focus on the optimization of recursive Chudnovsky-type algorithms over the projective line, introduced in [**BBP20**]. These algorithms are a specialization of the algorithm from Theorem 2.2 to the rational function field.

DEFINITION 2.4. Let q be a prime power and n be a positive integer. A recursive Chudnovsky-type algorithm $\mathcal{U}_{q,n}^{\mathcal{P}_n}(Q)$ over the projective line is an algorithm $\mathcal{U}_{q,n}^{F,\mathcal{P}}(\mathcal{D}, Q)$ satisfying the assumptions of Theorem 2.2 such that:

- $F/\mathbb{F}_q$ is the rational function field $\mathbb{F}_q(x)$,
- Q is a place of degree n of $\mathbb{F}_q(x)$,
- $\mathcal{D} = (n-1)P_\infty$, where P_∞ is the place at infinity of $\mathbb{F}_q(x)$,
- $\mathcal{P}_n$ is a set of places of degrees lower than n such that
$$\sum_{P \in \mathcal{P}_n} \deg P = 2n - 1,$$
- the multiplication in $F_P \simeq \mathbb{F}_{q^d}$, where $d = \deg P$, is computed by $\mathcal{U}_{q,d}^{\mathcal{P}_d}(P)$, where $P \in \mathcal{P}_n$.

Note that if $P \in \mathcal{P}_n$ is a rational place, the algorithm $\mathcal{U}_{q,1}^{\mathcal{P}_1}(P)$ consists in only a bilinear multiplication in $\mathbb{F}_q$. Such an algorithm verifies the criteria of Proposition 2.3. The bilinear complexity of these algorithms is given by the following.

PROPOSITION 2.5. *Let $\mathcal{U}_{q,n}^{\mathcal{P}_n}(Q)$ be a recursive Chudnovsky-type algorithm over the projective line. Its bilinear complexity is given by*
$$\mu_b(\mathcal{U}_{q,n}^{\mathcal{P}_n}(Q)) = \sum_{P \in \mathcal{P}_n} \mu_b(\mathcal{U}_{q,d}^{\mathcal{P}_d}(P)),$$
where $d = \deg P$.

Note that the evaluation at P_∞ is defined specifically in this context, since P_∞ is in the support of $\mathcal{D}$.

DEFINITION 2.6. Let k be a positive integer and P_∞ be the place at infinity of $\mathbb{F}_q(x)$. Let $\mathcal{D} = kP_\infty$, and let $f = \sum_{i=0}^{k} f_i x^i \in \mathcal{L}(\mathcal{D})$. We define the evaluation at P_∞ to be for all $f \in \mathcal{L}(\mathcal{D})$,
$$f(P_\infty) := f_k.$$

EXAMPLE 2.7. Consider the multiplication in $\mathbb{F}_{4^4}$ over $\mathbb{F}_4$. Let $P_0, P_1, P_\omega, P_{\omega^2}$ and P_∞ be the rational places of $\mathbb{F}_4[x]$. Let P^2 be a place of degree 2, and Q be a place of degree 4. Then, we can construct a recursive Chudnovsky-type algorithm over the projective line with $\mathcal{P}_4 = \{P_0, P_1, P_\omega, P_{\omega^2}, P_\infty, P^2\}$. This algorithm uses the algorithm $\mathcal{U}_{4,2}^{\mathcal{P}_2}(P^2)$, defined with $\mathcal{P}_2 = \{P_0, P_1, P_\infty\}$. The diagram of its construction is given in Table 1. As well as the Baum-Shokrollahi experiment [**BS91**], this algorithm has an optimal bilinear complexity $\mu_b(\mathcal{U}_{4,4}^{\mathcal{P}_4}(Q)) = 8$.

3. Optimization of scalar complexity

3.1. Some general results. First, we discuss how to adapt the work of [**BBD21**] to Chudnovsky-type algorithms using places of arbitrary degrees.

Let $\mathcal{U}_{q,n} = \mathcal{U}_{q,n}^{F,\mathcal{P}}(\mathcal{D}, Q)$ be an algorithm as defined in Theorem 2.2, and verifying the criteria of Proposition 2.3, for the multiplication in $\mathbb{F}_{q^n}$. Following [**BBD21**], we consider that the basis of $\mathbb{F}_{q^{\deg P_1}} \times \cdots \times \mathbb{F}_{q^{\deg P_N}}$ is always the canonical basis.

TABLE 1. Diagram of the construction of $\mathcal{U}_{4,4}^{\mathcal{P}}(Q)$.

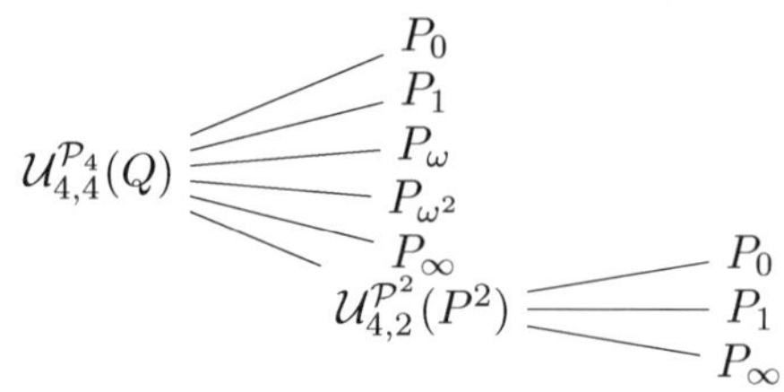

The basis $\mathcal{B}_Q$ of $F_Q = \mathbb{F}_{q^n}$ is defined by $\mathcal{B}_Q = Ev_Q(\mathcal{B}_\mathcal{D})$, where $\mathcal{B}_\mathcal{D}$ is the basis of $\mathcal{L}(\mathcal{D})$. We also denote by $\mathcal{B}_{2\mathcal{D}}$ the basis of the Riemann-Roch space $\mathcal{L}(2\mathcal{D})$. Since we take $\mathcal{D}$ as an effective divisor, we have that $\mathcal{L}(\mathcal{D}) \subset \mathcal{L}(2\mathcal{D})$, and we take $\mathcal{B}_{2\mathcal{D}} = \mathcal{B}_\mathcal{D} \cup \mathcal{B}_\mathcal{D}^c$, where $\mathcal{B}_\mathcal{D}^c$ is a basis of the supplementary space of $\mathcal{L}(\mathcal{D})$ in $\mathcal{L}(2\mathcal{D})$. Let T_D (resp. $T_{2\mathcal{D}}$) be the matrix of $E_\mathcal{P} : \mathcal{L}(\mathcal{D}) \longrightarrow \mathbb{F}_{q^{\deg P_1}} \times \cdots \times \mathbb{F}_{q^{\deg P_N}}$ in the basis $\mathcal{B}_\mathcal{D}$ (resp. $Ev_\mathcal{P} : \mathcal{L}(2\mathcal{D}) \longrightarrow \mathbb{F}_{q^{\deg P_1}} \times \cdots \times \mathbb{F}_{q^{\deg P_N}}$, in the basis $\mathcal{B}_{2\mathcal{D}}$). Let C be the matrix of the map E_Q from the Riemann-Roch Space $\mathcal{L}(2\mathcal{D})$, in the basis $\mathcal{B}_{2\mathcal{D}}$, to the finite field $\mathbb{F}_{q^n}$, in the basis $\mathcal{B}_Q$ over $\mathbb{F}_q$. Using these matrices, Algorithm (2) is written

$$XY = CT_{2\mathcal{D}}^{-1}(T_\mathcal{D}(X) \underline{\odot} T_\mathcal{D}(Y)), \tag{3}$$

where X and Y are the two elements of $\mathbb{F}_{4^4}$ in the basis $\mathcal{B}_Q$ being multiplied, and $\underline{\odot}$ is the generalized Hadamard product. In the following, we consider the product of C and $T_{2\mathcal{D}}^{-1}$ as one matrix $CT_{2\mathcal{D}}^{-1}$.

Recall that the scalar complexity of the algorithm $\mathcal{U}_{q,n}$ is defined as its number of multiplications by a non-trivial constant (distinct from 0 or 1) in $\mathbb{F}_q$. In $\mathcal{U}_{q,n}$, the matrices $T_\mathcal{D}$ and $CT_{2\mathcal{D}}^{-1}$ provide some scalar multiplications of the algorithm. We therefore wish to obtain matrices with as many coefficients equal to zero or one as possible, to have a maximum number of trivial multiplications that do not count in the scalar complexity. Consequently, we focus on the number of zeros and ones in these matrices. We denote by $N_z(T_D)$ (resp. $N_z(CT_{2\mathcal{D}}^{-1})$) the number of zeros in $T_\mathcal{D}$ (resp. $CT_{2\mathcal{D}}^{-1}$)) and also denote by $N_1(T_\mathcal{D})$ (resp. $N_1(CT_{2D}^{-1})$) the number of ones in the matrices $T_\mathcal{D}$ (resp. $CT_{2\mathcal{D}}^{-1}$). Note that it is useful to distinguish between zeros and ones, since the coefficients equal to one must be counted in the number of additions used by the algorithm. Since the matrix $T_\mathcal{D}$ is used twice in the algorithm, and $CT_{2\mathcal{D}}^{-1}$ only once, we denote the total number of zeros by $N_z = 2N_z(T_D) + N_z(CT_{2\mathcal{D}}^{-1})$ and the total number of ones by $N_1 = 2N_1(T_D) + N_1(CT_{2\mathcal{D}}^{-1})$ If the algorithm $\mathcal{U}_{q,n}$ is an original CCMA, the evaluations are only at rational places and all the scalar multiplications are given by the matrices $T_\mathcal{D}$ and $CT_{2\mathcal{D}}^{-1}$. Hence, the scalar complexity is given by [**BBD19, BBD21**]:

$$\mu_s(\mathcal{U}_{q,n}) = 3n(n+g-1) - N_z - N_1. \tag{4}$$

In our study, we allow the evaluations to be at places of arbitrary degrees. Consequently, we have to count the scalar multiplications involved in an algorithm $\mathcal{U}_{q,d}(P)$ (not necessarily of type Chudnovsky), where $d = \deg P$, for the multiplication in the residue class field $F_P \simeq \mathbb{F}_{q^d}$, required to multiply the evaluations at P.

PROPOSITION 3.1. *Let $\mathcal{U}_{q,n}$ be a Chudnovsky-type algorithm from Theorem 2.2, with evaluations at places of arbitrary degrees. Then, its scalar complexity is such that*

$$\mu_s(\mathcal{U}_{q,n}) = 3n(n+g-1) - N_z - N_1 + \sum_{P\in\mathcal{P}} \mu_s(\mathcal{U}_{q,d}(P)), \tag{5}$$

where $d=\deg P$ *and* $\mathcal{U}_{q,d}(P)$ *is the algorithm used to multiply the evaluations at* P.

PROOF. Follows from (4), where we add the scalar complexity of the multiplications in $\mathbb{F}_{q^{\deg P}}$ for all places P used by the algorithm. □

Since we have to add the scalar complexity of the algorithms $\mathcal{U}_{q,d}(P)$, for all $P \in \mathcal{P}$ and $d = \deg P$, the use of non rational places looks heavier for the scalar complexity. However, this is not necessarily the case. First, because a function field of lower genus can be used, which implies the use of smaller matrices, and also because the matrix $T_{\mathcal{D}}$ might contain more zeros.

PROPOSITION 3.2. *Let $\mathcal{U}_{q,n}$ be a Chudnovsky-type algorithm as defined in Theorem 2.2, satisfying Proposition 2.3. Consider $\mathcal{P}' \subset \mathcal{P}$ constructed by taking places that are in $\mathcal{P}$ by growing degrees as long as the sum of their degrees remains lower than or equal to $n+g-1$. Then, the number of zeros of $T_{\mathcal{D}}$ verifies*

$$N_z(T_{\mathcal{D}}) \leq n\big(n+g-1+\sum_{P\in\mathcal{P}\setminus\mathcal{P}'} (\deg P - 1)\big).$$

PROOF. By Proposition 2.3, the divisor $\mathcal{D}$ is taken effective and of degree $n+g-1$. Thus, a function f in $\mathcal{L}(\mathcal{D})$ has at most $\deg \mathcal{D} = n+g-1$ zeros. Consider $f \in \mathcal{B}_{\mathcal{D}}$, then a column of $T_{\mathcal{D}}$ is given by the evaluations of f at the places in $\mathcal{P}$. Moreover, the evaluation at a place of degree d gives d coefficients in $\mathbb{F}_q$, and such an evaluation can give $d-1$ zeros without vanishing. As the ratio $(d-1)/d$ is increasing, the column defined by the evaluations of f would have the largest number of zeros if f has (at most) $n+g-1$ zeros at the places of smallest possible degree, i.e. at the places in $\mathcal{P}'$, and $\deg P - 1$ coefficients equal to zero for each other place. Then, a column of $T_{\mathcal{D}}$ is given by the evaluations of a function in $\mathcal{L}(\mathcal{D})$ and has at most $n+g-1+\sum_{P\in\mathcal{P}\setminus\mathcal{P}'}(\deg P - 1)$ coefficients equal to zeros. The bound is obtained by counting this maximal number of zeros for all the n columns of $T_{\mathcal{D}}$. □

Another important result for the strategy of optimization is that for given $\mathcal{P}$, $\mathcal{D}$ and Q, since the basis $\mathcal{B}_{\mathcal{D}}$ of $\mathcal{L}(\mathcal{D})$ is fixed, and that the basis $\mathcal{B}_{2\mathcal{D}} = \mathcal{B}_{\mathcal{D}} \cup \mathcal{B}^c_{\mathcal{D}}$ of $\mathcal{L}(2\mathcal{D})$ is an extension of the basis of $\mathcal{L}(D)$, the algorithm does not depend on the choice of $\mathcal{B}^c_{\mathcal{D}}$. This result is established in [**BBD19**, Proposition 1], and is also true when places of higher degrees are used (same proof holds).

3.2. Optimization of the complexity of a recursive Chudnovsky-type algorithm over the projective line. Consider a recursive Chudnovsky-type algorithm over the projective line $\mathcal{U}^{\mathcal{P}_n}_{q,n}(Q)$. The bilinear complexity of such an algorithm is known by Proposition 2.5. In this section, we introduce our strategy to optimize its total complexity.

PROPOSITION 3.3. *Let $\mathcal{U}_{q,n}^{\mathcal{P}_n}(Q)$ be a recursive Chudnovsky-type algorithm over the projective line. Then, the scalar complexity of $\mathcal{U}_{q,n}^{\mathcal{P}_n}(Q)$ is given by*

$$\mu_s(\mathcal{U}_{q,n}^{\mathcal{P}_n}(Q)) = 3n(2n-1) - N_z - N_1 + \sum_{P\in\mathcal{P}_n} \mu_s(\mathcal{U}_{q,d}^{\mathcal{P}_d}(P)),$$

where $d = \deg P$.

PROOF. Follows immediately from Proposition 3.1, where the Chudnovsky-type algorithm $\mathcal{U}_{q,d}^{\mathcal{P}_d}(P)$ over the projective line is used to multiply in $F_P \simeq \mathbb{F}_{q^d}$. □

By the results obtained in [**BBD19, BBD21**] and the previous section, the optimization of the scalar complexity of $\mathcal{U}_{q,n}^{\mathcal{P}_n}(Q)$ only depends on the basis of $\mathcal{L}(D)$, when $\mathcal{P}_n$, $\mathcal{D}$ and the place Q of degree n are fixed. Hence, the main focus is to find a basis $\mathcal{B}_\mathcal{D}$ of $\mathcal{L}(\mathcal{D})$ such that the matrices $T_\mathcal{D}$ and $CT_{2\mathcal{D}}^{-1}$ have the most possible zeros and ones. The basis of $\mathbb{F}_{q^n}$ will be defined in accordance with this basis.

The strategy proposed in [**BBD19**] and significantly completed in [**BBD21**] is to construct a first basis $\mathcal{B}_\mathcal{D}$, and apply the linear group $GL_n(\mathbb{F}_q)$ to look for the best possible bases. It is effective, but expensive. With a recursive algorithm over the projective line, one can construct directly some bases of $\mathcal{L}(\mathcal{D})$ that improve the total complexity of the algorithm without using the action of the linear group.

3.2.1. *Optimization strategy.* Now, let us introduce the heart of the strategy of optimization of Chudnovsky-type algorithms over the projective line. We want to sculpt $\mathcal{B}_\mathcal{D}$ to obtain the minimum scalar multiplications and additions. For this purpose, we want to get as many zeros as possible before processing the number of ones. More precisely, we will focus on obtaining the most possible zeros and then ones in the matrix $T_\mathcal{D}$. Two reasons for that: we do not have information on how the choice of the basis of $\mathcal{L}(\mathcal{D})$ affects the matrix $CT_{2\mathcal{D}}^{-1}$, and moreover the matrix $T_\mathcal{D}$ counts twice. For all these reasons, our goal is finally to sculpt $\mathcal{B}_\mathcal{D}$ such that $T_\mathcal{D}$ has a maximal number of zeros, and then a maximal number of ones.

Recall that $\mathcal{D} = (n-1)P_\infty$, and hence $\mathcal{L}(\mathcal{D})$ is the space of polynomials over $\mathbb{F}_q$ of degrees at most $n-1$. Moreover, the places of $\mathbb{F}_q(x)$ are given by the irreducible polynomials over $\mathbb{F}_q[x]$ and the place at infinity. The idea is to take the vectors of the basis $\mathcal{B}_\mathcal{D}$ as products of irreducible polynomials associated to places in $\mathcal{P}_n$. Therefore, the evaluation of such a vector will vanish at the places used to define it. This translates into zeros in the matrix $T_\mathcal{D}$. We define such bases as $\mathcal{P}_n-$bases of $\mathcal{L}(\mathcal{D})$.

DEFINITION 3.4 ($\mathcal{P}_n-$basis of $\mathcal{L}(\mathcal{D})$). Let q be a prime power and $n > 1$ be an integer. Let $\mathcal{P}_n = \{P_1, \ldots, P_N\}$ be a set of distinct places of $\mathbb{F}_q(x)$ such that $\sum_{P\in\mathcal{P}_n} \deg P = 2n-1$. If P_j is not the place at infinity, let $P_j(x)$ be the monic irreducible polynomial associated to the place P_j. In the case of $P_j = P_\infty$, let $P_j(x) = 1$. We say that $\mathcal{B} = \{V_1, \ldots, V_n\}$ is a $\mathcal{P}_n-$basis of $\mathcal{L}(\mathcal{D})$ if every vector V_i of the basis $\mathcal{B}$ is defined as $V_i(x) = \prod_{P_j\in\mathcal{V}_i} P_j(x)$, where $\mathcal{V}_i$ is a subset of $\mathcal{P}_n$. The set $\mathcal{V}_i$ is called the support of V_i.

We can notice that since $\mathcal{L}(\mathcal{D})$ is the space of polynomials of degree at most $n-1$, the vectors defining the basis shall be of degree at most $n-1$. The following proposition gives a lower bound for the number of zeros in the matrices $T_\mathcal{D}$ constructed using a $\mathcal{P}_n-$basis of $\mathcal{L}(\mathcal{D})$.

PROPOSITION 3.5. *Let $\mathcal{B} = \{V_1, \dots, V_n\}$ be a $\mathcal{P}_n-$basis of $\mathcal{L}(\mathcal{D})$, and let $T_\mathcal{D}$ be a matrix obtained using $\mathcal{B}$. Then,*

$$N_z(T_\mathcal{D}) \geq \sum_{i=1}^{n} \deg V_i.$$

PROOF. Every column of $T_\mathcal{D}$ is given by the evaluation of a vector V_i at the places in $\mathcal{P}_n$. Every such vector is the product of some $P_{i,j}(x)$, where $P_{i,j} \in \mathcal{V}_i \subset \mathcal{P}_n$ and V_i can be written as the product $V_i(x) = P_{i,1}(x) \dots P_{i,N_i}(x)$, where $\sum_{j=1}^{N_i} \deg P_{i,j} = \deg V_i$. In particular, $V_i(P_{i,j}) = 0$, and it gives $\deg P_{i,j}$ zeros in the $i-$th column of $T_\mathcal{D}$, for all $j = 1, \dots, N_i$. Then, the $i-th$ column of $T_\mathcal{D}$ contains at least $\deg V_i$ zeros. This is the case for all the V_i in $\mathcal{B}_\mathcal{D}$, and then $N_z(T_\mathcal{D}) \geq \sum_{i=1}^{n} \deg V_i$. □

The preferred configuration to maximize the number of zeros is when the vectors of the $\mathcal{P}_n-$basis $\mathcal{B}$ are of degree $n-1$, or as close to $n-1$ as possible.

COROLLARY 3.6. *If $P_\infty \in \mathcal{P}_n$ and for all i, $\deg V_i = n-1$ or $n-2$. Then,*

$$N_z(T_\mathcal{D}) \geq n(n-1).$$

PROOF. If $\deg V_i = n-1$, then V_i has $n-1$ zeros. If $\deg V_i = n-2$, then $V_i(P_\infty) = 0$ since this evaluation is the coefficient in x^{n-1} of V_i (by Definition 2.6), and $n-2$ zeros at the places (distinct from P_∞) defining $\mathcal{V}_i$. Finally, the n vectors of the basis have at least $n-1$ zeros. □

3.2.2. *Generic optimization.* Thus far, we have seen that using $\mathcal{P}_n-$bases gives an information on the number of zeros in the matrix $T_\mathcal{D}$, and then can be used to improve the complexity of Chudnovsky-type algorithms over the projective line. It remains to be proven that such a basis always exists. An efficient way to obtain such a basis is to construct $\mathcal{B} = \{V_1, \dots, V_n\}$ such that for all i, $\deg V_i = i-1$. Secondly, we can construct the matrix $T_\mathcal{D}$ and maximize its number of ones by multiplying the vectors of the basis by a constant in $\mathbb{F}_q$. For each column in $T_\mathcal{D}$, suppose that $a \in \mathbb{F}_q$ is the non-zero scalar that occurs the most in the column. Then, we multiply the corresponding vector of the basis by a^{-1}. The generic construction of a $\mathcal{P}_n-$basis of $\mathcal{L}(\mathcal{D})$ is given in the following algorithm.

Algorithm 1 Construction of a generic $\mathcal{P}_n-$basis of $\mathcal{L}(\mathcal{D})$ and the associated matrix $T_\mathcal{D}$.

INPUT: q, n, $\mathcal{P}_n = \{P_1, \dots, P_N\}$ be a set of places such that $\sum_{P \in \mathcal{P}_n} \deg P = 2n-1$.

OUTPUT: $B_\mathcal{D}$, $T_\mathcal{D}$.

(1) For $i = 1, \dots, n$, construct $V_i(x) = \prod_{P_j \in \mathcal{V}_i} P_j(x)$, such that $\deg V_i = i-1$, and $\mathcal{V}_i$ is a subset of $\mathcal{P}_n$.
(2) Construct $T_\mathcal{D}$. For each column of $T_\mathcal{D}$, if $a \in \mathbb{F}_q$ is the scalar that occurs the most, multiply both the corresponding vector of $\mathcal{B}_\mathcal{D}$ and the associated columns of $T_\mathcal{D}$ by a^{-1}.

The natural strategy to construct Chudnovsky-type algorithms over the projective line is to include in $\mathcal{P}_n$ all places by increasing degrees, until the sum of their degrees is equal to $2n-1$ (if the sum gets bigger than $2n-1$, remove a place of the appropriate degree, see [**BBP20**, Section 4.2]).

PROPOSITION 3.7. *If $\mathcal{P}_n$ is constructed including places by increasing degrees, then Algorithm 1 is correct.*

PROOF. If $\mathcal{P}_n$ is constructed by taking places of increasing degrees, then it contains places of every degrees until some integer k, except in the case of $q = 2$ and the only degree 2 place of $\mathbb{F}_2(x)$ has been removed from $\mathcal{P}_n$. We can assume that P_∞ is always in $\mathcal{P}_n$. Note that the polynomial associated to P_∞ is defined by $P_\infty(x) = 1$ (see Definition 3.4). Suppose that $\mathcal{P}_n$ contains places of every degree until some integer k. Then, there exists $P_j(x)$ of degree j for $j = 1, \ldots, k$. Set $V_1(x) = 1$. Then, for $i = 2$ to $n-1$, we construct the polynomials $V_i(x)$ by taking the product of all monic polynomials by increasing degrees until the degree is equal to $i-1$ (if the degree gets greater than $i-1$, divide this product by an appropriate monic irreducible polynomial, that is in the support of V_i at this moment of the construction). If $q = 2$ and $\mathcal{P}_n$ does not contain the place of degree 2, we can divide V_i by the two irreducible polynomials of degree 1. Moreover, the sum of all $P_i(x)$ is of degree $2n-2 \geq n$, and then we can construct $V_i(x)$ of degree d for all $d = 0, \ldots, n-1$. Finally, since there exists some products of the $P_i(x)$ for any degree d, one can obtain n vectors $V_1(x), \ldots, V_n(x)$ of degrees $0, \ldots, n-1$ respectively, such that for all j the function $V_j(x)$ is the product of some distinct $P_i(x)$. Then, $\mathcal{B} = \{V_1(x), \ldots, V_n(x)\}$ is a $\mathcal{P}_n$−basis of $\mathcal{L}(\mathcal{D})$. □

Moreover, Algorithm 1 is ending in polynomial time.

PROPOSITION 3.8. *Algorithm 1 is running in time $\mathcal{O}(n^3 \log n \log\log n)$.*

PROOF. Step 1. For the n vectors, we take at most n products of $P_i(x)$. We roughly consider that we have at most n^2 products of polynomials whose degrees are bounded by n. Each product can be computed with $\mathcal{O}(n \log n \log\log n)$ operations by Schönhage-Strassen ([**SS71**], [**vzGG03**, Theorem 8.23.]), thus this step can be completed in time $\mathcal{O}(n^3 \log n \log\log n)$.

Step 2. The matrix $T_\mathcal{D}$ is constructible using $\mathcal{O}(n^3)$ operations in the base field. Indeed, the coefficients of the matrix are obtained by computing the evaluations of the polynomials in the basis of $\mathcal{L}(\mathcal{D})$ at the places in $\mathcal{P}_n$. The evaluation of such a polynomial $V(x)$, that is of degree at most $n-1$, at a place P of degree $d < n$ is obtained by the modular reduction of $V(x)$ modulo $P(x)$. More precisely, let $v(x) = V(x) \pmod{P(x)} = \sum_{i=0}^{d-1} v_i x^i$. The coefficients of $\{v_0, \ldots, v_{d-1}\}$ are exactly the evaluation of $V(x)$ at P in the basis $\{1, \alpha, \ldots, \alpha^{d-1}\}$, for α a root of $P(x)$. Such a computation gives d coefficients of the matrix, and can be computed using the Euclidean Algorithm for polynomials, that uses $\mathcal{O}(nd)$ operations by [**vzGG03**, Theorem 3.11.]. The matrix $T_\mathcal{D}$ having $\mathcal{O}(n^2)$ coefficients, it means that all of them can be computed using $\mathcal{O}(n^3)$ operations. Then, for the n columns, we count each occurrence of non-zero scalars in the $2n-1$ coefficients. Then, we have to multiply the n vectors of the basis, and the n columns of $T_\mathcal{D}$ by a scalar. This last part is computed in $\mathcal{O}(n^2)$. Finally, this step uses $\mathcal{O}(n^3)$ operations in $\mathbb{F}_q$. □

REMARK 3.9. In [**HvdH19**], it is proven that if there exists a Linnik constant with $L < 1 + 2^{-1162}$, the product of two degree n polynomials over $\mathbb{F}_q$ can be computed in $\mathcal{O}(n \log q \log(n \log q))$, uniformly in q. Considering that q is fixed, the running time of Algorithm 1 becomes $\mathcal{O}(n^3 \log n)$.

EXAMPLE 3.10. Table 2 gives an illustration of the improvement of the complexity of Chudnovsky-type algorithms over the projective line for small extensions

TABLE 2. Total complexity of Chudnovsky-type algorithms over $\mathbb{F}_2(x)$

n	2	3	4	5	6	$\cdots$	54
Non optimized	7	23	54	91	129		10152
Generic optimization	7	20	49	77	99		8703

of $\mathbb{F}_2$, and for the extension of degree 54. The non optimized algorithm uses the canonical basis $\{1, x, \ldots, x^{n-1}\}$ of $\mathcal{L}(\mathcal{D})$, while the generic optimization uses the basis provided by Algorithm 1. Details are given in Section 4.3.

3.2.3. *Non-generic optimization.* Even though we obtained a first improvement of the total complexity, this generic process does not provide the $\mathcal{P}_n$−basis of $\mathcal{L}(\mathcal{D})$ giving the best total complexity. For instance, we should take the vectors in the basis of the highest possible degree, since it ensures more zeros in the matrix. If the degree of the extension is low, we can check all possible $\mathcal{P}_n$−bases of $\mathcal{L}(\mathcal{D})$. Nevertheless, there are less than $2^{\#\mathcal{P}_n}$ possible vectors for the basis, and hence less than $\binom{2^{\#\mathcal{P}_n}}{n}$ possible $\mathcal{P}_n$−bases to try.

REMARK 3.11. Even if this complete optimization is too heavy to be used generically for large extension degree, it is still way more efficient than the optimization of [**BBD19, BBD21**] involving the action of the linear group. Considering the optimization in the extension of degree 13 of $\mathbb{F}_{16}$, the linear group is of cardinality $\simeq 10^{203}$, while our strategy would consist in looking through $\lessapprox 10^{72}$ possible $\mathcal{P}_n$−bases.

Depending on the time and resources that can be used, we can still improve the complexity of the algorithm. Instead of looking through all possible $\mathcal{P}_n$−bases, we can focus on the bases including only vectors of degree $n-1$ or $n-2$. Between these vectors, one can only look through the ones whose evaluations give a maximum number of zeros. An example of such optimization is given in Section 4.2.

This ends our strategy to optimize a Chudnovsky-type algorithm $\mathcal{U}_{q,n}^{\mathcal{P}_n}(Q)$ over the projective line, when the parameters $\mathcal{P}_n$ and Q are fixed. Finally, one can look for the best parameters, i.e. include in $\mathcal{P}_n$ in priority the places P such that the multiplication in F_P with $\mathcal{U}_{q,\deg P}^{\mathcal{P}_{\deg P}}(P)$ has the lowest complexity, or similarly for the place Q of degree n. A full optimization process is given in Section 4.1.

4. Examples

In this section, we provide several examples of optimizations of Chudnovsky-type algorithms over the projective line. All the computations were done using Magma Computational Algebra System [**BCP97**].

4.1. Multiplication in $\mathbb{F}_{256}$ over $\mathbb{F}_4$. We now illustrate the strategy introduced in the previous section to the multiplication in the extension of degree 4 of $\mathbb{F}_4$. The construction of a recursive Chudnovsky-type algorithm over the projective line to multiply in this extension has already been given in Example 2.7. More precisely, consider $\mathbb{F}_4 = \frac{\mathbb{F}_2[x]}{(x^2+x+1)}$. Hence, the elements of $\mathbb{F}_4$ are $\{0, 1, \omega, \omega^2\}$, where ω is a root of $x^2 + x + 1$. Let $\mathbb{F}_4(x)$, be the rational function field over $\mathbb{F}_4$. This function field has 5 rational places, that we denote by $P_0, P_1, P_\omega, P_{\omega^2}$ and

P_∞. These places are given by the irreducible polynomials $x, x+1, x+\omega, x+\omega^2$ and the place at infinity respectively. There exist 6 places of degree 2, which we denote by $P_1^2, \ldots, P_6^2$, and 60 places of degree 4. As in Example 2.7, we take $\mathcal{P}_4 = \{P_0, P_1, P_\omega, P_{\omega^2}, P_\infty, P^2\}$, where P^2 is one of the six places of degree 2, to obtain an algorithm of optimal bilinear complexity. Consequently, our algorithm requires to use the algorithm $\mathcal{U}_{4,2}^{\mathcal{P}_2}(P^2)$. We first focus on optimizing this algorithm, in order to take P^2 such that the complexity of $\mathcal{U}_{4,2}^{\mathcal{P}_2}(P^2)$ is minimal.

4.1.1. *Optimization of* $\mathcal{U}_{4,2}^{\mathcal{P}_2}(P^2)$. As seen in Example 2.7, the Chudnovsky-type algorithm over the projective line for the multiplication in the quadratic extension of $\mathbb{F}_4$ is defined using the ordered set $\mathcal{P}_2 = \{P_0, P_1, P_\infty\}$. Actually, the canonical basis $\{f_1, f_2\} = \{1, x\}$ of $\mathcal{L}(\mathcal{D})$ is already optimal. In fact, the matrix $T_\mathcal{D}$ is then given by

$$T_\mathcal{D} = \begin{pmatrix} f_1(P_0) & f_2(P_0) \\ f_1(P_1) & f_2(P_1) \\ f_1(P_\infty) & f_2(P_\infty) \end{pmatrix} = \begin{pmatrix} 1 & 0 \\ 1 & 1 \\ 0 & 1 \end{pmatrix}.$$

It has a maximal number of zero with respect to Proposition 3.2, and all its non-zero coefficients are equal to one. Hence, this matrix is optimal in terms of scalar complexity. Thus we do not need to search for a better basis of $\mathcal{L}(\mathcal{D})$. It remains to find for which places of degree 2 we obtain the more competitive algorithms. Hence, we compute $CT_{2\mathcal{D}}^{-1}$ for all the 6 possible places of degree 2 of $\mathbb{F}_4(x)$. We obtain

- $N_z(CT_{2\mathcal{D}}^{-1}) = 2$ with $P_1^2 = (x^2+x+\omega)$ and $P_2^2 = (x^2+x+\omega^2)$,
- $N_z(CT_{2\mathcal{D}}^{-1}) = 1$ with all other places.

Moreover, using P_1^2 or P_2^1, we have $N_1(CT_{2\mathcal{D}}^{-1}) = 3$. Hence, we can pick either P_1^2 or P_2^2 as the place of degree 2 in $\mathcal{P}_4$. By Proposition 3.3, we obtain

$$\mu_s(\mathcal{U}_{4,2}^{\mathcal{P}_2}(P_1^2)) = 1,$$

and the number of additions is given by

$$a(\mathcal{U}_{4,2}^{\mathcal{P}_2}(P_1^2)) = 4.$$

4.1.2. *Optimization of* $\mathcal{U}_{4,4}^{\mathcal{P}_4}(Q)$. Recall that $\mathcal{P}_4 = \{P_0, P_1, P_\omega, P_{\omega^2}, P_\infty, P^2\}$. By the previous section, one shall pick $P^2 = P_1^2 = (x^2+x+\omega)$ or $P_2^2 = (x^2+x+\omega^2)$. In the following, we choose $P^2 = P_1^2$.

We want to construct a good basis $\mathcal{B}_\mathcal{D}$ of $\mathcal{L}(\mathcal{D})$, with $\mathcal{D} = 3P_\infty$. Hence the Riemann-Roch space $\mathcal{L}(\mathcal{D})$ is the space of polynomials of degrees at most 3 over $\mathbb{F}_4$. Note that this time P_∞ is in $\mathcal{P}_4$. By Definition 2.6, a function f in $\mathcal{L}(\mathcal{D})$ has a zero at P_∞ if and only if f is a polynomial of degree at most 2. For all other places P in $\mathcal{P}_4$, let $P(x)$ be the corresponding polynomial. Then, a function f in $\mathcal{L}(\mathcal{D})$ has a zero at P if and only if $P(x) \mid f$. By Corollary 3.6, we shall construct the vectors of the basis as polynomials of degrees $n-1$ or $n-2$, which are the product of the polynomials defining the places in $\mathcal{P}_4$. Moreover, we want this vectors to vanish on rational places. Hence, we construct possible vectors for the basis of $\mathcal{L}(\mathcal{D})$ as

- the product of two irreducible polynomials of degree one $P_i(x)P_j(x)$, for $i, j \in \{0, 1, \omega, \omega^2\}$, then this vector has zeros at the places P_i, P_j and P_∞,
- the product of three irreducible polynomials of degree one $P_i(x)P_j(x)P_k(x)$, for $i, j, k \in \{0, 1, \omega, \omega^2\}$ then this vector has zeros at the places P_i, P_j and P_k.

Consequently, we can take the vectors of the basis $\mathcal{B}_{\mathcal{D}}$ as the product of three distinct elements of $\{1, x, x+1, x+\omega, x+\omega^2\}$ until a basis is found. This gives $\binom{5}{3} = 10$ possible vectors for the basis of $\mathcal{L}(\mathcal{D})$. Then, there are $\binom{10}{4} = 210$ possible combinations of these vectors to build the basis. Moreover, we want the vectors in each combination to be relatively prime in terms of polynomials, so that they do not all vanish at the same place. By computation, There are 150 possibilities left. We now consider that for given parameters, we only have to look through 150 possibilities to construct a $\mathcal{P}_4$−basis of $\mathcal{L}(\mathcal{D})$. In the strategy of [**BBD19**] and [**BBD21**], it was required to look through $\mid GL_4(\mathbb{F}_4) \mid = 2961100800$ possibilities. By Corollary 3.6, there are at least 12 zeros in the matrices $T_{\mathcal{D}}$ obtained using these bases. Nevertheless there can be at most 16 zeros by Proposition 3.2. We obtained exactly one basis of $\mathcal{L}(\mathcal{D})$ such that $N_z(T_{\mathcal{D}}) = 16$. This basis is given by

$$\mathcal{B}_{\mathcal{D}} = \{(x+\omega)(x+1), x(x+1), (x+\omega^2)(x+\omega), x(x+\omega^2)(x+\omega)\}.$$

The corresponding evaluation matrix $T_{\mathcal{D}}$ is then

$$T_{\mathcal{D}} = \begin{pmatrix} \omega & 0 & 1 & 0 \\ 0 & 1 & 0 & 0 \\ \omega & 1 & 0 & 0 \\ 0 & 0 & 1 & 1 \\ 0 & \omega & \omega^2 & 0 \\ \omega & 0 & 0 & \omega^2 \\ 0 & 0 & 0 & 1 \end{pmatrix}$$

where the rows are given by the evaluations at the places in $\mathcal{P}_4$, with the following order: $P_0, P_\omega, P_{\omega^2}, P_1, P_1^2$ and P_∞. Notice that the evaluation at P_1^2 takes two rows, in the basis $\{1, \alpha\}$, where α is a root of $P_1^2(x)$. Following Step 2 of Algorithm 1, one shall try to increase the number of ones in this matrix. In particular, the first column only contains 0 and ω. Hence, we modify the basis by multiplying the first vector by $\omega^{-1} = \omega^2$, we obtain

$$\mathcal{B}_{\mathcal{D}} = \{\omega^2(x+\omega)(x+1), x(x+1), (x+\omega^2)(x+\omega), x(x+\omega^2)(x+\omega)\}$$

and

$$T_{\mathcal{D}} = \begin{pmatrix} 1 & 0 & 1 & 0 \\ 0 & 1 & 0 & 0 \\ 1 & 1 & 0 & 0 \\ 0 & 0 & 1 & 1 \\ 0 & \omega & \omega^2 & 0 \\ 1 & 0 & 0 & \omega^2 \\ 0 & 0 & 0 & 1 \end{pmatrix}.$$

The last step is now to find a place Q that gives the best scalar complexity. Finally, we compute the matrices $CT_{2\mathcal{D}}^{-1}$ using the basis of $\mathcal{L}(2\mathcal{D})$ given by $\mathcal{B}_{2\mathcal{D}} = \mathcal{B}_{\mathcal{D}} \cup \{x^4, x^5, x^6\}$ for all the 60 places Q of degree 4 in $\mathbb{F}_4(x)$. There are 3 places such that $CT_{2\mathcal{D}}^{-1}$ has a maximum number of zeros $N_z(CT_{2\mathcal{D}}^{-1}) = 12$. Between those matrices, two have 4 coefficients equal to one, and the last one, that is defined using $Q = (x^4 + \omega x^2 + \omega x + \omega^2)$, has 6 coefficients equal to one. The matrix is in this latest case given by

$$CT_{2\mathcal{D}}^{-1} = \begin{pmatrix} \omega^2 & 0 & \omega & 0 & 1 & 0 & 1 \\ 1 & 0 & 0 & \omega^2 & \omega^2 & 1 & \omega^2 \\ 0 & \omega^2 & 0 & 0 & \omega & 0 & 1 \\ 0 & \omega & \omega & 0 & 0 & \omega & 1 \end{pmatrix}.$$

Finally, the algorithm $\mathcal{U}_{4,4}^{\mathcal{P}_4}(Q)$ is obtained with these parameters. The finite field $\mathbb{F}_{4^4}$ is represented as $\mathbb{F}_4[x]/(Q(x)) = \mathbb{F}_4[\beta]$, for β a root of $Q(x)$. Its basis over $\mathbb{F}_4$ is given by $\mathcal{B}_Q = Ev_Q(\mathcal{B}_\mathcal{D})$ and hence by

$$\mathcal{B}_Q = \{\beta^{43}, \beta^{198}, \beta^{108}, \beta^{109}\}.$$

4.1.3. *Comparison with other algorithms.* The matrices obtained contain $N_z(T_\mathcal{D}) = 16$ and $N_z(CT_{2\mathcal{D}}^{-1}) = 12$ zeros, and $N_1(T_\mathcal{D}) = 9$ and $N_1(CT_{2\mathcal{D}}^{-1}) = 6$ ones. Finally, we can compute the scalar complexity of $\mathcal{U}_{4,4}^{\mathcal{P}_4}(Q)$, including the scalar complexity of $\mathcal{U}_{4,2}^{\mathcal{P}_2}(P_1^2)$. By Proposition 3.3, we obtain,

$$\mu_s(\mathcal{U}_{4,4}^{\mathcal{P}_4}(Q)) = 17,$$

and

$$a(\mathcal{U}_{4,4}^{\mathcal{P}_4}(Q)) = 4 + 2 \times 5 + 12 = 26.$$

Originally, the Baum-Shokrollahi experiment [**BS91**] introduced an algorithm for the extension of degree 4 of $\mathbb{F}_4$ with optimal bilinear complexity. This algorithm is an original CCMA over the function field defined by the Fermat curve $u^3+v^3 = 1$. It also uses 51 scalar multiplications and 52 additions. In [**BBD19, BBD21**], the same algorithm is optimized (BS Optimized) with a good choice of the basis of $\mathbb{F}_{4^4}$ to obtain only 19 scalar multiplications and 43 additions. In this paper, the proposed algorithm is constructed over the rational function field, and only requires 17 scalar multiplications and 26 additions, for the same bilinear complexity.

At last, we want to compare our algorithm to well-known methods of polynomial interpolation. The generalized Karatsuba algorithm computes the product of two 4-terms polynomials using 9 (bilinear) multiplications and 24 additions (see [**WP06**, Appendix]). Once this product is computed, the modulo $Q(x)$ reduction still needs to be performed. For the comparison, we define $\mathbb{F}_{4^4}$ as in our construction, using $Q(x) = x^4 + \omega x^2 + \omega x + \omega^2$. The reduction then uses 9 additions and 8 scalar multiplications. The comparison between these methods is given in Table 3. We can see that the total complexity of our algorithm is equal to Karatsuba's to the nearest 1.

REMARK 4.1. Other experiments have similar performances, for example for the degree 3 extension of $\mathbb{F}_2$, regardless of the polynomial used to define the extension (see Table 6 and Table 7).

TABLE 3. Comparison of algorithms for the multiplication in $\mathbb{F}_{4^4}$

Algorithm	$\mu_b(\mathcal{U})$	$\mu_s(\mathcal{U})$	$a(\mathcal{U})$	$\mu(\mathcal{U})$
Baum-Shokrollahi [**BS91**]	8	51	52	111
BS Optimized [**BBD21**]	8	19	43	70
Our construction	8	17	26	51
Karatsuba [**WP06**] + Reduction	9	8	33	50

4.2. The degree 13 extension of $\mathbb{F}_{16}$. Let the finite field $\mathbb{F}_{16}$ be defined as $\mathbb{F}_2(\omega)$, where ω is a root of x^4+x+1. In [**Bal02**], Ballet constructed a Chudnovsky-Chudnovsky Multiplication Algorithm with quasi-optimal bilinear complexity for the multiplication in the extension of degree $n = 13$ of $\mathbb{F}_{16}$. This algorithm is defined using the hyperelliptic curve given by the plane equation $y^2 + y = x^5$ of genus 2 that has 33 rational points. The algorithm uses 27 bilinear multiplications, that is still the best known bilinear complexity for the multiplication in this extension. The calculation of the number of operations of such an algorithm in Magma gives 833 scalar multiplications and 840 additions.

We can define a Chudnovsky-type algorithm over the projective line for the multiplication in $\mathbb{F}_{16^{13}}$ over $\mathbb{F}_{16}$. The rational function field over $\mathbb{F}_{16}$ has 17 rational places and 120 places of degree 2. We construct the set $\mathcal{P}_{16}$ with the 17 rational places and 4 places of degree 2. Then, the sum of the degrees of the places in $\mathcal{P}_{16}$ is equal to $17+2\times 4 = 25 = 2n-1$. As in the previous example, we start by including in $\mathcal{P}_{16}$ the places P^2 of degree 2 such that the algorithm $\mathcal{U}_{16,2}^{\mathcal{P}_2}(P^2)$ has the best complexity. There are 8 places P^2 of degree 2 such that $\mu_s(\mathcal{U}_{16,2}^{\mathcal{P}_2}(P^2)) = 1$ and $a(\mathcal{U}_{16,2}^{\mathcal{P}_2}(P^2)) = 4$. We include 4 of them in $\mathcal{P}_{16}$. In the following, we consider that the places of degree 2 in $\mathcal{P}_{16}$ are given by $(x^2+x+\omega^7)$, $(x^2+x+\omega^{14})$, $(x^2+x+\omega^{13})$ and $(x^2+x+\omega^{11})$. Consider the place $Q = (x^{13}+x^4+x^3+x+1)$ of $\mathbb{F}_{16}(x)$ of degree 13, and $\mathcal{D} = 12P_\infty$. We can now construct the algorithm $\mathcal{U}_{16,13}^{\mathcal{P}_{16}}(Q)$. Without any optimization, we use the canonical basis of $\mathcal{L}(\mathcal{D})$ given by $\{1, x, \ldots, x^{12}\}$. The algorithm then uses 29 bilinear multiplications, 686 scalar multiplications and 815 additions.

4.2.1. *Generic optimization.* With Algorithm 1, we can construct a $\mathcal{P}_{16}$−basis of $\mathcal{L}(\mathcal{D})$. This basis is given by

$$
\begin{aligned}
V_1 &= 1,\\
V_2 &= x,\\
V_3 &= \omega^{11}x^2 + \omega^{12}x,\\
V_4 &= \omega^{13}x^3 + \omega^3x^2 + \omega x,\\
V_5 &= \omega^{13}x^4 + \omega^9x^3 + \omega^{11}x^2 + \omega^4x,\\
V_6 &= \omega^{12}x^5 + \omega^{10}x^4 + \omega^3x^3 + x^2 + \omega^7x,\\
V_7 &= \omega^{13}x^6 + \omega^5x^5 + x^4 + \omega^3x^3 + \omega^{14}x^2 + \omega^{13}x,\\
V_8 &= \omega^{12}x^7 + \omega^7x^6 + \omega^{11}x^5 + \omega x^4 + \omega^3x^3 + \omega^6x^2 + \omega^3x,\\
V_9 &= \omega^{11}x^8 + \omega^2x^7 + \omega^9x^6 + \omega^8x^5 + \omega^{12}x^4 + \omega^6x^3 + \omega^7x^2 + \omega^9x,\\
V_{10} &= \omega^{14}x^9 + \omega^{13}x^8 + \omega x^7 + \omega^3x^6 + \omega x^5 + \omega^{12}x^4 + \omega^4x^3 + \omega^{10}x^2 + \omega^5x,\\
V_{11} &= \omega^7x^{10} + \omega^{11}x^9 + \omega^7x^8 + \omega^5x^7 + \omega^6x^6 + \omega^{11}x^5 + \omega^5x^4\\
&\quad + \omega^2x^3 + \omega x^2 + \omega^7x,\\
V_{12} &= \omega^5x^{11} + \omega^7x^{10} + \omega^8x^9 + \omega^{14}x^8 + \omega^{11}x^7 + \omega^4x^6 + \omega^7x^5\\
&\quad + \omega^6x^4 + \omega^{11}x^3 + \omega^6x^2 + x,\\
V_{13} &= x^{12} + \omega^9x^{11} + \omega^8x^{10} + \omega^4x^9 + \omega^9x^8 + \omega^{13}xx^7 + \omega^4x^6 + \omega^{12}x^5\\
&\quad + \omega^4x^4 + \omega^5x^3 + \omega^3x^2 + \omega^6x.
\end{aligned}
$$

Using this basis, the complexity of the algorithm is reduced to 614 scalar multiplications and 705 additions.

4.2.2. *Non-generic optimization.* In Remark 3.11, we saw that there are too many possible $\mathcal{P}_{16}$−bases for an exhaustive search. Nevertheless, we can still improve our algorithm. Using the proof of Proposition 3.2, a column of the matrix $T_{\mathcal{D}}$ contains at most $n-1+4=16$ zeros, since $\mathcal{P}_{16}$ contains 4 places of degree 2. Moreover, this equality is possible if and only if the corresponding vector of the basis of $\mathcal{L}(\mathcal{D})$ vanishes only on rational places. Thus, we consider the set $\mathcal{S}=\{1,x,x+\omega,x+\omega^2,\dots,x+\omega^{15}\}$. By corollary 3.6, we want to construct products of these elements of degree 11 or 12, such that such a function vanishes on 12 rational places of $\mathbb{F}_{16}(x)$. Such a polynomial is the product of 12 elements of $\mathcal{S}$. Hence, there are $\binom{17}{12}=6188$ possibilities. Moreover, the evaluation of each of these vectors gives at most 16 zeros. For all these functions f, we compute $Ev_{\mathcal{P}}(f)$, the vector of the evaluations of f at the places in $\mathcal{P}_{16}$. There are 49 of them containing 16 zeros. Finally, it remains to find a basis using 13 of these vectors. There are still $\binom{49}{13}=262596783764$ possibilities. This is very few compared to an exhaustive search of a $\mathcal{P}_{16}$−basis ($\lessapprox 10^{72}$ possibilities), but still too much. We finally randomly search a basis using these vectors, and apply Algorithm 1, Step 2 to reduce the number of scalar multiplications. We obtain the following.

$$
\begin{aligned}
V_1 &= \omega^4x^{12}+\omega^5x^{10}+\omega^8x^9+\omega^6x^8+\omega^2x^6+\omega^{10}x^5+\omega^3x^4+\omega x^3+\omega^9x^2\\
&\quad+\omega^{12}x+1,\\
V_2 &= \omega^4x^{12}+\omega^5x^{10}+\omega^8x^9+\omega^3x^8+\omega^2x^6+\omega^{10}x^5+\omega^9x^4+\omega x^3\\
&\quad+\omega^{12}x^2+\omega^6x,\\
V_3 &= \omega x^{12}+\omega^5x^{10}+\omega^2x^9+\omega^6x^8+\omega^8x^6+\omega^{10}x^5+\omega^3x^4+\omega^4x^3\\
&\quad+\omega^9x^2+\omega^{12}x,\\
V_4 &= \omega^9x^{11}+\omega^9x^{10}+\omega^{10}x^9+\omega^7x^7+\omega^7x^6+\omega^{14}x^5+x^4+\omega^{12}x^2+\omega^6x,\\
V_5 &= \omega^2x^{12}+\omega^{10}x^{10}+\omega^4x^9+\omega^9x^8+\omega x^6+\omega^5x^5+\omega^{12}x^4+\omega^8x^3\\
&\quad+\omega^6x^2+\omega^3x,\\
V_6 &= \omega^{13}x^{11}+\omega^{13}x^{10}+\omega^3x^9+\omega^{13}x^7+\omega^{13}x^6+\omega^8x^5+\omega^3x^3+\omega^8x^2+\omega^3x,\\
V_7 &= x^{12}+x^9+\omega^{10}x^8+x^6+\omega^5x^4+x^3+\omega^{10}x^2+\omega^5x,\\
V_8 &= \omega^8x^{12}+\omega^{10}x^{10}+\omega x^9+\omega^{12}x^8+\omega^4x^6+\omega^5x^5+\omega^6x^4+\omega^2x^3+\omega^3x^2\\
&\quad+\omega^9x+1,\\
V_9 &= \omega^{10}x^{12}+x^{10}+\omega^5x^9+\omega^8x^8+\omega^5x^6+x^5+\omega^4x^4+\omega^{10}x^3+\omega^2x^2+\omega x,\\
V_{10} &= x^{11}+x^{10}+x^9+\omega^{10}x^7+\omega^{10}x^6+\omega^5x^5+\omega^5x^4+x^2+\omega^5x,\\
V_{11} &= \omega^{12}x^{11}+\omega^{12}x^{10}+\omega^5x^9+\omega^{11}x^7+\omega^{11}x^6+\omega^7x^5+x^4+\omega^6x^2+\omega^3x,\\
V_{12} &= \omega^2x^{11}+\omega^2x^{10}+\omega^{10}x^9+x^7+x^6+\omega^6x^5+\omega^8x^4+\omega^9x^3+\omega^2x^2+\omega^6x,\\
V_{13} &= \omega^{14}x^{11}+\omega^{14}x^{10}+\omega^9x^9+\omega^{14}x^7+\omega^{14}x^6+\omega^4x^5+\omega^9x^3+\omega^4x^2+\omega^9x.
\end{aligned}
$$

Using this basis, the Chudnovsky-type algorithm $\mathcal{U}^{\mathcal{P}_{16}}_{16,13}(Q)$ now uses 423 scalar multiplications and 487 additions.

Karatsuba algorithm is more expensive in terms of bilinear complexity, using 66 bilinear multiplications instead of 29 with our method. It also uses 277 additions

([**WP06**, Appendix]). As in the previous section, we compute the reduction modulo $Q(x)$ with 66 additions. We notice that there is no scalar multiplication. This is due to the choice of $Q(x)$ which has all its coefficients in $\mathbb{F}_2$. This kind of situation is more favorable to Karatsuba's technique than to our method for the scalar complexity. On the other hand, we can see that our method is clearly more efficient than the CCMA method using a curve of genus 2. The complexities of all these algorithms are summarized in Table 4.

TABLE 4. Comparison of algorithms for the multiplication in $\mathbb{F}_{16^{13}}$

Algorithm	$\mu_b(\mathcal{U})$	$\mu_s(\mathcal{U})$	$a(\mathcal{U})$	$\mu(\mathcal{U})$
CCMA [**Bal02**]	27	833	840	1700
Our construction				
Non optimized	29	686	815	1530
Generic optimization	29	614	705	1348
Non-generic optimization	29	423	487	939
Karatsuba [**WP06**] + Reduction	66	0	338	404

4.3. Generic optimization over $\mathbb{F}_2$. For this last example, we fix the base field to be $\mathbb{F}_2$. We want to construct and optimize generically Chudnovsky-type algorithms over the projective line to reach large extensions. In the following, each set of places $\mathcal{P}_n$ is constructed by taking all places of growing degrees until the sum is equal to $2n-1$. Recall that if at some point the sum is bigger than $2n-1$ we can remove from $\mathcal{P}_n$ a place to obtain exactly $2n-1$. Note that since we consider extensions of $\mathbb{F}_2$, there are no scalar multiplication. Moreover, the number of additions depends on the place of degree n used to define the extension. For this reason, we return a list of values for the number of additions, following the order of the places given by Magma. We give the results for the extensions of degrees until 6 for a recursive Chudnovsky-type over the projective line first non-optimized (Table 5), then generically optimized (Table 6), and compared to the Karatsuba Algorithm ([**WP06**, Appendix]) with the polynomial reduction (Table 7).

Using all places of degrees lower than or equal to 6 of $\mathbb{F}_2(x)$, one can define a Chudnovsky-type algorithm over the projective line for the multiplication in the extension of degree 54 of $\mathbb{F}_2$. The set $\mathcal{P}_{54}$ then contains all of these places. Considering $Q(x) = x^{54} + x^{34} + x^{32} + x^{31} + x^{30} + x^{29} + x^{27} + x^{25} + x^{21} + x^{18} + x^{17} + x^{16} + x^{15} + x^{13} + x^7 + x^4 + x^2 + x + 1$, we obtain the results of Table 8.

REMARK 4.2. In this paper, we focused on constructing the matrices that gives the less possible operations when applied canonically. We did not focus on how to compute the multiplication by those matrices. For instance, if a non trivial α appears more than once in a column of a matrix, we can compute the multiplication by α once and for all, thus reducing the number of scalar multiplications.

REMARK 4.3. This strategy of optimization is specialized to Chudnovsky-type algorithms over the rational function field $\mathbb{F}_q(x)$, where the places are fully defined by polynomials over $\mathbb{F}_q$. Nevertheless, one can consider the generalization of this

TABLE 5. Non optimized generic Chudnovsky-type algorithms over the projective line

Extension degree	$\mu_b(\mathcal{U})$	$a(\mathcal{U})$	$\min\{\mu(\mathcal{U})\}$
2	3	$[4]$	7
3	6	$[18, 17]$	23
4	11	$[44, 44, 43]$	54
5	15	$[81, 85, 76, 78, 78, 79]$	91
6	18	$[118, 125, 115, 112, 112, 126, 112, 115, 111]$	129

TABLE 6. Generically optimized Chudnovsky-type algorithms over the projective line

Extension degree	$\mu_b(\mathcal{U})$	$a(\mathcal{U})$	$\min\{\mu(\mathcal{U})\}$
2	3	$[4]$	7
3	6	$[14, 14]$	20
4	11	$[41, 41, 38]$	49
5	15	$[65, 68, 68, 62, 63, 66]$	77
6	18	$[88, 93, 95, 95, 89, 93, 81, 89, 85]$	99

TABLE 7. Karatsuba [**WP06**] + Reduction

Extension degree	$\mu_b(\mathcal{U})$	$a(\mathcal{U})$	$\min\{\mu(\mathcal{U})\}$
2	3	$[6]$	9
3	6	$[19, 19]$	25
4	9	$[32, 30, 35]$	49
5	15	$[57, 60, 58, 60, 63, 59]$	72
6	18	$[82, 78, 71, 67, 78, 79, 79, 83, 75]$	85

TABLE 8. Comparison of algorithms for the multiplication in $\mathbb{F}_{2^{54}}$

Algorithm	$\mu_b(\mathcal{U})$	$a(\mathcal{U})$	$\mu(\mathcal{U})$
Our Construction			
Non optimized	303	9849	10152
Generic optimization	303	8400	8703
Karatsuba [**WP06**]+ Reduction	630	4512	5142

strategy to optimize algorithms over a function field $F/\mathbb{F}_q$ of genus $g > 0$, by using local uniformizers of the places instead of monic irreducible polynomials.

REMARK 4.4. This work, together with [**BBD19, BBD21**], are the very first works on the scalar optimization of Chudnovsky-type algorithms, and it reduces significantly the number of algebraic operations used by these algorithms. Concerning practical efficiency, this is a first step before being able to explore and realize efficient implementations of the formulae given by the method. It would then be relevant to realize timings of implementations of our algorithms, and possibly to compare it for instance with the specific algorithms over $\mathbb{F}_4$ presented by Harvey, Lecerf and van der Hoeven in [**HvdHL16**]. But this work of comparison with these algorithms is sufficiently important to require a further work of its own. More precisely, it requires to translate the algorithms we obtained in terms of computer instructions, for example using multiplications, additions, but also shifts. Furthermore, it would also be interesting to compare our results with other algorithms of evaluation and interpolation over rational points (other than Karatsuba's), that are closer to our method, like the Toom-Cook methods optimized by Bodrato [**Bod07**] and Bodrato and Zanoni in [**BZ07**]. But even this comparison requires a non-trivial translation of our method, which can only be done later.

Acknowledgment

The authors are deeply grateful to the anonymous referees for their comments, that helped to improve and complete this article.

References

[Bal02] Stéphane Ballet, *Quasi-optimal algorithms for multiplication in the extensions of* $\mathbb{F}_{16}$ *of degree 13, 14 and 15*, J. Pure Appl. Algebra **171** (2002), no. 2-3, 149–164, DOI 10.1016/S0022-4049(01)00137-2. MR1904474

[BBD19] Stéphane Ballet, Alexis Bonnecaze, and Thanh-Hung Dang, *On the scalar complexity of* Chudnovsky^2 *multiplication algorithm in finite fields*, Algebraic informatics, Lecture Notes in Comput. Sci., vol. 11545, Springer, Cham, 2019, pp. 64–75, DOI 10.1007/978-3-030-21363-3_6. MR3976187

[BBD21] Stéphane Ballet, Alexis Bonnecaze, and Thanh-Hung Dang, *Optimization of the scalar complexity of Chudnovsky^2 multiplication algorithms in finite fields*, Cryptogr. Commun. **13** (2021), no. 4, 495–517, DOI 10.1007/s12095-021-00494-y. MR4298229

[BBP20] Stéphane Ballet, Alexis Bonnecaze, and Bastien Pacifico, *Multiplication in finite fields with Chudnovsky-type algorithms on the projective line*, 2020, ⟨hal-02911546⟩ https://doi.org/10.48550/arxiv.2007.16082.

[BCP97] Wieb Bosma, John Cannon, and Catherine Playoust, *The Magma algebra system. I. The user language*, J. Symbolic Comput. **24** (1997), no. 3-4, 235–265, DOI 10.1006/jsco.1996.0125. Computational algebra and number theory (London, 1993). MR1484478

[BCP+21] S. Ballet, J. Pieltant, M. Rambaud, H. Randriambololona, R. Rolland, and J. Chaumine, *On the tensor rank of multiplication in finite extensions of finite fields and related issues in algebraic geometry* (Russian, with Russian summary), Uspekhi Mat. Nauk **76** (2021), no. 1(457), 31–94, DOI 10.4213/rm9928. MR4223937

[BCS97] Peter Bürgisser, Michael Clausen, and M. Amin Shokrollahi, *Algebraic complexity theory*, Grundlehren der mathematischen Wissenschaften [Fundamental Principles of Mathematical Sciences], vol. 315, Springer-Verlag, Berlin, 1997. With the collaboration of Thomas Lickteig, DOI 10.1007/978-3-662-03338-8. MR1440179

[Bod07] Marco Bodrato, *Towards optimal Toom-Cook multiplication for univariate and multivariate polynomials in characteristic 2 and 0*, Arithmetic of finite fields, Lecture Notes in Comput. Sci., vol. 4547, Springer, Berlin, 2007, pp. 116–133, DOI 10.1007/978-3-540-73074-3_10. MR2373888

[BS91] Ulrich Baum and Mohammad Amin Shokrollahi, *An optimal algorithm for multiplication in* $\mathbf{F}_{256}/\mathbf{F}_4$, Appl. Algebra Engrg. Comm. Comput. **2** (1991), no. 1, 15–20, DOI 10.1007/BF01810851. MR1209240

[BZ07] Marco Bodrato and Alberto Zanoni, *Integer and polynomial multiplication: towards optimal Toom-Cook matrices*, ISSAC 2007, ACM, New York, 2007, pp. 17–24, DOI 10.1145/1277548.1277552. MR2396179

[CC88] D. V. Chudnovsky and G. V. Chudnovsky, *Algebraic complexities and algebraic curves over finite fields*, J. Complexity **4** (1988), no. 4, 285–316, DOI 10.1016/0885-064X(88)90012-X. MR974928

[Fü09] Martin Fürer, *Faster integer multiplication*, SIAM J. Comput. **39** (2009), no. 3, 979–1005, DOI 10.1137/070711761. MR2538847

[HvdH19] David Harvey and Joris van der Hoeven, *Polynomial multiplication over finite fields in time $O(n \log n)$*, Journal of the ACM, Volume 69, Issue 2, April 2022, Article No.: 12, pp. 1–40, `https://doi.org/10.1145/3505584`.

[HvdHL16] David Harvey, Joris van der Hoeven, and Grégoire Lecerf, *Fast polynomial multiplication over $\mathbb{F}_{2^{60}}$*, Proceedings of the 2016 ACM International Symposium on Symbolic and Algebraic Computation, ACM, New York, 2016, pp. 255–262. MR3565722

[Kar63] Anatolii Karatsuba, *Multiplication of multidigit number on automata*, Soviet Physics Doklady **7** (1963), 595–596.

[SS71] A. Schönhage and V. Strassen, *Schnelle Multiplikation grosser Zahlen* (German, with English summary), Computing (Arch. Elektron. Rechnen) **7** (1971), 281–292, DOI 10.1007/bf02242355. MR292344

[Sti08] Henning Stichtenoth, *Algebraic function fields and codes*, 2nd ed., Graduate Texts in Mathematics, vol. 254, Springer-Verlag, Berlin, 2009. MR2464941

[STV92] Igor E. Shparlinski, Michael A. Tsfasman, and Serge G. Vladut, *Curves with many points and multiplication in finite fields*, Coding theory and algebraic geometry (Luminy, 1991), Lecture Notes in Math., vol. 1518, Springer, Berlin, 1992, pp. 145–169, DOI 10.1007/BFb0087999. MR1186422

[vzGG03] Joachim von zur Gathen and Jürgen Gerhard, *Modern computer algebra*, 2nd ed., Cambridge University Press, Cambridge, 2003. MR2001757

[WP06] Andre Weimerskirch and Christof Paar, *Generalizations of the Karatsuba Algorithm for efficient implementations.*, IACR Cryptology ePrint Archive. (2006).

Institut Mathématiques de Marseille, Aix Marseille Univ, CNRS, I2M, Marseille, France

Email address: `Stephane.Ballet@univ-amu.fr`

Institut Mathématiques de Marseille, Aix Marseille Univ, CNRS, I2M, Marseille, France

Email address: `Alexis.Bonnecaze@univ-amu.fr`

Institut Matématiques de Marseille, Aix Marseille Univ, CNRS, I2M, Marseille, France

Email address: `Bastien.Pacifico@univ-amu.fr`

Contemporary Mathematics
Volume **779**, 2022
https://doi.org/10.1090/conm/779/15669

On the constant $D(q)$ defined by Homma

Peter Beelen, Maria Montanucci, and Lara Vicino

ABSTRACT. Let $\mathcal{X}$ be a projective, irreducible, nonsingular algebraic curve over the finite field $\mathbb{F}_q$ with q elements and let $|\mathcal{X}(\mathbb{F}_q)|$ and $g(\mathcal{X})$ be its number of rational points and genus respectively. The Ihara constant $A(q)$ has been intensively studied during the last decades, and it is defined as the limit superior of $|\mathcal{X}(\mathbb{F}_q)|/g(\mathcal{X})$ as the genus of $\mathcal{X}$ goes to infinity. In 2012 Homma defined an analogue $D(q)$ of $A(q)$, where the nonsingularity of $\mathcal{X}$ is dropped and $g(\mathcal{X})$ is replaced with the degree of $\mathcal{X}$. We will call $D(q)$ Homma's constant. In this paper, upper and lower bounds for the value of $D(q)$ are found.

1. Introduction

Let p be a prime and let $q = p^e$ be a prime power. Let $\mathcal{X}$ be a projective, nonsingular, geometrically irreducible curve of genus g. The interaction between the genus g of $\mathcal{X}$ and the number $|\mathcal{X}(\mathbb{F}_q)|$ of its rational points has been subject of intense studies during the last years. It is well known that the Weil bound

$$|\mathcal{X}(\mathbb{F}_q)| \leq q + 1 + 2g\sqrt{q}$$

is not sharp if g is large compared to q. Put

$$N_q(g) := \max |\mathcal{X}(\mathbb{F}_q)|, \tag{1.1}$$

where the maximum is taken over all curves $\mathcal{X}/\mathbb{F}_q$ with genus g. The *Ihara constant* is defined by

$$A(q) := \limsup_{g \to \infty} \frac{N_q(g)}{g}. \tag{1.2}$$

This is a measure of the asymptotic behaviour of the number of rational points on curves over $\mathbb{F}_q$ when the genus becomes large. Ihara's constant $A(q)$ has been intensively studied during the last decades. For any q, we have $A(q) \leq \sqrt{q} - 1$ (see [**4**]), and if q is a square we have (see [**13**, **21**]) $A(q) = \sqrt{q} - 1$.

For any q, using class field theory, Serre [**17**] showed that $A(q) > c\log(q)$ for some constant $c > 0$ independent of q. In particular $A(q) > 0$ for all q. For $q = p^{2m+1}$, with $m > 0$, the currently best-known lower bound is $A(q) \geq 2(1/(p^m - 1) + 1/(p^{m+1} - 1))^{-1}$, see [**2**]. The exact value of $A(q)$ is however unknown when q is not a square.

2020 *Mathematics Subject Classification.* Primary 14G15, 14H50; Secondary 11G20, 14H25.
Key words and phrases. Algebraic curve, rational point, finite field.
The first and second authors were supported by The Danish Council for Independent Research (DFF-FNU), project *Correcting on a Curve*, Grant No. 8021-00030B.

If the curve $\mathcal{X}$ is seen as a projective curve $\mathcal{X} \subseteq \mathbb{P}^n(\mathbb{F}_q)$ of degree $d > 0$ and it is not necessarily required to be nonsingular, a different question can be addressed: how large can $|\mathcal{X}(\mathbb{F}_q)|$ be with respect to d?

In a series of papers [**10**–**12**] it has been shown that if $\mathcal{X}$ is a (possibly reducible) plane curve without $\mathbb{F}_q$-linear components, then

$$|\mathcal{X}(\mathbb{F}_q)| \leq (d-1)q + 1, \tag{1.3}$$

except for curves isomorphic over $\mathbb{F}_4$ to the curve defined by

$$\mathcal{K}: (X+Y+Z)^4 + (XY+YZ+ZX)^2 + XYZ(X+Y+Z) = 0,$$

which satisfies $|\mathcal{K}(\mathbb{F}_4)| = 14$. The bound (1.3) was originally conjectured by Sziklai [**19**], and he found that some curves actually achieve this bound.

The natural question on whether the bound (1.3) is valid for curves in higher dimensional projective space $n \geq 3$ was analyzed by Homma in [**9**]. There, it is obtained that (1.3) is also true when $n \geq 3$ and $\mathcal{X}$ has no $\mathbb{F}_q$-linear components, unless $d = q = 4$ and $\mathcal{X}$ is $\mathbb{F}_q$-isomorphic to the plane curve $\mathcal{K}$.

In the same paper [**9**], an analogue of Ihara constant $A(q)$ (1.2) is given when replacing the genus g with the degree d. First, we replace $N_q(g)$ as defined in (1.1), with $M_q(d) := \max |\mathcal{X}(\mathbb{F}_q)|$ where this time the maximum is taken over all irreducible curves of a fixed degree d in a projective space of some dimension over $\mathbb{F}_q$. Here the dimension is not fixed and therefore allowed to be arbitrarily large. Then the analogue of $A(q)$ is defined as

$$D(q) := \limsup_{d \to \infty} \frac{M_q(d)}{d}, \tag{1.4}$$

which measures the asymptotic behavior of the number of rational points of projective curves over $\mathbb{F}_q$ when d becomes large. In [**9**] it was observed that since the bound (1.3) is valid for curves in any projective space $\mathbb{P}^n(\mathbb{F}_q)$, $n \geq 2$, with the exception already mentioned above, one may conclude that $D(q) \leq q$. In the same paper also the lower bound $D(q) \geq A(q)/2$ was derived, but the exact value of $D(q)$ remains unknown for all q.

In this paper, new upper and lower bounds for the value of $D(q)$, which we from now on will call Homma's constant, are found by a refinement of Homma's methods and by using towers of algebraic function fields. Our main results are summarized in the following theorem.

THEOREM 1.5. *Let $q = p^e$ be a prime power and let $D(q)$ be Homma's constant as defined in* (1.4). *Then*

(1) $D(q) \leq q - 1$,
(2) $D(q) \geq 1$ *provided that* $q > 2$,
(3) $D(q^2) \geq \frac{q}{q+1} A(q^2) = \frac{q^2-q}{q+1}$.

Note that the lower bound $D(q) \geq 1$ is interesting for small values of q only, since otherwise Homma's lower bound $D(q) \geq A(q)/2$ is better. The values $q \leq 31$ for which the lower bound $D(q) \geq 1$ is currently the best known are listed in Remark 4.6.

The paper is organized as follows. We start by slightly improving Homma's upper bound on $D(q)$ in Section 2 by refining his argument, thus proving Item 1 of Theorem 1.5. Next we prove Item 2 of Theorem 1.5 in Section 3 by explicitly constructing a sequence of curves whose degrees are close to their number of rational points. Finally, the main part of the paper is devoted to proving Item 3 of Theorem 1.5 in the final section.

2. An upper bound for $D(q)$: the proof of Item 1 in Theorem 1.5

The upper bound $D(q) \leq q$ obtained by Homma in [**9**, Proposition 5.4] was deduced from the bound (1.3), but in the same paper the following theorem was given.

THEOREM 2.1 ([**9**, Theorem 3.2]). *Let $\mathcal{X}$ be a nondegenerate irreducible curve of degree d in $\mathbb{P}^n(\mathbb{F}_q)$. Then*

$$|\mathcal{X}(\mathbb{F}_q)| \leq \frac{(q-1)(q^{n+1}-1)}{q(q^n-1)-n(q-1)}d. \tag{2.2}$$

Here the word *nondegenerate* means that $\mathcal{X}$ is not contained in any hyperplane of $\mathbb{P}^n(\mathbb{F}_q)$. At this point, using this result, we are ready to prove Item 1 in Theorem 1.5.

Indeed for a fixed value of q, considering equation (2.2) and dividing both sides by d gives

$$\frac{|\mathcal{X}(\mathbb{F}_q)|}{d} \leq \frac{(q-1)(q^{n+1}-1)}{q(q^n-1)-n(q-1)} = \frac{(q-1)\dfrac{(q^{n+1}-1)}{q^{n+1}}}{\dfrac{q(q^n-1)}{q^{n+1}} - \dfrac{n(q-1)}{q^{n+1}}}. \tag{2.3}$$

This observation can be used to improve the upper bound for $D(q)$. Note that by taking the $\limsup_{d\to\infty} M_q(d)/d$ as in (1.4), we are by definition of $D(q)$ considering curves of increasing degree. However, the dimension of the projective spaces containing the curves will be increasing as d increases. Indeed, if for a family of curves $(\mathcal{X}_i)_{i\geq 0}$, with degrees d_i tending to infinity as i tends to infinity, there exists an n such that for all i, $\mathcal{X}_i \subseteq \mathbb{P}^n$, then $|\mathcal{X}_i(\mathbb{F}_q)| \leq |\mathbb{P}^n(\mathbb{F}_q)| = (q^{n+1}-1)/(q-1)$, implying that $|\mathcal{X}_i(\mathbb{F}_q)|/d_i$ tends to zero as i tends to infinity.

Now let $(\mathcal{X}_i)_{i\geq 0}$, be a family of curves with degrees d_i tending to infinity such that $\limsup_{i\to\infty} |\mathcal{X}_i(\mathbb{F}_q)|/d_i > 0$. Further assume for each i that $\mathcal{X}_i$ is a nondegenerate curve contained in $\mathbb{P}^{n_i}$. We have seen that n_i tends to infinity as i tends to infinity. But then we obtain from equation (2.3):

$$D(q) \leq \lim_{i\to\infty} \frac{(q-1)\dfrac{(q^{n_i+1}-1)}{q^{n_i+1}}}{\dfrac{q(q^{n_i}-1)}{q^{n_i+1}} - \dfrac{n_i(q-1)}{q^{n_i+1}}} = q-1.$$

This proves Item 1 of Theorem 1.5.

3. A lower bound for $D(q)$: the proof of Item 2 in Theorem 1.5

For a prime power $q = p^e$ strictly larger than two, consider the tower of function fields $\mathcal{T} = (T_m)_{m\geq 1}$ over $\mathbb{F}_q$ defined recursively as

$$T_1 = \mathbb{F}_q(x_1) \qquad \text{and} \qquad T_{i+1} = T_i(x_{i+1}) \qquad \text{with} \qquad x_{i+1}^{q-1} = -1 + (x_i+1)^{q-1}.$$

The tower $\mathcal{T}$ is similar to an asymptotically good tower considered in [**18**, Proposition 7.3.3], but the variation we consider is actually not asymptotically good. It is not hard to see that the place of T_1 corresponding to the zero of x_1 is totally ramified in the tower. In particular, the equation $x_{i+1}^{q-1} = -1 + (x_i + 1)^{q-1}$ is absolutely irreducible when viewed as a polynomial in $T_i[x_{i+1}]$. This implies in particular that the ideal $I_\ell := \langle x_2^{q-1} + 1 - (x_1+1)^{q-1}, \ldots, x_\ell^{q-1} + 1 - (x_{\ell-1}+1)^{q-1}\rangle \subseteq \mathbb{F}_q[x_1, \ldots, x_\ell]$ is a prime ideal. Since we want to deal with projective curves, the following proposition is essential.

PROPOSITION 3.1. *Let $\ell > 1$ be an integer and define $I'_\ell := \langle x_2^{q-1} + z^{q-1} - (x_1 + z)^{q-1}, \ldots, x_\ell^{q-1} + z^{q-1} - (x_{\ell-1} + z)^{q-1}\rangle \subseteq \mathbb{F}_q[x_1, \ldots, x_\ell, z]$. Then I'_ℓ is a homogeneous prime ideal and the homogenization of the prime ideal $I_\ell := \langle x_2^{q-1} + 1 - (x_1 + 1)^{q-1}, \ldots, x_\ell^{q-1} + 1 - (x_{\ell-1} + 1)^{q-1}\rangle \subseteq \mathbb{F}_q[x_1, \ldots, x_\ell]$.*

PROOF. For convenience, let us write $g_i := x_{i+1}^{q-1} + 1 - (x_i + 1)^{q-1}$ and $g'_i := x_{i+1}^{q-1} + z^{q-1} - (x_i + z)^{q-1}$. We have already seen that the ideal I_ℓ is a prime ideal. Now let $>_{\text{deglex}}$ denote the degree-lexicographic ordering with $x_\ell >_{\text{deglex}} \cdots >_{\text{deglex}} x_1$ as a monomial order in $\mathbb{F}_q[x_1, \ldots, x_\ell]$. Since under this monomial ordering the leading terms of the g_i are co-prime, the set $\{g_1, \ldots, g_{\ell-1}\}$ is a Gröbner basis of I_ℓ. Then from [**3**, §8.4, Theorem 4] $\{g'_1, \ldots, g'_{\ell-1}\}$ is a Gröbner basis for the homogenization of I_ℓ. Hence I'_ℓ is the homogenization of the prime ideal I_ℓ and in particular I'_ℓ is a homogeneous prime ideal. □

Now consider the projective curve $\mathcal{X}_\ell \subset \mathbb{P}^\ell$ defined over $\mathbb{F}_q$ given by the homogeneous equations

$$x_{i+1}^{q-1} = -z^{q-1} + (x_i + z)^{q-1} \quad \text{for } i = 1, \ldots, \ell - 1. \tag{3.2}$$

Proposition 3.1 implies that $\mathcal{X}_\ell \subset \mathbb{P}^\ell$ is indeed an irreducible projective curve. It actually implies that $\mathcal{X}_\ell$ is a complete intersection, which in turn implies that $\deg(\mathcal{X}_\ell) = \deg(g'_1) \cdots \deg(g'_{\ell-1}) = (q-1)^{\ell-1}$.

Now we consider the number of $\mathbb{F}_q$-rational points on $\mathcal{X}_\ell$. To estimate this number, we consider the number of projective points $[x_1 : x_2 : \cdots : x_\ell : 0]$ satisfying equation (3.2). Substituting $z = 0$ in equation (3.2), we obtain that

$$x_{i+1}^{q-1} = x_i^{q-1} \quad \text{for } i = 1, \ldots, \ell - 1.$$

Choosing $x_1 = 1$, we see that any solution is defined over $\mathbb{F}_q$ and that there are exactly $(q-1)^{\ell-1}$ points at the infinity on $\mathcal{X}_\ell$. In particular, $|\mathcal{X}_\ell(\mathbb{F}_q)| \geq (q-1)^{\ell-1}$. Hence

$$D(q) \geq \limsup_{\ell\to\infty} \frac{|\mathcal{X}_\ell(\mathbb{F}_q)|}{\deg(\mathcal{X}_\ell)} \geq \frac{(q-1)^{\ell-1}}{(q-1)^{\ell-1}} = 1.$$

This completes the proof of Item 2 of Theorem 1.5.

4. A lower bound for $D(q^2)$: the proof of Item 3 in Theorem 1.5

In order to prove Item 3 in Theorem 1.5 we use a tower of function fields over $\mathbb{F}_{q^2}$ constructed recursively by Garcia and Stichtenoth in [**6**] as follows:

$$F_1 = \mathbb{F}_{q^2}(x_1) \qquad \text{and} \qquad F_{i+1} = F_i(x_{i+1}) \qquad \text{with} \qquad x_{i+1}^q + x_{i+1} = \frac{x_i^q}{x_i^{q-1} + 1}.$$

This tower is optimal in the sense that if $N_1(F_i)$ denotes the number of rational places and $g(F_i)$ the genus of F_i, then $\lim_{m\to\infty} N_1(F_m)/g(F_m) = q - 1 = A(q^2)$.

Indeed, any zero of the function $x_1 - \alpha$ in F_1 for $\alpha \in \mathbb{F}_{q^2} \setminus \{\alpha \mid \alpha^q + \alpha = 0\}$ splits completely in the extension F_m/F_1, implying that $N_1(F_m) \geq (q-1)q^m$. Moreover, in [**6**, Remark 3.8], the genus $g(F_m)$ of F_m is computed for all $m \geq 1$. It is given by

$$g(F_m) = \begin{cases} (q^{m/2}-1)^2 & \text{if } \; m \equiv 0 \pmod 2, \\ (q^{\frac{m+1}{2}}-1)(q^{\frac{m-1}{2}}-1) & \text{if } \; m \equiv 1 \pmod 2. \end{cases}$$

Hence optimality of the tower follows. For computing the genus $g(F_m)$, it is proven that the pole P_∞ of $x_1 \in F_1$ is totally ramified in all extensions F_m/F_1, $m \geq 2$, see also [**15**, Proposition 1.1]. We denote by $P_\infty^{(m)}$ the unique extension of P_∞ in F_m. Note that $P_\infty^{(m)}$ is a rational place, since P_∞ is totally ramified in F_m/F_1.

Even though it is in general a difficult challenge to compute the Weierstrass semigroups at places in a tower, Pellikaan, Stichtenoth, and Torres [**15**] computed the Weierstrass semigroup at $P_\infty^{(m)}$ for all $m \geq 1$. The nice property proven by the authors in [**15**] is that the semigroups at $P_\infty^{(m)}$ can be computed from the one at $P_\infty^{(m-1)}$, following a recursive procedure. Indeed from [**15**, Theorem 3.1]

$$H(P_\infty^{(m)}) = \begin{cases} \mathbb{Z}_{\geq 0} & \text{if } \; m = 1 \\ qH(P_\infty^{(m-1)}) \cup \mathbb{Z}_{\geq c_m} & \text{if } \; m > 1 \end{cases} \tag{4.1}$$

where $c_m := q^m - q^{\lceil \frac{m}{2} \rceil}$ is the conductor of $H(P_\infty^{(m)})$.

Let $\{\gamma_1, \ldots, \gamma_\ell\}$ be a set of generators of $H(P_\infty^{(m)})$, so that

$$H(P_\infty^{(m)}) = \langle \gamma_1, \ldots, \gamma_\ell \rangle,$$

and $0 < \gamma_1 < \cdots < \gamma_\ell$. Note that equation (4.1) implies that $\gamma_1 = q^{m-1}$, being the smallest positive element of $H(P_\infty^{(m)})$. This implies that $H(P_\infty^{(m)}) \cap \mathbb{Z}_{<c_m+q^{m-1}}$ is a generating set and that therefore we may assume that

$$\gamma_\ell \leq c_m + q^{m-1} - 1. \tag{4.2}$$

By definition of the Weierstrass semigroup $H(P_\infty^{(m)})$, there exist functions f_1, $\ldots$, $f_\ell \in F_m$ such that

$$(f_i)_\infty = \gamma_i P_\infty^{(m)}, \; i = 1, \ldots, \ell.$$

In [**16**], the functions f_1, $\ldots$, f_ℓ are used to define a birational morphism between a nonsingular projective curve $\mathcal{X}$ and a curve $\mathcal{X}'$, with only one point at infinity. Since we use the language of function fields, we need to reformulate the results from [**16**] slightly. Intuitively, we simply use the functions f_1, $\ldots$, f_n to define a map from the set of places of F_m to an algebraic curve $\mathcal{X}_m$. However, this map, which we denote by φ_m, is easiest to describe when first extending the constant field of F_m to $\overline{\mathbb{F}}_q$, the algebraic closure of $\mathbb{F}_q$, since then all places are rational:

$$\varphi_m : \mathbb{P}(\overline{\mathbb{F}}_q F_m) \longrightarrow \mathbb{P}^\ell$$

defined by

$$\begin{aligned} \varphi_m(Q) &= [1 : f_1(Q) : \cdots : f_\ell(Q)], && \text{if } Q \neq P_\infty^{(m)}, \\ \varphi_m(Q) &= [0 : \cdots : 0 : 1], && \text{otherwise.} \end{aligned}$$

Note that [**7**, Theorem 4.2.2] implies that indeed the image of the map φ_m is a projective curve $\mathcal{X}_m$. Since f_1, $\ldots$, f_ℓ are defined over $\mathbb{F}_{q^2}$, so is $\mathcal{X}_m$. Therefore we will from now on consider the curve $\mathcal{X}_m$ as a curve defined over $\mathbb{F}_{q^2}$. Moreover, [**16**, Theorem 15] states among other things that the function field of $\mathcal{X}_m$, when considered over the field $\mathbb{F}_{q^2}$, is exactly F_m, that apart from possibly $\varphi_m(P_\infty^{(m)})$, the curve has no singularities and that $P_\infty^{(m)}$ is the only place of F_m centered at $\varphi_m(P_\infty^{(m)})$. In particular φ_m induces a bijection between $\mathbb{P}(\overline{\mathbb{F}}_q F_m) \setminus \{P_\infty^{(m)}\}$ and $\mathcal{X}_m \setminus \{\varphi_m(P_\infty^{(m)})\}$.

REMARK 4.3. The curve $\mathcal{X}_m$ is a non-degenerate curve in $\mathbb{P}^\ell$. Indeed if this was not the case, then there would exist a combination $a_1 + a_2 f_1 + \cdots + a_{\ell+1} f_\ell$, for some $a_i \in \overline{\mathbb{F}}_q$ not all equal to zero, such that $a_1 + a_2 f_1 + \cdots + a_{\ell+1} f_\ell \equiv 0$, which is impossible by the linear independence of $\{1, f_1, \ldots, f_\ell\}$ over $\overline{\mathbb{F}}_q$ given by [**18**, Proposition 3.6.1].

Now we investigate the degree and number of $\mathbb{F}_{q^2}$-rational points on $\mathcal{X}_m$. The number of rational points is easy to bound, since the rational places of F_m are in bijection with the points on $\mathcal{X}_m$ defined over $\mathbb{F}_{q^2}$. Indeed, the place $P_\infty^{(m)}$ corresponds to the projective point $[0 : \cdots : 0 : 1]$, while the remaining rational points of $\mathcal{X}_m$ are non-singular and hence each corresponds to a unique rational place of F_m. This shows that

$$|\mathcal{X}_m(\mathbb{F}_{q^2})| = N_1(F_m) \geq (q-1)q^m. \tag{4.4}$$

The inequality $N_1(F_m) \geq (q-1)q^m$ was already mentioned before.

At this point we need to derive some information on the degree $\deg(\mathcal{X}_m)$ of the curve $\mathcal{X}_m$. The following inequality holds:

$$\deg(\mathcal{X}_m) \leq \gamma_\ell \leq c_m + q^{m-1} - 1. \tag{4.5}$$

This can be proven as follows. First of all, the last inequality is simply equation (4.2). Now recall that the degree can also be seen as the the maximum number of intersection points with a hyperplane. The points of intersection of the curve $\mathcal{X}_m$ and a hyperplane of equation $a_0x_0 + \cdots + a_\ell x_\ell = 0$ in $\mathbb{P}^\ell$ correspond, by the definition of φ_m, to the places that are zeros of the function $\sum_{i=0}^\ell a_i f_i \in \mathscr{L}(\gamma_\ell \bar{P}_\infty^{(m)})$. Here $\mathscr{L}(\gamma_\ell \bar{P}_\infty^{(m)})$ denotes the Riemann–Roch space of the divisor $\gamma_\ell \bar{P}_\infty^{(m)}$. Since the pole divisor of $\sum_{i=0}^\ell a_i f_i$ has degree at most γ_ℓ the same is true for its zero divisor. Hence the number of intersection points is at most γ_ℓ.

Now combining equations (4.4) and (4.5), we obtain:

$$D(q^2) \geq \limsup_{m\to\infty} \frac{|\mathcal{X}_m(\mathbb{F}_{q^2})|}{\deg(\mathcal{X}_m)} \geq \limsup_{m\to\infty} \frac{(q-1)q^m}{c_m + q^{m-1} - 1} = \frac{q^2-q}{q+1}.$$

Since $A(q^2) = q - 1$, Item 3 of Theorem 1.5 follows.

REMARK 4.6. Thereom 1.5 (3) improves Homma's lower bound $D(q^2) \geq A(q^2)/2$ for any values of q. The bound $D(q) \geq 1$ is instead interesting for small values of $q > 2$, since then Homma's lower bound $D(q) \geq A(q)/2$ is weaker. The following table provides for those small values of q the best known lower bound for $A(q)/2$. For all other values of q, except possibly when q is a prime, $A(q) \geq 2$.

q	$A(q)/2 \geq$	reference
3	0.2464	**[5]**
4	0.5	**[13, 21]**
5	0.3636	**[1, 20]**
7	0.4615	**[8]**
8	0.75	**[22]**
11	0.5714	**[8]**
13	0.6	**[14]**
17	0.8	**[14]**
19	0.8	**[8]**
23	0.9230	**[8]**
29	0.9523	**[8]**
31	0.9523	**[8]**

References

[1] Bruno Angles and Christian Maire, *A note on tamely ramified towers of global function fields*, Finite Fields Appl. **8** (2002), no. 2, 207–215, DOI 10.1006/ffta.2000.0336. MR1894514

[2] Alp Bassa, Peter Beelen, Arnaldo Garcia, and Henning Stichtenoth, *Towers of function fields over non-prime finite fields* (English, with English and Russian summaries), Mosc. Math. J. **15** (2015), no. 1, 1–29, 181, DOI 10.17323/1609-4514-2015-15-1-1-29. MR3427409

[3] David Cox, John Little, and Donal O'Shea, *Ideals, varieties, and algorithms: An introduction to computational algebraic geometry and commutative algebra*, Undergraduate Texts in Mathematics, Springer-Verlag, New York, 1992, DOI 10.1007/978-1-4757-2181-2. MR1189133

[4] S. G. Vlèduts and V. G. Drinfel′d, *The number of points of an algebraic curve* (Russian), Funktsional. Anal. i Prilozhen. **17** (1983), no. 1, 68–69. MR695100

[5] Iwan Duursma and Kit-Ho Mak, *On lower bounds for the Ihara constants $A(2)$ and $A(3)$*, Compos. Math. **149** (2013), no. 7, 1108–1128, DOI 10.1112/S0010437X12000796. MR3078640

[6] Arnaldo Garcia and Henning Stichtenoth, *On the asymptotic behaviour of some towers of function fields over finite fields*, J. Number Theory **61** (1996), no. 2, 248–273, DOI 10.1006/jnth.1996.0147. MR1423052

[7] David M. Goldschmidt, *Algebraic functions and projective curves*, Graduate Texts in Mathematics, vol. 215, Springer-Verlag, New York, 2003, DOI 10.1007/b97844. MR1934359

[8] L. L. Hall-Seelig, *New lower bounds for the Ihara function $A(q)$ for small primes*, J. Number Theory **133** (2013), no. 10, 3319–3324, DOI 10.1016/j.jnt.2013.04.002. MR3071814

[9] Masaaki Homma, *A bound on the number of points of a curve in a projective space over a finite field*, Theory and applications of finite fields, Contemp. Math., vol. 579, Amer. Math. Soc., Providence, RI, 2012, pp. 103–110, DOI 10.1090/conm/579/11523. MR2975736

[10] Masaaki Homma and Seon Jeong Kim, *Around Sziklai's conjecture on the number of points of a plane curve over a finite field*, Finite Fields Appl. **15** (2009), no. 4, 468–474, DOI 10.1016/j.ffa.2009.02.008. MR2535590

[11] Masaaki Homma and Seon Jeong Kim, *Sziklai's conjecture on the number of points of a plane curve over a finite field II*, Finite fields: theory and applications, Contemp. Math., vol. 518, Amer. Math. Soc., Providence, RI, 2010, pp. 225–234, DOI 10.1090/conm/518/10208. MR2648551

[12] Masaaki Homma and Seon Jeong Kim, *Sziklai's conjecture on the number of points of a plane curve over a finite field III*, Finite Fields Appl. **16** (2010), no. 5, 315–319, DOI 10.1016/j.ffa.2010.05.001. MR2678619

[13] Yasutaka Ihara, *Some remarks on the number of rational points of algebraic curves over finite fields*, J. Fac. Sci. Univ. Tokyo Sect. IA Math. **28** (1981), no. 3, 721–724 (1982). MR656048

[14] Wen-Ching W. Li and Hiren Maharaj, *Coverings of curves with asymptotically many rational points*, J. Number Theory **96** (2002), no. 2, 232–256. MR1932454

[15] Ruud Pellikaan, Henning Stichtenoth, and Fernando Torres, *Weierstrass semigroups in an asymptotically good tower of function fields*, Finite Fields Appl. **4** (1998), no. 4, 381–392, DOI 10.1006/ffta.1998.0217. MR1648573

[16] Keith Saints and Chris Heegard, *Algebraic-geometric codes and multidimensional cyclic codes: a unified theory and algorithms for decoding using Gröbner bases*, IEEE Trans. Inform. Theory **41** (1995), no. 6, 1733–1751, DOI 10.1109/18.476246. Special issue on algebraic geometry codes. MR1391032

[17] Jean-Pierre Serre, *Sur le nombre des points rationnels d'une courbe algébrique sur un corps fini* (French, with English summary), C. R. Acad. Sci. Paris Sér. I Math. **296** (1983), no. 9, 397–402. MR703906

[18] Henning Stichtenoth, *Algebraic function fields and codes*, 2nd ed., Graduate Texts in Mathematics, vol. 254, Springer-Verlag, Berlin, 2009. MR2464941

[19] Peter Sziklai, *A bound on the number of points of a plane curve*, Finite Fields Appl. **14** (2008), no. 1, 41–43, DOI 10.1016/j.ffa.2007.09.004. MR2381474

[20] Alexandre Temkine, *Hilbert class field towers of function fields over finite fields and lower bounds for $A(q)$*, J. Number Theory **87** (2001), no. 2, 189–210, DOI 10.1006/jnth.2000.2596. MR1824142

[21] M. A. Tsfasman, S. G. Vlăduţ, and Th. Zink, *Modular curves, Shimura curves, and Goppa codes, better than Varshamov-Gilbert bound*, Math. Nachr. **109** (1982), 21–28, DOI 10.1002/mana.19821090103. MR705893

[22] Th. Zink, *Degeneration of Shimura surfaces and a problem in coding theory*, Fundamentals of computation theory (Cottbus, 1985), Lecture Notes in Comput. Sci., vol. 199, Springer, Berlin, 1985, pp. 503–511, DOI 10.1007/BFb0028834. MR821267

Department of Applied Mathematics and Computer Science, Technical University of Denmark, Kongens Lyngby 2800, Denmark
Email address: pabe@dtu.dk

Department of Applied Mathematics and Computer Science, Technical University of Denmark, Kongens Lyngby 2800, Denmark
Email address: marimo@dtu.dk

Department of Applied Mathematics and Computer Science, Technical University of Denmark, Kongens Lyngby 2800, Denmark
Email address: lavi@dtu.dk

Contemporary Mathematics
Volume **779**, 2022
https://doi.org/10.1090/conm/779/15670

How big is the image of the Galois representations attached to CM elliptic curves?

Francesco Campagna and Riccardo Pengo

ABSTRACT. Using an analogue of Serre's open image theorem for elliptic curves with complex multiplication, one can associate to each CM elliptic curve E defined over a number field F a natural number $\mathcal{I}(E/F)$ which describes how big the image of the Galois representation associated to E is. We show how one can compute $\mathcal{I}(E/F)$, using a closed formula that we obtain from the classical theory of complex multiplication.

1. Introduction

Fix an algebraic closure $\overline{\mathbb{Q}}$ of the field of rational numbers $\mathbb{Q}$. Let E be an elliptic curve defined over a number field $F \subseteq \overline{\mathbb{Q}}$, and let:

$$\rho_E : G_F \to \operatorname{Aut}_{\mathbb{Z}}(E_{\text{tors}}) \tag{1.1}$$

be the representation of the absolute Galois group $G_F := \operatorname{Gal}(\overline{F}/F)$ associated to its action on the torsion points $E_{\text{tors}} := E(\overline{F})_{\text{tors}}$ of the elliptic curve E.

If E does not have complex multiplication (CM), *i.e.* $\operatorname{End}_{\overline{F}}(E) \cong \mathbb{Z}$, Serre's open image theorem [**17**, Théorème 3] implies that the index:

$$\mathcal{I}(E/F) := |\operatorname{Aut}_{\mathbb{Z}}(E_{\text{tors}}) \colon \rho_E(G_F)|$$

is finite. One is naturally led to investigate the dependence of $\mathcal{I}(E/F)$ on E and F. For instance, one can ask whether there exists an explicit, closed formula for $\mathcal{I}(E/F)$, whose terms can be effectively computed starting from a Weierstraß equation of E. At the time of writing, and to the best of the authors' knowledge, no such formula is available in the literature. The previous question can then be weakened, by asking whether there exists an upper bound for $\mathcal{I}(E/F)$, which can be effectively computed in terms of E. An affirmative answer to this second question has been provided by Lombardo in [**12**]. In fact, it has even been conjectured that there should exist such an upper bound which does not depend on E, but only on

---

2020 *Mathematics Subject Classification.* Primary 11G05, 14K22, 11F80, 11G15; Secondary 11Y40.

Key words and phrases. Elliptic curves, Complex multiplication, Galois representations.

The first author was supported by ANR-20-CE40-0003 Jinvariant. Moreover, he wishes to thank the Max Planck Institute for Mathematics in Bonn for its financial support, great work conditions and an inspiring atmosphere. The second author performed this work within the framework of the LABEX MILYON (ANR-10-LABX-0070) of Université de Lyon, within the program "Investissements d'Avenir" (ANR-11-IDEX-0007) operated by the French National Research Agency (ANR). Both authors thank the IRN GANDA for support.

the field of definition F. This conjecture is explicitly mentioned for $F = \mathbb{Q}$ in the introduction to the recent work of Rouse, Sutherland and Zureick-Brown [**16**], and is known to hold true under the assumption of Serre's uniformity conjecture, by previous work of Zywina (see [**26**, Theorem 1.4]).

On the other hand, if E has complex multiplication by an order $\mathcal{O}$ in an imaginary quadratic field K, *i.e.* $\mathrm{End}_{\overline{F}}(E) \cong \mathcal{O}$, the index of the image of ρ_E inside $\mathrm{Aut}_{\mathbb{Z}}(E_{\mathrm{tors}})$ is infinite. Nevertheless, as we recall in Section 2, one can formulate an analogue of Serre's open image theorem for E, by replacing $\mathrm{Aut}_{\mathbb{Z}}(E_{\mathrm{tors}})$ with a smaller subgroup $\mathcal{G}(E/F) \subseteq \mathrm{Aut}_{\mathbb{Z}}(E_{\mathrm{tors}})$, explicitly defined in (2.4), which is closed and of infinite index inside $\mathrm{Aut}_{\mathbb{Z}}(E_{\mathrm{tors}})$. As a consequence, the index:

$$\mathcal{I}(E/F) := |\mathcal{G}(E/F) \colon \rho_E(G_F)|$$

is finite, and, as above, one can ask whether it can be expressed by means of an explicit and closed formula. The main goal of this paper is to show how to use the classical theory of complex multiplication to give the following affirmative answer to this question.

THEOREM 1.1. *Let $\mathcal{O}$ be an order in an imaginary quadratic field $K \subseteq \overline{\mathbb{Q}}$. Let E be an elliptic curve that has complex multiplication by $\mathcal{O}$ and is defined over a number field $F \subseteq \overline{\mathbb{Q}}$. Denote by $K^{\mathrm{ab}} \subseteq \overline{\mathbb{Q}}$ the maximal abelian extension of K contained in $\overline{\mathbb{Q}}$, and by $FK \subseteq \overline{\mathbb{Q}}$ and $FK^{\mathrm{ab}} \subseteq \overline{\mathbb{Q}}$ the composita of F with K and K^{ab} respectively. Then:*

$$\mathcal{I}(E/F) = [(FK) \cap K^{\mathrm{ab}} \colon H_{\mathcal{O}}] \cdot \frac{|\mathcal{O}^{\times}|}{[F(E_{\mathrm{tors}}) : FK^{\mathrm{ab}}]} \tag{1.2}$$

where $H_{\mathcal{O}} \subseteq K^{\mathrm{ab}}$ is the ring class field *of K relative to the order $\mathcal{O}$ (see [**9**, § 9]), and $F(E_{\mathrm{tors}}) \subseteq \overline{\mathbb{Q}}$ is the field obtained by adjoining to F all the coordinates of all the points lying in E_{tors}.*

Note that the right-hand side of (1.2) makes sense because the field extension $K \subseteq H_{\mathcal{O}}$ is abelian, and, whenever $\mathrm{End}_{\overline{F}}(E) \cong \mathcal{O}$, one knows that $FK^{\mathrm{ab}} \subseteq F(E_{\mathrm{tors}})$ [**5**, § 4.1 and Remark 3.8], and $H_{\mathcal{O}} = K(j(E)) \subseteq FK$ [**9**, Theorem 11.1], where $j(E) \in F$ denotes the *j-invariant* of the elliptic curve E. Moreover, the classical theory of complex multiplication implies that the degree of the field extension $FK^{\mathrm{ab}} \subseteq F(E_{\mathrm{tors}})$ is finite and divides $|\mathcal{O}^{\times}|$. We explain this in more detail in Section 3, which is mainly devoted to the proof of Theorem 1.1.

As an immediate consequence of Theorem 1.1, one has the divisibility:

$$\mathcal{I}(E/F) \,\big|\, [(FK) \cap K^{\mathrm{ab}} \colon H_{\mathcal{O}}] \cdot |\mathcal{O}^{\times}| \tag{1.3}$$

which shows that $\mathcal{I}(E/F)$ can be bounded solely in terms of F, for every CM elliptic curve $E_{/F}$. This improves the upper bounds for $\mathcal{I}(E/F)$ previously proved by Lombardo [**13**, Theorem 6.6] and Bourdon and Clark [**3**, Corollary 1.5]. Moreover, Theorem 1.1 applied to any elliptic curve $E_{/\mathbb{Q}}$ which has complex multiplication by an imaginary quadratic order $\mathcal{O}$ shows that $\mathcal{I}(E/\mathbb{Q}) = |\mathcal{O}^{\times}|$. In the case $\mathcal{O} = \mathbb{Z}[i]$, this strengthens the conclusion of [**14**, Theorem 1.3].

The foregoing discussion shows that $\mathcal{I}(E/F)$ is very well understood in the CM case. However, it may not appear immediately clear how to apply (1.2) to compute $\mathcal{I}(E/F)$ in concrete examples. We explain how to do so in Section 4. In fact, after rewriting (1.2) appropriately (see Proposition 4.1), we obtain an algorithm that takes as inputs a number field F and a CM elliptic curve $E_{/F}$, and outputs

$\mathcal{I}(E/F)$. More precisely, we rephrase Equation (1.2) in terms of a finite extension $L \supseteq FK$ such that $F(E_{\text{tors}}) = LK^{\text{ab}}$. We prove in Proposition 4.2 that one can always take $L = F(E[I])$ to be the *I-division field* generated by the coordinates of the points $P \in E[I]$ belonging to the *I-torsion subgroup*:

$$E[I] := \bigcap_{\alpha \in I} \ker \left(E(\overline{F}) \xrightarrow{[\alpha]_E} E(\overline{F}) \right)$$

where $I \subseteq \mathcal{O}$ is any ideal such that $|\mathbb{Z}/(I \cap \mathbb{Z})| > \max(2, |\mathcal{O}^\times|/2)$, and the map:

$$[\cdot]_E \colon \mathcal{O} \xrightarrow{\sim} \operatorname{End}_{\overline{F}}(E)$$

is the normalised isomorphism described in Lemma 2.1. In practice, if $j(E) \neq 0$ one usually takes $L = F(E[3])$ in order to ease the computational burden. We devote Section 5 to the application of this algorithm to some explicit examples of elliptic curves E that have complex multiplication by imaginary quadratic orders $\mathcal{O}$ of class number two.

2. Analogues of Serre's open image theorem for CM elliptic curves

Let E be an elliptic curve defined over a number field $F \subseteq \overline{\mathbb{Q}}$. Then, the absolute Galois group G_F naturally acts both on the set $E_{\text{tors}} = \varinjlim_N E[N]$, and on the adelic Tate module $\mathcal{T}(E) := \varprojlim_N E[N]$. The first action gives rise to the Galois representation ρ_E appearing in (1.1), whereas the action on $\mathcal{T}(E)$ induces another Galois representation $\varrho_E \colon G_F \to \operatorname{Aut}_{\widehat{\mathbb{Z}}}(\mathcal{T}(E))$. As done in [**17**, § 4.1, Remarque (1)], one can construct an isomorphism:

$$\nu \colon \operatorname{Aut}_{\widehat{\mathbb{Z}}}(\mathcal{T}(E)) \xrightarrow{\sim} \operatorname{Aut}_{\widehat{\mathbb{Z}}}(E_{\text{tors}}) = \operatorname{Aut}_{\mathbb{Z}}(E_{\text{tors}})$$

such that $\rho_E = \nu \circ \varrho_E$. As a consequence, one can indifferently study the Galois representation ρ_E, as done in this paper, or its twin ϱ_E, as done in some of our references.

If E does not have complex multiplication, *i.e.* if $\operatorname{End}_F(E) \cong \operatorname{End}_{\overline{F}}(E) \cong \mathbb{Z}$, then the celebrated "open image theorem", proved by Serre in [**17**, Théorème 3], shows that the image of the Galois representation ρ_E is a subgroup of finite index inside $\operatorname{Aut}_{\mathbb{Z}}(E_{\text{tors}}) \cong \operatorname{GL}_2(\widehat{\mathbb{Z}})$, where $\widehat{\mathbb{Z}} := \varprojlim_N (\mathbb{Z}/N\mathbb{Z})$ denotes the profinite completion of $\mathbb{Z}$. On the other hand, if the elliptic curve E has complex multiplication, the image of ρ_E is not open inside $\operatorname{Aut}_{\mathbb{Z}}(E_{\text{tors}})$. However, one can formulate a CM analogue of Serre's open image theorem by replacing $\operatorname{Aut}_{\mathbb{Z}}(E_{\text{tors}})$ with an appropriate closed subgroup $\mathcal{G}(E/F) \subseteq \operatorname{Aut}_{\mathbb{Z}}(E_{\text{tors}})$, which we now describe.

Suppose now that $\operatorname{End}_{\overline{F}}(E) \ncong \mathbb{Z}$. Then the endomorphism ring $\operatorname{End}_{\overline{F}}(E)$ can be canonically identified with an order inside an imaginary quadratic field, as the following classical lemma shows.

LEMMA 2.1. *Let F be a number field, and $E_{/F}$ be an elliptic curve such that $\operatorname{End}_{\overline{F}}(E) \ncong \mathbb{Z}$, where $\overline{F}$ denotes a fixed algebraic closure of F. Then, there exists an imaginary quadratic field K and an order $\mathcal{O} \subseteq K$ such that $\operatorname{End}_{\overline{F}}(E) \cong \mathcal{O}$. Moreover, for each embedding $\iota \colon K \hookrightarrow \overline{F}$, there exists a unique isomorphism:*

$$[\cdot]_{E,\iota} \colon \mathcal{O} \xrightarrow{\sim} \operatorname{End}_{\overline{F}}(E)$$

such that $[\alpha]_{E,\iota}^*(\omega) = \iota(\alpha) \cdot \omega$ for every $\alpha \in \mathcal{O}$ and every invariant differential ω defined over $E_{\overline{F}}$, where $[\alpha]_{E,\iota}^*(\omega)$ denotes the pull-back of ω along the endomorphism $[\alpha]_{E,\iota}$.

Proof. See [**21**, Chapter III, Corollary 9.4] for the existence of K and $\mathcal{O}$. Moreover, the existence of $[\cdot]_{E,\iota}$ follows from [**20**, Chapter II, Proposition 1.1], after fixing an embedding $\overline{F} \hookrightarrow \mathbb{C}$. Finally, observe that for any two isomorphisms $[\cdot], [\cdot]' \colon \mathcal{O} \xrightarrow{\sim} \mathrm{End}_{\overline{F}}(E)$, there exists an automorphism $\sigma \colon \mathcal{O} \xrightarrow{\sim} \mathcal{O}$ with the property that $[\alpha] = [\sigma(\alpha)]'$ for every $\alpha \in \mathcal{O}$. Hence, if these isomorphisms satisfy the requirements of the lemma, we see that $\iota(\alpha - \sigma(\alpha)) \cdot \omega = 0$ for every $\alpha \in \mathcal{O}$ and every invariant differential ω. Thus, we have that $\sigma = \mathrm{Id}_{\mathcal{O}}$, which allows us to conclude. □

Now, suppose at first that $E_{/F}$ is an elliptic curve with the property that $\mathrm{End}_F(E) \cong \mathrm{End}_{\overline{F}}(E) \cong \mathcal{O}$ for some order $\mathcal{O}$ inside an imaginary quadratic field K. Then by [**19**, Chapter II, Proposition 30] we necessarily have $K \subseteq F$ and one can easily show (using for instance [**20**, Chapter II, Theorem 2.2]) that the absolute Galois group G_F of F acts as $\mathcal{O}$-module automorphisms on E_{tors}. Thus, we have:

$$\rho_E(G_F) \subseteq \mathrm{Aut}_{\mathcal{O}}(E_{\mathrm{tors}}) =: \mathcal{G}(E/F) \tag{2.1}$$

where $\mathcal{G}(E/F)$ is an abelian group canonically isomorphic to $\widehat{\mathcal{O}}^\times$, the unit group of the profinite completion $\widehat{\mathcal{O}} := \varprojlim_N (\mathcal{O}/N\mathcal{O})$. In particular, the field extension $F \subseteq F(E_{\mathrm{tors}})$ is abelian. Note also that $\mathrm{Aut}_{\mathcal{O}}(E_{\mathrm{tors}})$ is closed inside $\mathrm{Aut}_{\mathbb{Z}}(E_{\mathrm{tors}})$, since we have:

$$\mathrm{Aut}_{\mathcal{O}}(E_{\mathrm{tors}}) = \bigcap_{N \in \mathbb{N}} \mathrm{res}_N^{-1}(\mathrm{Aut}_{\mathcal{O}}(E[N]))$$

where $\mathrm{res}_N : \mathrm{Aut}_{\mathbb{Z}}(E_{\mathrm{tors}}) \to \mathrm{Aut}_{\mathbb{Z}}(E[N])$ denotes the natural restriction map. On the other hand, $\mathrm{Aut}_{\mathcal{O}}(E_{\mathrm{tors}})$ is not open inside $\mathrm{Aut}_{\mathbb{Z}}(E_{\mathrm{tors}}) \cong \mathrm{GL}_2(\widehat{\mathbb{Z}})$, because the latter does not contain any abelian subgroup of finite index. However, the subgroup $\rho_E(G_F)$ is open in $\mathrm{Aut}_{\mathcal{O}}(E_{\mathrm{tors}})$, as shown in [**17**, § 4.5] using the classical theorems of complex multiplication. Since $\mathrm{Aut}_{\mathcal{O}}(E_{\mathrm{tors}}) \cong \widehat{\mathcal{O}}^\times$ is a profinite group, this in particular implies that the index of $\rho_E(G_F)$ inside $\mathrm{Aut}_{\mathcal{O}}(E_{\mathrm{tors}})$ is finite. We can regard this result as an analogue of Serre's open image theorem for those CM elliptic curves whose field of definition contains the field K.

Assume now that the elliptic curve $E_{/F}$ has the properties that $\mathrm{End}_F(E) \cong \mathbb{Z}$ and $\mathrm{End}_{\overline{F}}(E) \cong \mathcal{O}$, for some order $\mathcal{O}$ inside an imaginary quadratic field K. Again by [**19**, Chapter II, Proposition 30], under these assumptions we must have $K \not\subseteq F$. Since not all the geometric endomorphisms of E are defined over the base field, in this case the Galois group G_F does not respect the $\mathcal{O}$-module structure on E_{tors}. More precisely, since we fixed an embedding $\mathcal{O} \subseteq K \subseteq \overline{\mathbb{Q}} = \overline{F}$, there exists a unique isomorphism $[\cdot]_E : \mathcal{O} \xrightarrow{\sim} \mathrm{End}_{\overline{F}}(E)$ such that for every $\alpha \in \mathcal{O}$ and every invariant differential ω on the elliptic curve $E_{\overline{F}}$, the equality $[\alpha]_E^*(\omega) = \alpha\omega$ holds. Then an automorphism $\sigma \in G_F$ acts on $[\alpha]_E(P)$ as:

$$\sigma\left([\alpha]_E(P)\right) = [\sigma(\alpha)]_E(\sigma(P)) \tag{2.2}$$

as follows from [**20**, Chapter II, Theorem 2.2]. We then see that for every $\sigma \in G_F$ and each fixed $\tau \in G_F$ restricting to the unique non-trivial element in $\mathrm{Gal}(FK/F)$, exactly one among σ and $\sigma\tau$ acts $\mathcal{O}$-linearly on E_{tors}. We deduce that:

$$\rho_E\left(G_F\right) \subseteq \langle \mathrm{Aut}_{\mathcal{O}}(E_{\mathrm{tors}}), \rho_E(\tau) \rangle := \mathcal{G}(E/F) \tag{2.3}$$

and one can easily show that the group $\mathcal{G}(E/F)$ does not actually depend on τ, thus justifying the notation. Indeed, if both $\tau, \tau' \in G_F$ restrict to the unique non-trivial element of $\mathrm{Gal}(FK/F)$, one has that $\tau\tau' \in \mathrm{Gal}(\overline{F}/FK)$. This implies that

$\rho_E(\tau\tau') \in \mathrm{Aut}_{\mathcal{O}}(E_{\mathrm{tors}})$, which gives $\langle \mathrm{Aut}_{\mathcal{O}}(E_{\mathrm{tors}}), \rho_E(\tau)\rangle = \langle \mathrm{Aut}_{\mathcal{O}}(E_{\mathrm{tors}}), \rho_E(\tau')\rangle$ as wanted. Moreover, $\rho_E(\tau)$ normalises $\mathrm{Aut}_{\mathcal{O}}(E_{\mathrm{tors}})$, as follows from (2.2) and the fact that $\rho_E(\tau)^2 \in \mathrm{Aut}_{\mathcal{O}}(E_{\mathrm{tors}})$. Hence, we see that $\mathrm{Aut}_{\mathcal{O}}(E_{\mathrm{tors}})$ is a normal subgroup of $\mathcal{G}(E/F)$ with index $|\mathcal{G}(E/F) : \mathrm{Aut}_{\mathcal{O}}(E_{\mathrm{tors}})| = 2$. As a consequence, $\mathcal{G}(E/F)$ is closed inside $\mathrm{Aut}_{\mathbb{Z}}(E_{\mathrm{tors}})$, and so it is a profinite group. On the other hand, $\mathcal{G}(E/F)$ is not open inside $\mathrm{Aut}_{\mathbb{Z}}(E_{\mathrm{tors}})$, because it contains the abelian group $\mathrm{Aut}_{\mathcal{O}}(E_{\mathrm{tors}})$ as a finite-index subgroup. Thus, $\rho_E(G_F)$ cannot be open inside $\mathrm{Aut}_{\mathbb{Z}}(E_{\mathrm{tors}})$. Nevertheless, $\rho_E(G_F)$ is open inside the closed subgroup $\mathcal{G}(E/F)$, as the following lemma shows.

LEMMA 2.2. *Let $E_{/F}$ be an elliptic curve with complex multiplication by an order $\mathcal{O}$ in an imaginary quadratic field $K \not\subseteq F$, and let $E' := E_{FK}$ denote the base-change of E to the compositum FK. Then $\rho_E(G_F)$ is open in $\mathcal{G}(E/F)$, and the following equality:*

$$\mathcal{I}(E/F) := |\mathcal{G}(E/F) : \rho_E(G_F)| = |\mathrm{Aut}_{\mathcal{O}}(E'_{\mathrm{tors}}) : \rho_{E'}(G_{FK})| =: \mathcal{I}(E/FK)$$

holds.

PROOF. Since $\mathrm{Aut}_{\mathcal{O}}(E_{\mathrm{tors}})$ is closed and of finite index in $\mathcal{G}(E/F)$, it is also open in the same group. Moreover, the subgroup $\rho_{E'}(G_{FK}) \subseteq \mathrm{Aut}_{\mathcal{O}}(E'_{\mathrm{tors}})$ is open by [**17**, § 4.5, Corollaire], and clearly the equalities $\rho_{E'}(G_{FK}) = \rho_E(G_{FK})$ and $\mathrm{Aut}_{\mathcal{O}}(E'_{\mathrm{tors}}) = \mathrm{Aut}_{\mathcal{O}}(E_{\mathrm{tors}})$ hold. Thus we see that $\rho_E(G_{FK})$ is an open subgroup of $\rho_E(G_F)$ and we conclude that the latter is open in $\mathcal{G}(E/F)$. In particular, $\rho_E(G_F)$ is a closed subgroup of finite index inside $\mathcal{G}(E/F)$.

To prove the equality of indices, we use the fact that $FK \subseteq F(E_{\mathrm{tors}})$, by [**4**, Lemma 3.15]. Since ρ_E induces an injective map $\mathrm{Gal}(F(E_{\mathrm{tors}})/F) \hookrightarrow \mathcal{G}(E/F)$, we have $|\rho_E(G_F) : \rho_E(G_{FK})| = 2$. Now, the computation:

$$|\mathcal{G}(E/F) : \rho_E(G_F)| = \frac{1}{2}\,|\mathcal{G}(E/F) : \rho_E(G_{FK})| = |\mathrm{Aut}_{\mathcal{O}}(E_{\mathrm{tors}}) : \rho_E(G_{FK})|$$

allows us to conclude. □

We summarise our discussion so far. Given a number field F and an elliptic curve $E_{/F}$ with complex multiplication by an order $\mathcal{O}$ in an imaginary quadratic field K, we define, following (2.1) and (2.3):

$$\mathcal{G}(E/F) := \begin{cases} \mathrm{Aut}_{\mathcal{O}}(E_{\mathrm{tors}}) & \text{if } K \subseteq F, \\ \langle \mathrm{Aut}_{\mathcal{O}}(E_{\mathrm{tors}}), \rho_E(\tau)\rangle & \text{if } K \not\subseteq F \end{cases} \tag{2.4}$$

where, if $K \not\subseteq F$, we let $\tau \in G_F$ be any automorphism that restricts to the unique non-trivial element of $\mathrm{Gal}(FK/F)$. Then, in the previous discussion, we have shown that $\mathcal{G}(E/F)$ is a profinite group, which contains $\rho_E(G_F)$ as an open subgroup. Moreover, if we define the *CM index* $\mathcal{I}(E/F)$ to be:

$$\mathcal{I}(E/F) := |\mathcal{G}(E/F) : \rho_E(G_F)| \tag{2.5}$$

then by Lemma 2.2 we have that $\mathcal{I}(E/F) = \mathcal{I}(E/FK)$ is finite.

3. A formula for the index

The aim of this section is to provide a proof of Theorem 1.1. We place ourselves in the setting of the theorem, by fixing an order $\mathcal{O}$ inside an imaginary quadratic field $K \subseteq \overline{\mathbb{Q}}$ and an elliptic curve E which has complex multiplication by $\mathcal{O}$ and is defined over a number field $F \subseteq \overline{\mathbb{Q}}$. We explained in Lemma 2.2 that the equality

$\mathcal{I}(E/F) = \mathcal{I}(E/FK)$ holds true, hence we will assume without loss of generality that $K \subseteq F$. This in particular implies that $H_{\mathcal{O}} \subseteq F$, where, as in Theorem 1.1, $H_{\mathcal{O}}$ denotes the ring class field of K relative to the order $\mathcal{O}$.

The formula (1.2) appearing in Theorem 1.1 is a byproduct of the first main theorem of complex multiplication (see [**11**, Chapter 10, Theorem 8]). The latter asserts the existence of a unique continuous group homomorphism $\mu\colon \mathbb{A}_F^\times \to K^\times$ such that, for every $s \in \mathbb{A}_F^\times$ and every complex uniformisation $\xi\colon \mathbb{C} \twoheadrightarrow E(\mathbb{C})$ with $\Lambda := \ker(\xi) \subseteq K$, the following diagram:

$$\begin{array}{ccc} K/\Lambda & \xrightarrow{(\mu(s)\,\mathrm{N}_{F/K}(s^{-1}))\cdot} & K/\Lambda \\ \downarrow{\scriptstyle \xi} & & \downarrow{\scriptstyle \xi} \\ E_{\mathrm{tors}} & \xrightarrow{[s,F]} & E_{\mathrm{tors}} \end{array}$$

commutes. Here $\mathrm{N}_{F/K}\colon \mathbb{A}_F^\times \to \mathbb{A}_K^\times$ denotes the idelic norm map, whereas the notation $[\cdot, F]\colon \mathbb{A}_F^\times \twoheadrightarrow \mathrm{Gal}(F^{\mathrm{ab}}/F)$ stands for the global Artin map, and the upper horizontal arrow is given by the idelic multiplication map (see [**11**, Page 100]). In particular, the action of the idèle $\mu(s)\,\mathrm{N}_{F/K}(s^{-1}) \in \mathbb{A}_K^\times$ on the set of lattices contained in K, described in [**11**, Chapter 8, Theorem 10], fixes Λ. Since Λ is an invertible fractional ideal of $\mathcal{O}$, this implies that $\mu(s)\,\mathrm{N}_{F/K}(s^{-1})$ fixes also $\mathcal{O}$. Thus, the finite idèle $(\mu(s)\,\mathrm{N}_{F/K}(s^{-1}))_{\mathsf{fin}}$ lies in the subgroup $\widehat{\mathcal{O}}^\times \subseteq \mathbb{A}_K^\times$. Hence, the association $s \mapsto (\mu(s)\,\mathrm{N}_{F/K}(s^{-1}))_{\mathsf{fin}}$ defines a continuous group homomorphism $\theta_E\colon \mathbb{A}_F^\times \to \widehat{\mathcal{O}}^\times$, which makes the following diagram:

$$\begin{array}{ccc} \mathbb{A}_F^\times & \xrightarrow{\theta_E} & \widehat{\mathcal{O}}^\times \\ {\scriptstyle [\cdot,F]|_{F(E_{\mathrm{tors}})}}\downarrow & & \downarrow{\scriptstyle \wr} \\ \mathrm{Gal}(F(E_{\mathrm{tors}})/F) & \overset{\rho_E}{\hookrightarrow} & \mathrm{Aut}_{\mathcal{O}}(E_{\mathrm{tors}}) \end{array} \tag{3.1}$$

commute. We are now ready to prove Theorem 1.1.

PROOF OF THEOREM 1.1. Define ψ_E to be the group homomorphism:

$$\psi_E \;:\; \mathrm{Aut}_{\mathcal{O}}(E_{\mathrm{tors}}) \cong \widehat{\mathcal{O}}^\times \xrightarrow{\mathfrak{a}_{\mathcal{O}}} \mathrm{Gal}(K^{\mathrm{ab}}/H_{\mathcal{O}})$$

where $\mathfrak{a}_{\mathcal{O}}\colon \widehat{\mathcal{O}}^\times \twoheadrightarrow \mathrm{Gal}(K^{\mathrm{ab}}/H_{\mathcal{O}})$ is the composition of the natural embedding $\widehat{\mathcal{O}}^\times \hookrightarrow \mathbb{A}_K^\times$ with the map $\mathbb{A}_K^\times \twoheadrightarrow G_K^{\mathrm{ab}}$ given by $s \mapsto [s^{-1}, K]$. It is easy to show that ψ_E fits in a short exact sequence:

$$1 \to \mathrm{Aut}_F(E) \to \mathrm{Aut}_{\mathcal{O}}(E_{\mathrm{tors}}) \xrightarrow{\psi_E} \mathrm{Gal}(K^{\mathrm{ab}}/H_{\mathcal{O}}) \to 1 \tag{3.2}$$

because $\ker(\mathfrak{a}_{\mathcal{O}}) = \ker([\cdot, K]) \cap \widehat{\mathcal{O}}^\times = K^\times \cap \widehat{\mathcal{O}}^\times = \mathcal{O}^\times$. Then, we can form the following square:

$$\begin{array}{ccc} \mathrm{Gal}(F(E_{\mathrm{tors}})/F) & \overset{\rho_E}{\hookrightarrow} & \mathrm{Aut}_{\mathcal{O}}(E_{\mathrm{tors}}) \\ \downarrow & & \downarrow{\scriptstyle \psi_E} \\ \mathrm{Gal}(K^{\mathrm{ab}}/F \cap K^{\mathrm{ab}}) & \overset{\iota}{\hookrightarrow} & \mathrm{Gal}(K^{\mathrm{ab}}/H_{\mathcal{O}}) \end{array} \tag{3.3}$$

where the map on the left is defined by the composition:

$$\mathrm{Gal}(F(E_{\mathrm{tors}})/F) \twoheadrightarrow \mathrm{Gal}(FK^{\mathrm{ab}}/F) \xrightarrow{\sim} \mathrm{Gal}(K^{\mathrm{ab}}/F \cap K^{\mathrm{ab}})$$

of a restriction map and a natural isomorphism coming from Galois theory. We claim that (3.3) commutes. Indeed, extending (3.3) by diagram (3.1) gives the following square:

$$\begin{array}{ccc} \mathbb{A}_F^\times & \xrightarrow{\theta_E} & \widehat{\mathcal{O}}^\times \\ {\scriptstyle [\cdot,F]|_{K^{\mathrm{ab}}}}\downarrow & & \downarrow{\scriptstyle \mathfrak{a}_\mathcal{O}} \\ \mathrm{Gal}(K^{\mathrm{ab}}/F\cap K^{\mathrm{ab}}) & \overset{\iota}{\hookrightarrow} & \mathrm{Gal}(K^{\mathrm{ab}}/H_\mathcal{O}) \end{array} \tag{3.4}$$

which commutes because, for every $s \in \mathbb{A}_F^\times$, one has:

$$\mathfrak{a}_\mathcal{O}(\theta_E(s)) = [(\mu(s)\cdot \mathrm{N}_{F/K}(s^{-1}))^{-1}, K] = [\mathrm{N}_{F/K}(s), K] = \iota([s,F]|_{K^{\mathrm{ab}}})$$

using the fact that $K^\times \cdot (K \otimes_\mathbb{Q} \mathbb{R})^\times \subseteq \ker([\cdot, K])$, as explained in [**1**, Chapter IX, Theorem 3], and the functoriality of class field theory [**15**, Chapter VI, Proposition 5.2]. Thus (3.3) commutes, because (3.4) does, and the vertical maps in the commutative diagram (3.1) are surjective.

Now, (3.2) and (3.3) induce the following commutative diagram:

$$\begin{array}{ccccc} \mathrm{Gal}(F(E_{\mathrm{tors}})/FK^{\mathrm{ab}}) & \hookrightarrow & \mathrm{Gal}(F(E_{\mathrm{tors}})/F) & \twoheadrightarrow & \mathrm{Gal}(K^{\mathrm{ab}}/F\cap K^{\mathrm{ab}}) \\ \downarrow{\scriptstyle \iota'} & & \downarrow{\scriptstyle \rho_E} & (3.3) & \downarrow{\scriptstyle \iota} \\ \mathrm{Aut}_F(E) & \hookrightarrow & \mathrm{Aut}_\mathcal{O}(E_{\mathrm{tors}}) & \underset{\psi_E}{\twoheadrightarrow} & \mathrm{Gal}(K^{\mathrm{ab}}/H_\mathcal{O}) \end{array} \tag{3.5}$$

whose rows are exact. This shows in particular that the degree of the extension $FK^{\mathrm{ab}} \subseteq F(E_{\mathrm{tors}})$ is finite and divides $|\mathrm{Aut}_F(E)| = |\mathcal{O}^\times|$. Finally, we have:

$$\begin{aligned} \mathcal{I}(E/F) &= |\mathrm{coker}(\rho_E)| = |\mathrm{coker}(\iota)| \cdot |\mathrm{coker}(\iota')| \\ &= [F\cap K^{\mathrm{ab}} \colon H_\mathcal{O}] \cdot \frac{|\mathcal{O}^\times|}{[F(E_{\mathrm{tors}}) : FK^{\mathrm{ab}}]} \end{aligned}$$

by the snake lemma, which allows us to conclude. □

An immediate consequence of Theorem 1.1 is the following improvement of the bounds provided by [**13**, Theorem 6.6] and [**3**, Corollary 1.5].

COROLLARY 3.1. *Let $\mathcal{O}$ be an order inside an imaginary quadratic field K. For every number field $F \subseteq \overline{\mathbb{Q}}$, and every elliptic curve $E_{/F}$ with complex multiplication by $\mathcal{O}$, the index $\mathcal{I}(E/F)$ divides $[(FK)\cap K^{\mathrm{ab}} \colon H_\mathcal{O}] \cdot |\mathcal{O}^\times|$.*

Moreover, Theorem 1.1 can be rephrased in a simpler fashion, if one assumes that $|\mathcal{O}^\times| = 2$, which holds for every order $\mathcal{O}$ of discriminant $\Delta_\mathcal{O} < -4$.

COROLLARY 3.2. *Let $\mathcal{O}$ be an order inside an imaginary quadratic field K, and suppose that $\Delta_\mathcal{O} < -4$. Let E be an elliptic curve with complex multiplication by $\mathcal{O}$, defined over a number field $F \subseteq \overline{\mathbb{Q}}$. Then, the following equality:*

$$\frac{\mathcal{I}(E/F)}{[(FK)\cap K^{\mathrm{ab}} : H_\mathcal{O}]} = \begin{cases} 2, & \textit{if } F(E_{\mathrm{tors}}) = FK^{\mathrm{ab}} \\ 1, & \textit{otherwise} \end{cases} \tag{3.6}$$

holds.

The dichotomy provided by (3.6) reflects a property of CM elliptic curves introduced by Shimura in [**18**, Pages 216-218], and studied in [**5**, § 5]. In particular, Corollary 3.2 generalises [**5**, Corollary 5.8], which was proved by different means.

REMARK 3.3. Setting $F = \mathbb{Q}(j(E))$ in Theorem 1.1 we see that $\mathcal{I}(E/F) \in \{1, |\mathcal{O}^\times|\}$. However, this does not allow to describe explicitly the image $\rho_E(G_F)$ as a subgroup of $\mathrm{Aut}_{\mathbb{Z}}(E_{\mathrm{tors}}) \cong \mathrm{GL}_2(\widehat{\mathbb{Z}})$, since the latter can vary amongst infinitely many possible subgroups, as it happens already for $F = \mathbb{Q}$ (see [**5**, Theorem 6.3]). On the other hand, the image of $\rho_E(G_F)$ under the projections $\mathrm{GL}_2(\widehat{\mathbb{Z}}) \twoheadrightarrow \mathrm{GL}_2(\mathbb{Z}_\ell)$ for $\ell \in \mathbb{N}$ a prime, belongs, up to conjugation, to a finite list of subgroups which has been explicitly determined by Lozano-Robledo [**14**].

To conclude this section, we observe that Theorem 1.1 implies that the index $\mathcal{I}(E/F)$ is invariant under appropriate twisting of the elliptic curve E, as specified by the following corollary.

COROLLARY 3.4. *Let $\mathcal{O}$ be an order inside an imaginary quadratic field K, and set $d := |\mathcal{O}^\times|$. Let $E_{/F}$ be an elliptic curve defined over a number field $F \subseteq \overline{\mathbb{Q}}$ such that* $\mathrm{End}_F(E) \cong \mathcal{O}$. *Suppose that E is the twist of another elliptic curve $E'_{/F}$ by $\sqrt[d]{\alpha}$, for some $\alpha \in F^\times$ such that $L := F(\sqrt[d]{\alpha}) \subseteq FK^{\mathrm{ab}}$. Then $\mathcal{I}(E/F) = \mathcal{I}(E'/F)$.*

PROOF. First of all, note that the extension $F \subseteq L$ is well defined, because $K \subseteq F$ by the hypothesis $\mathrm{End}_F(E) \cong \mathcal{O}$, and thus the group of d-th roots of unity $\mathcal{O}^\times$ is also contained in F. Then, one has:

$$\rho_E(\sigma) = \rho_{E'}(\sigma) \cdot \chi_\alpha(\sigma) \tag{3.7}$$

for every $\sigma \in G_F$, where $\rho_E \colon G_F \to \mathcal{G}(E/F) \cong \widehat{\mathcal{O}}^\times$ and $\rho_{E'} \colon G_F \to \mathcal{G}(E'/F) \cong \widehat{\mathcal{O}}^\times$ are the Galois representations associated to E and E'. Moreover, the map:

$$\chi_\alpha \colon G_F \to \mathcal{O}^\times \subseteq \widehat{\mathcal{O}}^\times$$

is the Kummer character attached to the extension $F \subseteq L$. In particular, for every $\sigma \in G_F$ the unit $\chi_\alpha(\sigma) \in \mathcal{O}^\times$ is defined by the equality $\sigma(\sqrt[d]{\alpha}) = \chi_\alpha(\sigma) \cdot \sqrt[d]{\alpha}$.

Now, for every $\sigma \in \mathrm{Gal}(\overline{\mathbb{Q}}/LF(E'_{\mathrm{tors}}))$, we have that $\rho_{E'}(\sigma) = \chi_\alpha(\sigma) = 1$, hence (3.7) implies that $\rho_E(\sigma) = 1$. Thus, the inclusion $F(E_{\mathrm{tors}}) \subseteq LF(E'_{\mathrm{tors}})$ holds. On the other hand, if $\tau \in \mathrm{Gal}(\overline{\mathbb{Q}}/F(E_{\mathrm{tors}}))$, the hypothesis $L \subseteq FK^{\mathrm{ab}}$ and the inclusion $FK^{\mathrm{ab}} \subseteq F(E_{\mathrm{tors}})$ imply that τ fixes L. Therefore $\rho_E(\tau) = \chi_\alpha(\tau) = 1$, and (3.7) gives that $\rho_{E'}(\tau) = 1$. Hence, the opposite inclusion $LF(E'_{\mathrm{tors}}) \subseteq F(E_{\mathrm{tors}})$ holds. Thus, we have that $F(E_{\mathrm{tors}}) = LF(E'_{\mathrm{tors}}) = F(E'_{\mathrm{tors}})$, where the last equality follows from the hypothesis $L \subseteq FK^{\mathrm{ab}}$ and the inclusion $FK^{\mathrm{ab}} \subseteq F(E'_{\mathrm{tors}})$. Finally, using Theorem 1.1, one gets that $\mathcal{I}(E/F) = \mathcal{I}(E'/F)$, as we wanted to prove. □

4. How to compute the index in practice

In this section we show how one can concretely compute the index $\mathcal{I}(E/F)$ for any given CM elliptic curve E defined over a number field F. Thanks to Lemma 2.2, we can and will assume throughout this section, without loss of generality, that the number field F contains the CM field K.

The starting point of our discussion is the formula (1.2) provided by Theorem 1.1. Let us observe that (1.2), albeit completely explicit, involves the degree of the finite extension $FK^{\mathrm{ab}} \subseteq F(E_{\mathrm{tors}})$ which *a priori* can not be implemented in a computer, because FK^{ab} is an infinite algebraic extension of $\mathbb{Q}$. Nevertheless, the following result shows how one can rewrite (1.2) as an equality involving only finite abelian groups and number fields.

PROPOSITION 4.1. *Let $\mathcal{O}$ be an order inside an imaginary quadratic field $K \subseteq \overline{\mathbb{Q}}$. Fix a number field $F \subseteq \overline{\mathbb{Q}}$ and an elliptic curve $E_{/F}$ such that* $\mathrm{End}_F(E) \cong \mathcal{O}$. *Then, we have:*

$$\mathcal{I}(E/F) = \frac{|\mathcal{O}^\times| \cdot [L \cap K^{\mathrm{ab}} \colon K]}{|\mathrm{Pic}(\mathcal{O})| \cdot [L \colon F]} \tag{4.1}$$

for every finite extension $F \subseteq L$ such that $F(E_{\mathrm{tors}}) = LK^{\mathrm{ab}}$ is the compositum of L and K^{ab} inside $\overline{\mathbb{Q}}$.

PROOF. Combining Theorem 1.1 with the equality:

$$[F(E_{\mathrm{tors}}) \colon FK^{\mathrm{ab}}] = [LK^{\mathrm{ab}} \colon FK^{\mathrm{ab}}] = \frac{[L \colon F]}{[L \cap K^{\mathrm{ab}} \colon F \cap K^{\mathrm{ab}}]} = \frac{[L \colon F][F \cap K^{\mathrm{ab}} \colon K]}{[L \cap K^{\mathrm{ab}} \colon K]}$$

allows us to conclude, because $[F \cap K^{\mathrm{ab}} \colon K] = [F \cap K^{\mathrm{ab}} \colon H_{\mathcal{O}}] \cdot |\mathrm{Pic}(\mathcal{O})|$. □

Using Proposition 4.1, we can now reduce the computation of $\mathcal{I}(E/F)$ to the following steps:

- S.1 compute $|\mathcal{O}^\times|$ and $|\mathrm{Pic}(\mathcal{O})|$;
- S.2 find a finite extension $F \subseteq L$ such that $F(E_{\mathrm{tors}}) = LK^{\mathrm{ab}}$, and compute $[L \colon F]$;
- S.3 compute $[L \cap K^{\mathrm{ab}} \colon K]$, *i.e.* the degree of the maximal abelian sub-extension of $K \subseteq L$.

To achieve S.1 one can use for instance the algorithms described in [**7**, § 5.3] for the computation of $|\mathrm{Pic}(\mathcal{O})|$, and the fact that $|\mathcal{O}^\times| = 2$ unless $\mathcal{O} = \mathbb{Z}[i]$, for which $|\mathcal{O}^\times| = 4$, or $\mathcal{O} = \mathbb{Z}\left[\frac{1+\sqrt{-3}}{2}\right]$, for which $|\mathcal{O}^\times| = 6$. Moreover, once S.2 has been carried out, and the extension $F \subseteq L$ is known, one can deal with the last step S.3 in (at least) two different ways:

- one can use the isomorphism:

$$\mathrm{Gal}(L \cap K^{\mathrm{ab}}/K) \cong \mathrm{Gal}(L'/K)^{\mathrm{ab}} \tag{4.2}$$

  where $K \subseteq L' \subseteq L$ denotes the maximal sub-extension of $K \subseteq L$ which is Galois over K, and the notation S^{ab} stands for the abelianization of a finite group S (*i.e.* its maximal abelian quotient). In order to compute the right hand side of (4.2), note that, if $G := \mathrm{Gal}(\widetilde{L}/K)$ denotes the Galois group of the Galois closure $\widetilde{L}$ of the extension $K \subseteq L$, and $H^G \subseteq G$ denotes the normal closure of the subgroup $H := \mathrm{Gal}(\widetilde{L}/L)$ inside G, then we have $\mathrm{Gal}(L'/K) \cong G/H^G$. Since both G and H can be computed as subgroups of the symmetric group $\mathfrak{S}_n$ on $n = [L \colon K]$ letters (see [**7**, § 6.3]), the abelian group $(G/H^G)^{\mathrm{ab}}$ can also be explicitly computed, for instance using the functions NORMALCLOSURE and MAXIMALABELIANQUOTIENT in GAP [**10**];
- one can compute $[L \cap K^{\mathrm{ab}} \colon K]$ as the index $|\mathrm{Cl}_{\mathfrak{m}}(K) \colon \mathrm{T}_{\mathfrak{m}}(L/K)|$ of the norm group $\mathrm{T}_{\mathfrak{m}}(L/K)$ inside the ray class group $\mathrm{Cl}_{\mathfrak{m}}(K)$ modulo the relative discriminant $\mathfrak{m} := \delta_{L/K}$ of $K \subseteq L$ (see [**15**, Chapter VI, § 7]). This norm group $\mathrm{T}_{\mathfrak{m}}(L/K)$ can be computed using an adaptation of [**8**, Algorithm 4.4.5] to the non-Galois case. More precisely:

- in the fourth step of the aforementioned algorithm, one can proceed even if the polynomials T_j do not have the same degree, by taking as f the greatest common divisor of their degrees. Indeed, $\mathrm{T}_{\mathfrak{m}}(L/K)$ is by definition generated by the classes of $\mathfrak{p}^{f(\mathfrak{P}/\mathfrak{p})}$, where $\mathfrak{p} := \mathfrak{P} \cap \mathcal{O}_K$ and $\mathfrak{P}$ varies amongst the prime ideals of $\mathcal{O}_L$ coprime with $\mathfrak{m} \cdot \mathcal{O}_L$, and the inertia degrees $f(\mathfrak{P}/\mathfrak{p})$ correspond exactly to the degrees of the polynomials T_j mentioned above;
- in the second step of the same algorithm, one should always output the matrix M even if $\det(M) \neq [L\colon K]$. In fact, $\det(M)$ will be precisely the index of the norm group inside $\mathrm{Cl}_{\mathfrak{m}}(K)$, *i.e.* one has that $[L \cap K^{\mathrm{ab}}\colon K] = \det(M)$.

Note that this modification does indeed work (assuming the validity of the Generalised Riemann Hypothesis), because $\mathrm{T}_{\mathfrak{m}}(L/K) = \mathrm{T}_{\mathfrak{m}}(L \cap K^{\mathrm{ab}}/K)$ by [**1**, Chapter XIV, Theorem 7].

Thus, in order to have a complete procedure for the computation of the CM index $\mathcal{I}(E/F)$, we only need to prove that one can always find a finite extension $F \subseteq L$ such that $F(E_{\mathrm{tors}}) = LK^{\mathrm{ab}}$ as in $\boxed{\text{S.2}}$. The next proposition shows that one can take L to be essentially any division field.

PROPOSITION 4.2. *Let $\mathcal{O}$ be an order inside an imaginary quadratic field K and let $E_{/F}$ be an elliptic curve defined over a number field $F \subseteq \overline{\mathbb{Q}}$ such that* $\mathrm{End}_F(E) \cong \mathcal{O}$. *Fix an ideal $I \subseteq \mathcal{O}$ and let $L := F(E[I])$ be the I-division field associated to E. Then $F(E_{\mathrm{tors}}) = LK^{\mathrm{ab}}$ whenever $|\mathbb{Z}/(I \cap \mathbb{Z})| > 2$ if $j(E) \neq 0$, and $|\mathbb{Z}/(I \cap \mathbb{Z})| > 3$ otherwise.*

PROOF. The inclusion $LK^{\mathrm{ab}} \subseteq F(E_{\mathrm{tors}})$ is clear, and the other containment can be proved as in [**5**, Proposition 5.7]. More precisely, fix an embedding $\overline{\mathbb{Q}} \hookrightarrow \mathbb{C}$ and a complex uniformisation $\xi\colon \mathbb{C} \twoheadrightarrow E(\mathbb{C})$, such that $\ker(\xi) = \Lambda$ for some lattice $\Lambda \subseteq K$. Then [**18**, Theorem 5.4] shows that, for every field automorphism $\sigma\colon \mathbb{C} \to \mathbb{C}$ which fixes FK^{ab}, there exists a complex uniformisation $\xi'\colon \mathbb{C} \twoheadrightarrow E(\mathbb{C})$ such that $\sigma(\xi(z)) = \xi'(z)$ for every $z \in K$. This implies in particular that there exists $\varepsilon \in \mathcal{O}^\times$ such that $\sigma(P) = [\varepsilon]_E(P)$ for every $P \in E_{\mathrm{tors}}$. If now σ fixes also the division field $L = F(E[I])$, one must have $\varepsilon = 1$ by our assumptions on I. We conclude that σ fixes the entire $F(E_{\mathrm{tors}})$, which in turn implies that $F(E_{\mathrm{tors}}) \subseteq LK^{\mathrm{ab}}$ as we wanted to show. □

Using Proposition 4.2, we see that $\boxed{\text{S.1}}$, $\boxed{\text{S.2}}$ and $\boxed{\text{S.3}}$ indeed describe a procedure to compute the index $\mathcal{I}(E/F)$ for any CM elliptic curve defined over any number field F. In practice, in $\boxed{\text{S.2}}$ it is convenient to choose a "small" division field $L = F(E[I])$, for instance by using $I = 3\mathcal{O}$ (when $j(E) \neq 0$), which gives $[L\colon F] \leq 8$. However, if one already knows an elliptic curve $E'_{/F}$ with $j(E') = j(E)$ and such that $F(E'_{\mathrm{tors}}) = FK^{\mathrm{ab}}$, then the subsequent Proposition 4.3, whose proof is analogous to that of Corollary 3.4, shows that one can take L to be a Kummer extension of F with degree $[L : F] \leq |\mathcal{O}^\times| \leq 6$. Since computations involving division fields of elliptic curves are typically hard, taking such an L is certainly more advantageous in this situation.

PROPOSITION 4.3. *Let $\mathcal{O}$ be an order inside an imaginary quadratic field K, and set $d := |\mathcal{O}^\times|$. Let $E_{/F}$ be an elliptic curve defined over a number field $F \subseteq \overline{\mathbb{Q}}$ such that* $\mathrm{End}_F(E) \cong \mathcal{O}$. *Suppose that there exists another elliptic curve $E'_{/F}$ such* *that $F(E'_{\mathrm{tors}}) = FK^{\mathrm{ab}}$, and that E is the twist of E' by $\sqrt[d]{\alpha}$, for some $\alpha \in F^\times$. Then $F(E_{\mathrm{tors}}) = LK^{\mathrm{ab}}$, where $L = F(\sqrt[d]{\alpha})$.*

PROOF. If $\sigma \in \mathrm{Gal}(\overline{\mathbb{Q}}/LF(E'_{\mathrm{tors}}))$, we see from the twisting formula (3.7) that $\rho_{E'}(\sigma) = \chi_\alpha(\sigma) = \rho_E(\sigma) = 1$, hence $F(E_{\mathrm{tors}}) \subseteq LF(E'_{\mathrm{tors}})$. Vice versa, if one has $\tau \in \mathrm{Gal}(\overline{\mathbb{Q}}/F(E_{\mathrm{tors}}))$ then $\rho_E(\tau) = 1$ and $\rho_{E'}(\tau) = \chi_\alpha(\tau^{-1}) \in \mathcal{O}^\times$. However, (3.5) shows that $\rho_{E'}(G_F) \cap \mathcal{O}^\times = \{1\}$, because $F(E'_{\mathrm{tors}}) = FK^{\mathrm{ab}}$ by assumption. Hence $\rho_{E'}(\tau) = \chi_\alpha(\tau) = 1$, which gives $F(E_{\mathrm{tors}}) = LF(E'_{\mathrm{tors}}) = LK^{\mathrm{ab}}$, as we wanted. □

REMARK 4.4. Note that the condition $F(E'_{\mathrm{tors}}) = FK^{\mathrm{ab}}$ is invariant under base change along a finite extension $F \subseteq F'$. In particular, if $\mathrm{Pic}(\mathcal{O}) = \{1\}$, one can take as E' any base change to F of an elliptic curve $E_{/K}$ which has complex multiplication by $\mathcal{O}$. On the other hand, if $\mathrm{Pic}(\mathcal{O}) \neq \{1\}$, constructing such an elliptic curve is a non-trivial matter, as we will see in the next section.

5. Explicit examples

We now want to provide some examples of index computations for CM elliptic curves E defined over the corresponding field of moduli $\mathbb{Q}(j(E))$. A way of constructing such curves is to consider an elliptic curve $\mathcal{E}$ defined over the function field $\mathbb{Q}(j)$, with j-invariant $j(\mathcal{E}) = j$ and discriminant $\Delta_{\mathcal{E}} \in \mathbb{Q}(j)$, and then specialise the parameter to $j = j_0$ for some CM j-invariant $j_0 \in \overline{\mathbb{Q}}$ such that $\Delta_{\mathcal{E}}(j_0) \neq 0$. When we want to emphasize that the specialization at j_0 of the elliptic curve $\mathcal{E}$ has complex multiplication by some order $\mathcal{O}$, we say that $j_0 \in \overline{\mathbb{Q}}$ is *relative to the order* $\mathcal{O}$. With a view towards doing explicit calculations in the mostly popular computer algebra systems in computational number theory, we consider and compare the following choices of $\mathcal{E}$:

(1) the curve:

$$\mathcal{E}_{\mathrm{SAGE}} : y^2 = x^3 + (-3j^2 + 5184j)x - 2j^3 + 6912j^2 - 5971968j$$

implemented under the command ELLIPTICCURVE_FROM_J(J,FALSE) in the software SageMath [**24**]. We warn the reader that, without setting the second optional parameter equal to FALSE, the command ELLIPTICCURVE_FROM_J, applied to a rational number $j_0 \in \mathbb{Q}$, returns an elliptic curve $E_{/\mathbb{Q}}$ which has j-invariant $j(E) = j_0$, and minimal conductor among all its twists. This curve, in general, can be different from the specialization of $\mathcal{E}_{\mathrm{SAGE}}$ at $j = j_0$;

(2) the curve:

$$\mathcal{E}_{\mathrm{PARI}} : y^2 = x^3 + (-3j^2 + 5184j)x + 2j^3 - 6912j^2 + 5971968j$$

implemented under the command ELLFROMJ(J) in the software PARI/GP [**23**];

(3) the curve:

$$\mathcal{E}_{\mathrm{MAGMA}} : y^2 + xy = x^3 - \frac{36}{j - 1728}x - \frac{1}{j - 1728}$$

implemented under the command ELLIPTICCURVEFROMJINVARIANT(J) in the software MAGMA [**2**].

The above families are clearly all defined over $\mathbb{Q}(j)$, and their singular specializations occur only at the values $j_0 \in \{0, 1728\}$. Moreover, it is easily verified that

$\mathcal{E}_{\mathrm{PARI}}$ and $\mathcal{E}_{\mathrm{SAGE}}$ are isomorphic over $\mathbb{Q}(j,\sqrt{-1})$ while $\mathcal{E}_{\mathrm{SAGE}}$ and $\mathcal{E}_{\mathrm{MAGMA}}$ are isomorphic over $\mathbb{Q}\left(j,\sqrt{\frac{1728-j}{3}}\right)$.

Now, for every CM j-invariant $j_0 \in \overline{\mathbb{Q}}$ relative to an order of class number 2, we want to compute the index $\mathcal{I}(E_{j_0}/\mathbb{Q}(j_0))$ where E_{j_0} is the fiber over j_0 in any of the three families described above (one can check that all these fibers are non-singular). First of all, we show that for every CM invariant $j_0 \in \overline{\mathbb{Q}}$ the CM fibers E_{j_0} in the above families have the same index $\mathcal{I}(E_{j_0}/\mathbb{Q}(j_0))$.

Fix now a CM j-invariant $j_0 \in \overline{\mathbb{Q}} \setminus \{0, 1728\}$ relative to an order $\mathcal{O}$. Let moreover $(E_{j_0}, E'_{j_0}, E''_{j_0})$ be the specialisations of the families $(\mathcal{E}_{\mathrm{SAGE}}, \mathcal{E}_{\mathrm{PARI}}, \mathcal{E}_{\mathrm{MAGMA}})$ to $j = j_0$. If $H_{\mathcal{O}} = K(j_0)$ denotes the ring class field relative to the order $\mathcal{O}$ then by Lemma 2.2 we have $\mathcal{I}(E_{j_0}/\mathbb{Q}(j_0)) = \mathcal{I}(E_{j_0}/H_{\mathcal{O}})$ and similarly with the other two elliptic curves, so we assume that everything is base-changed to the ring class field. Since by the discussion above E_{j_0} and E'_{j_0} are twisted over $H_{\mathcal{O}}$ by $\alpha = -1$ and $H_{\mathcal{O}}(\sqrt{-1}) \subseteq K^{\mathrm{ab}}$ (being the compositum of two abelian extensions of K), Corollary 3.4 allows us to conclude that $\mathcal{I}(E_{j_0}/\mathbb{Q}(j_0)) = \mathcal{I}(E'_{j_0}/\mathbb{Q}(j_0))$. Furthermore, the elliptic curve $\mathcal{E}_{\mathrm{MAGMA}}$ admits a short Weierstraß form:

$$y^2 = x^3 - \left(\frac{27j}{j-1728}\right)x + \frac{54j}{j-1728}$$

whose discriminant is given by $\Delta_j := 6^{12} \cdot j^2/(j-1728)^3$. Thus, we see that:

$$H_{\mathcal{O}}(\sqrt{j_0 - 1728}) = H_{\mathcal{O}}(\sqrt{\Delta_{j_0}}) \subseteq H_{\mathcal{O}}(E''_{j_0}[2])$$

for every CM j-invariant $j_0 \in \overline{\mathbb{Q}}$, relative to the order $\mathcal{O}$. Since $H_{\mathcal{O}}(E''_{j_0}[2])$ is generated over $H_{\mathcal{O}}$ by the Weber functions evaluated at 2-torsion points, we have that $H_{\mathcal{O}}(E''_{j_0}[2]) \subseteq K^{\mathrm{ab}}$ (see [**5**, Theorem 4.7]). Thus $H_{\mathcal{O}}\left(\sqrt{(1728-j_0)/3}\right)$ is abelian over K, and Corollary 3.4 shows that $\mathcal{I}(E_{j_0}/\mathbb{Q}(j_0)) = \mathcal{I}(E''_{j_0}/\mathbb{Q}(j_0))$. Hence, we can conclude that the three families $\mathcal{E}_{\mathrm{PARI}}, \mathcal{E}_{\mathrm{SAGE}}$ and $\mathcal{E}_{\mathrm{MAGMA}}$, when specialised to the same CM j-invariant, have the same CM index. We will use in the rest of the paper, the elliptic curves E_{j_0} obtained by specialising the family $\mathcal{E}_{\mathrm{SAGE}}$. Note that, once the imaginary quadratic order $\mathcal{O}$ is fixed, the index $\mathcal{I}(E_{j_0}/\mathbb{Q}(j_0))$ does not depend on the particular j-invariant $j_0 \in \overline{\mathbb{Q}}$ relative to $\mathcal{O}$ to which one specializes the family $\mathcal{E}_{\mathrm{SAGE}}$, because all these j-invariants are conjugate under the action of the absolute Galois group $\mathrm{Gal}(\overline{\mathbb{Q}}/\mathbb{Q})$ (see [**9**, Proposition 13.2]).

Let us turn now to the computation of the index $\mathcal{I}(E_{j_0}/\mathbb{Q}(j_0))$, where we take $j_0 \in \overline{\mathbb{Q}}$ to be a CM j-invariant relative to an order of class number 2. The procedure described in Section 4 simplifies considerably in this case. Indeed, in general, for any imaginary quadratic order $\mathcal{O} \neq \mathbb{Z}\left[\frac{1+\sqrt{-3}}{2}\right]$ and any elliptic curve E with complex multiplication by $\mathcal{O}$ and defined over the ring class field $H_{\mathcal{O}}$, one has that:

$$\mathcal{I}(E/H_{\mathcal{O}}) = \frac{2}{[H_{\mathcal{O}}(E[3]) \colon H_{\mathcal{O}}(E[3]) \cap K^{\mathrm{ab}}]} \tag{5.1}$$

as one can see by combining Proposition 4.1 and Proposition 4.2. Moreover, since:

$$[H_{\mathcal{O}}(E[3]) \colon H_{\mathcal{O}}(E[3]) \cap K^{\mathrm{ab}}] = \begin{cases} 1, & \text{if the extension } K \subseteq H_{\mathcal{O}}(E[3]) \text{ is abelian;} \\ 2, & \text{otherwise (as follows from (5.1)),} \end{cases}$$

we see, using Lemma 2.2, that the computation of $\mathcal{I}(E_{j_0}/\mathbb{Q}(j_0))$ reduces to understanding whether or not the 3-division field of E_{j_0} is an abelian extension of

K. We implemented this computation in SAGEMATH (importing also the functions POLREDBEST and RNFISABELIAN from PARI/GP), as shown in Algorithm 5.1. We ran this algorithm for all the j-invariants relative to orders $\mathcal{O}$ of class number 2, whose discriminants $\Delta_{\mathcal{O}}$ are given by the following list:

$$\begin{aligned}\Delta_{\mathcal{O}} \in \{ &-15, -20, -24, -32, -35, -36, -40, -48, -51, -52,\\ &-60, -64, -72, -75, -88, -91, -99, -100, -112, -115,\\ &-123, -147, -148, -187, -232, -235, -267, -403, -427\}\end{aligned}$$

which can be obtained either by applying the algorithms described in [**25**], and implemented under the function DISCRIMINANTS_WITH_BOUNDED_CLASS_NUMBER of the SAGEMATH module SAGE.SCHEMES.ELLIPTIC_CURVES.CM, or by appealing to the classical result [**22**, Theorem 1], and then applying the class number formula [**9**, Theorem 7.24]. The results of this computation show that $\mathcal{I}(E_{j_0}/\mathbb{Q}(j_0)) = 1$ unless $\Delta_{\mathcal{O}} = -15$, in which case $\mathcal{I}(E_{j_0}/\mathbb{Q}(j_0)) = 2$.

To conclude, consider the order $\mathcal{O} = \mathbb{Z}[\sqrt{-5}]$ of discriminant $\Delta_{\mathcal{O}} = -20$, such that $\mathcal{I}(E_{j_0}/\mathbb{Q}(j_0)) = 1$ for every CM j-invariant $j_0 \in \overline{\mathbb{Q}}$ relative to $\mathcal{O}$. We now construct, by a suitable twist of $E := E_{j_0}$ over the Hilbert class field $H := H_{\mathcal{O}}$, another elliptic curve $E'_{/H}$ with complex multiplication by $\mathcal{O}$, with the property that $\mathcal{I}(E'/H) = 2$. To do so, we specialize $j_0 = 282880\sqrt{5} + 632000$, so that

Algorithm 5.1. SAGEMATH code to compute the index $\mathcal{I}(E_{j_0}/\mathbb{Q}(j_0))$, relative to the elliptic curve E_{j_0} obtained by specialising the family $\mathcal{E}_{\text{SAGE}}$ to a CM j-invariant j_0.

Input: DELTA $= \Delta_{\mathcal{O}}$, the discriminant of an imaginary quadratic order $\mathcal{O}$.

```
from sage.libs.pari.convert_sage import gen_to_sage
R.<x> = PolynomialRing(QQ)
K.<D> = NumberField(x^2-Delta)
F.<j> = K.extension(hilbert_class_polynomial(Delta))
E = EllipticCurve_from_j(j,F)
Fabs.<a> =
    NumberField(gen_to_sage(pari(F.absolute_polynomial()).polredbest(),
    {'x' : x}))
Eabs = E.base_extend(F.embeddings(Fabs)[0])
F3.<f3> = Eabs.division_field(3)
F3best.<f3best> =
    NumberField(gen_to_sage(pari(F3.absolute_polynomial()).polredbest(),
    {'x' : x}))
F3rel.<f3rel> = F3best.relativize(K.embeddings(F3best)[0])
if F3rel.is_galois_relative() == True:
    Index = gp.rnfisabelian(pari('y^2 +
        '+str(-Delta)).nfinit(),pari(F3rel.relative_polynomial())) + 1
else:
    Index = 1
```

Output: INDEX $= \mathcal{I}(E_{j_0}/\mathbb{Q}(j_0))$, for any CM j-invariant j_0 relative to $\mathcal{O}$

$E := E_{j_0}$ is given by:

$$E\colon y^2 = x^3 + 29736960(36023\sqrt{5} - 80550)x - 55826186240(16154216\sqrt{5} + 36121925) \tag{5.2}$$

and we follow the procedure described in the proof of [**5**, Theorem 5.11].

More precisely, observe that $H = \mathbb{Q}(\sqrt{-5}, i)$ and the ideal $3 \cdot \mathcal{O}$ factors as $3 \cdot \mathcal{O} = \mathfrak{p}_3 \cdot \overline{\mathfrak{p}}_3$, where $\mathfrak{p}_3 = (3, \sqrt{-5}+1)$ and $\overline{\mathfrak{p}}_3 = (3, \sqrt{-5}-1)$. By [**5**, Theorem 4.6] one has that $H_{\mathfrak{p}_3} = H_{\overline{\mathfrak{p}}_3} = H$, where $H_{\mathfrak{p}_3}$ and $H_{\overline{\mathfrak{p}}_3}$ denote respectively the ray class fields of K modulo $\mathfrak{p}_3$ and $\overline{\mathfrak{p}}_3$. This in particular implies, using [**20**, Chapter II, Theorem 5.6], that the x-coordinates of the points $P \in E[\mathfrak{p}_3] \cup E[\overline{\mathfrak{p}}_3]$ lie in H. Moreover, it follows from [**3**, Lemma 2.4] that $|E[\mathfrak{p}_3]| = |E[\overline{\mathfrak{p}}_3]| = 3$, which shows that each non-trivial $\mathfrak{p}_3$-torsion point has the same x-coordinate, and similarly for non-trivial $\overline{\mathfrak{p}}_3$-torsion points. From the factorization:

$$\begin{aligned}\phi_{E,3}(x) = 3\cdot&(x + 594880 + 59840i - 26048\sqrt{-5} + 266816\sqrt{5})\cdot\\ &(x + 594880 - 59840i + 26048\sqrt{-5} + 266816\sqrt{5})\cdot\\ &(x^2 - (1189760 + 533632\sqrt{5})x - 2668089262080 - 1193205432320\sqrt{5})\end{aligned}$$

of the 3-division polynomial $\phi_{E,3} \in H[x]$, one can verify that the number:

$$x_3 := -594880 - 59840i + 26048\sqrt{-5} - 266816\sqrt{5}$$

is the x-coordinate of all the non-trivial $\mathfrak{p}_3$-torsion points. Hence we have that $H(E[\mathfrak{p}_3]) = H(\sqrt{\alpha})$, where the number:

$$\alpha := 13956546560 \cdot (1190435 + 2307955i - 1032149\sqrt{-5} + 532379\sqrt{5})$$

is obtained by substituting x_3 in the right hand side of (5.2). It can be checked that the extension $K \subseteq H(\sqrt{\alpha})$ is not Galois, and in particular not abelian, which is compatible with the fact that $\mathcal{I}(E/\mathbb{Q}(j_0)) = 1$.

Thus, the twisted elliptic curve $E' := E^{(\alpha)}$, given by the global minimal Weierstraß model:

$$\begin{aligned}E'\colon y^2 - \left(\frac{1 - i + \sqrt{-5} + \sqrt{5}}{2}\right)xy - \left(\frac{1 + i + \sqrt{-5} + \sqrt{5}}{2}\right)y =\\ = x^3 + x^2 + \left(2i - \sqrt{5}\right)x - 1 + 2i\end{aligned} \tag{5.3}$$

has index $\mathcal{I}(E'/H) = 2$, as follows from (5.1). Indeed, the first point of [**5**, Proposition 5.1] implies that $H(E'[\mathfrak{p}_3]) = H_{\mathfrak{p}_3}$, which entails that $H(E'[3])$ coincides with the 3-ray class field of K, as can also be checked by direct computation. Note finally that $H(E'_{\mathrm{tors}}) = K^{\mathrm{ab}}$, as follows from Corollary 3.2.

REMARK 5.1. The interested reader can find at [**6**] a SAGEMATH notebook in which we implemented the computations carried out to find the elliptic curve E' appearing in (5.3).

Acknowledgments

We would like to thank François Brunault, Ian Kiming, Fabien Pazuki and Peter Stevenhagen for many useful discussions. We also thank the anonymous referees for their helpful comments and suggestions.

References

[1] E. Artin and J. Tate, *Class field theory*, W. A. Benjamin, Inc., New York-Amsterdam, 1968. MR0223335

[2] Wieb Bosma, John Cannon, and Catherine Playoust, *The Magma algebra system. I. The user language*, J. Symbolic Comput. **24** (1997), no. 3-4, 235–265, DOI 10.1006/jsco.1996.0125. Computational algebra and number theory (London, 1993). MR1484478

[3] Abbey Bourdon and Pete L. Clark, *Torsion points and Galois representations on CM elliptic curves*, Pacific J. Math. **305** (2020), no. 1, 43–88, DOI 10.2140/pjm.2020.305.43. MR4077686

[4] Abbey Bourdon, Pete L. Clark, and James Stankewicz, *Torsion points on CM elliptic curves over real number fields*, Trans. Amer. Math. Soc. **369** (2017), no. 12, 8457–8496, DOI 10.1090/tran/6905. MR3710632

[5] Francesco Campagna and Riccardo Pengo, *Entanglement in the family of division fields of elliptic curves with complex multiplication*, To appear in *Pacific Journal of Mathematics.*

[6] Francesco Campagna and Riccardo Pengo, *Finding explicitly a CM elliptic curve with small Galois image*, SAGEMATH notebook, available at: HTTPS://BIT.LY/3OYZOOB.

[7] Henri Cohen, *A course in computational algebraic number theory*, Graduate Texts in Mathematics, vol. 138, Springer-Verlag, Berlin, 1993, DOI 10.1007/978-3-662-02945-9. MR1228206

[8] Henri Cohen, *Advanced topics in computational number theory*, Graduate Texts in Mathematics, vol. 193, Springer-Verlag, New York, 2000, DOI 10.1007/978-1-4419-8489-0. MR1728313

[9] David A. Cox, *Primes of the form $x^2 + ny^2$*, 2nd ed., Pure and Applied Mathematics (Hoboken), John Wiley & Sons, Inc., Hoboken, NJ, 2013. Fermat, class field theory, and complex multiplication, DOI 10.1002/9781118400722. MR3236783

[10] The GAP Group, *GAP – Groups, Algorithms, and Programming*, 2021, Version 4.11.1.

[11] Serge Lang, *Elliptic functions*, 2nd ed., Graduate Texts in Mathematics, vol. 112, Springer-Verlag, New York, 1987. With an appendix by J. Tate, DOI 10.1007/978-1-4612-4752-4. MR890960

[12] Davide Lombardo, *Bounds for Serre's open image theorem for elliptic curves over number fields*, Algebra Number Theory **9** (2015), no. 10, 2347–2395, DOI 10.2140/ant.2015.9.2347. MR3437765

[13] Davide Lombardo, *Galois representations attached to abelian varieties of CM type* (English, with English and French summaries), Bull. Soc. Math. France **145** (2017), no. 3, 469–501, DOI 10.24033/bsmf.2745. MR3766118

[14] Álvaro Lozano-Robledo, *Galois representations attached to elliptic curves with complex multiplication*, To appear in Algebra & Number Theory.

[15] Jürgen Neukirch, *Algebraic number theory*, Grundlehren der mathematischen Wissenschaften [Fundamental Principles of Mathematical Sciences], vol. 322, Springer-Verlag, Berlin, 1999. Translated from the 1992 German original and with a note by Norbert Schappacher; With a foreword by G. Harder, DOI 10.1007/978-3-662-03983-0. MR1697859

[16] Jeremy Rouse, Andrew V. Sutherland, and David Zureick-Brown, *ℓ-adic images of Galois for elliptic curves over $\mathbb{Q}$*, 2021, `arXiv:2106.11141`.

[17] Jean-Pierre Serre, *Propriétés galoisiennes des points d'ordre fini des courbes elliptiques* (French), Invent. Math. **15** (1972), no. 4, 259–331, DOI 10.1007/BF01405086. MR387283

[18] Goro Shimura, *Introduction to the arithmetic theory of automorphic functions*, Publications of the Mathematical Society of Japan, vol. 11, Princeton University Press, Princeton, NJ, 1994. Reprint of the 1971 original; Kanô Memorial Lectures, 1. MR1291394

[19] Goro Shimura, *Abelian varieties with complex multiplication and modular functions*, Princeton Mathematical Series, vol. 46, Princeton University Press, Princeton, NJ, 1998, DOI 10.1515/9781400883943. MR1492449

[20] Joseph H. Silverman, *Advanced topics in the arithmetic of elliptic curves*, Graduate Texts in Mathematics, vol. 151, Springer-Verlag, New York, 1994, DOI 10.1007/978-1-4612-0851-8. MR1312368

[21] Joseph H. Silverman, *The arithmetic of elliptic curves*, 2nd ed., Graduate Texts in Mathematics, vol. 106, Springer, Dordrecht, 2009, DOI 10.1007/978-0-387-09494-6. MR2514094

[22] H. M. Stark, *On complex quadratic fields wth class-number two*, Math. Comp. **29** (1975), 289–302, DOI 10.2307/2005481. MR369313

[23] The PARI Group, *PARI/GP version 2.11.2*, Univ. Bordeaux, 2019.

[24] The Sage Developers, *Sagemath, the Sage Mathematics Software System (Version 9.0)*, 2020.

[25] Mark Watkins, *Class numbers of imaginary quadratic fields*, Math. Comp. **73** (2004), no. 246, 907–938, DOI 10.1090/S0025-5718-03-01517-5. MR2031415

[26] David Zywina, *Possible indices for the Galois image of elliptic curves over* $\mathbb{Q}$, 2015, arXiv:1508.07663.

Max Planck Institute for Mathematics, Vivatsgasse 7, 53111 Bonn, Germany
Email address: campagna@mpim-bonn.mpg.de

École normale supérieure de Lyon, Unité de Mathématiques Pures et Appliquées, 46 allée d'Italie, 69007 Lyon, France
Email address: riccardo.pengo@ens-lyon.fr

Contemporary Mathematics
Volume **779**, 2022
https://doi.org/10.1090/conm/779/15671

Multiradical isogenies

Wouter Castryck and Thomas Decru

ABSTRACT. We argue that for all integers $N \geq 2$ and $g \geq 1$ there exist "multiradical" isogeny formulae, that can be iteratively applied to compute $(N^k, \ldots, N^k)$-isogenies between principally polarized g-dimensional abelian varieties, for any value of $k \geq 2$. The formulae are complete: each iteration involves the extraction of $g(g+1)/2$ different Nth roots, whence the epithet multiradical, and by varying which roots are chosen one computes all $N^{g(g+1)/2}$ extensions to an $(N^k, \ldots, N^k)$-isogeny of the incoming $(N^{k-1}, \ldots, N^{k-1})$-isogeny. Our group-theoretic argumentation is heuristic, but it is supported by concrete formulae for several prominent families. As our main application, we illustrate the use of multiradical isogenies by implementing a hash function from $(3,3)$-isogenies between Jacobians of superspecial genus-2 curves, showing that it outperforms its $(2,2)$-counterpart by an asymptotic factor ≈ 9 in terms of speed.

1. Introduction

In a previous joint work with Vercauteren [**10**], we introduced the concept of *radical isogenies* between elliptic curves, which in low degree allow for a very fast computation of isogeny chains over finite fields, *e.g.*, of the type used in Charles, Goren and Lauter's hash function [**12**] and in the Couveignes–Rostovtsev–Stolbunov key exchange protocol [**14, 42**] and its descendant CSIDH [**11**].

The central observation was that for any integer $N \geq 2$ there exist explicit formulae which, upon input of an elliptic curve E — say given in long Weierstrass form — over a perfect field K with $\operatorname{char} K \nmid N$ and a point $P \in E$ of order N, produce the coordinates of an order-N point $P' \in E' = E/\langle P \rangle$ such that the isogeny $\varphi' : E' \to E'/\langle P' \rangle$ cyclically extends $\varphi : E \to E/\langle P \rangle$. This, of course, assumes that we have a defining equation for E' at hand, such as the one provided by Vélu [**45**]. Moreover, the formulae can be chosen to enjoy the following properties.

(1) Radicality. The formulae are algebraic expressions in the coefficients of E, the coordinates of P and a radical $\sqrt[N]{\mathfrak{r}_1}$, where $\mathfrak{r}_1$ is itself an algebraic expression in these coefficients and coordinates.

(2) Completeness. By varying the Nth root chosen, *i.e.*, by scaling $\sqrt[N]{\mathfrak{r}_1}$ with powers of a primitive Nth root of unity $\zeta_N \in \overline{K}$, we obtain generators

---

2020 *Mathematics Subject Classification.* Primary 14G50, Secondary 14K02, 14H40.

This work was supported by the Research Council KU Leuven grant C14/18/067, by CyberSecurity Research Flanders with reference code VR20192203, and by the Research Foundation Flanders (FWO) through the WOG Coding Theory and Cryptography.

for all N subgroups $G' \subseteq E'$ of order N which are such that $E' \to E'/G'$ cyclically extends φ.

(3) Good reduction. The formulae are naturally defined over $\mathbb{Z}[1/N]$, *i.e.*, they work over any perfect field K with $\operatorname{char} K \nmid N$.

(The last property is in fact conjectural [**10**, Conj. 1].) Concrete versions of our radical isogeny formulae for $N = 2, \ldots, 13$ can be found in the GitHub repository that accompanies [**10**]. For the sake of illustration, we have included the details of the case $N = 5$ in Section 4.

The current paper studies how radical isogenies generalize to principally polarized (p.p.) abelian varieties of any given dimension $g \geq 1$. That is, we are looking for formulae which, upon input of a g-dimensional p.p. abelian variety A over a perfect field K with $\operatorname{char} K \nmid N$ and points $P_1, \ldots, P_g \in A$ that generate an $(N, \ldots, N)$-subgroup[1] $G \subseteq A$, produce the coordinates of points $P_1', \ldots, P_g' \in A' = A/G$ generating an $(N, \ldots, N)$-subgroup $G' \subseteq A'$ such that the composition $A \to A' = A/G \to A'/G'$ is an $(N^2, \ldots, N^2)$-isogeny.

When aiming for universally applicable formulae, a major bottleneck is the lack of an analogue of the long Weierstrass form for p.p. abelian varieties of dimension $g \geq 2$. That is, we do not know of a set of defining equations from which every g-dimensional p.p. abelian variety A can be obtained by specializing coefficients. Moreover, in practical applications, we are mostly interested in instances of A that are described in a more implicit form, *e.g.*, as the Jacobian of some genus-g curve, or as a product of Jacobians of lower-genus curves. Things are complicated further by the fact that the isogenous p.p. abelian variety A' may be of a different type, *e.g.*, if A is a Jacobian, then this may not be the case for A'.

We therefore focus on smaller families, parametrized by the points s of some quasi-affine set $\mathcal{S}$. Concretely, we assume to have algebraic formulae at our disposal which can be evaluated at the coordinates of any point $s \in \mathcal{S}$, each time producing a g-dimensional p.p. abelian variety A_s together with points $P_{s,1}, \ldots, P_{s,g}$ that generate an $(N, \ldots, N)$-subgroup $G_s \subseteq A_s$. We furthermore assume that the family comes equipped with Vélu-like formulae providing an explicit description of the isogenous p.p. abelian variety $A_s' = A_s/G_s$. Several examples of such families can be found in Section 4 and Section 5.

CONJECTURE 1. *Under the above assumptions, there always exist accompanying formulae which, when evaluated at s, produce points $P_{s,1}', \ldots, P_{s,g}' \in A_s'$ generating a subgroup $G_s' \subseteq A_s'$ such that the composition $A_s \to A_s' \to A_s'/G_s'$ is an $(N^2, \ldots, N^2)$-isogeny. Moreover, these formulae can be chosen to enjoy the following properties:*

(1) *Multiradicality*. *They are algebraic expressions in the coordinates of s and radicals $\sqrt[N]{\mathfrak{r}_1}, \ldots, \sqrt[N]{\mathfrak{r}_{g(g+1)/2}}$, where in turn the radicands $\mathfrak{r}_i$ are algebraic expressions in the coordinates of s.*

(2) *Completeness*. *By varying the Nth roots chosen,* i.e., *by scaling them with powers of $\zeta_N \in \overline{K}$, we obtain generating sets for all $N^{g(g+1)/2}$ subgroups $G_s' \subseteq A_s'$ such that $A_s \to A_s' = A_s/G_s \to A_s'/G_s'$ is an $(N^2, \ldots, N^2)$-isogeny.*

[1] See Section 2.2 for a definition.

(3) ***Good reduction.*** *If the family $\mathcal{S}$ is defined over $\mathbb{Z}[1/M]$ for some multiple M of N, then so are our formulae,* i.e., *they work over any perfect field K with* $\operatorname{char} K \nmid M$.

Formulae of the above kind will be called *multiradical isogeny* formulae. We refer to Section 3 for a more extensive discussion of Conjecture 1, where we will provide a group-theoretic heuristic argument in favor of the existence of multiradical isogeny formulae. However, we stress that each of the above subclaims remains conjectural. We will also discuss an addendum to Conjecture 1, namely that one can always take the radicands $\mathfrak{r}_1, \ldots, \mathfrak{r}_{g(g+1)/2}$ to be representants of the Tate pairings $t_N(P_{s,i}, P_{s,j})$, $1 \leq i \leq j \leq g$, in the sense of Frey and Rück [**24**], as soon as these are well-defined.

Further support comes from concrete examples of multiradical isogeny formulae, which are discussed in Section 4 and Section 5. For arbitrary N and in arbitrary dimension g, we discuss fully split $(N, \ldots, N)$-isogenies from g-fold products of elliptic curves. Other examples focus on Jacobians of genus-2 curves, where we discuss non-split $(2, 2)$-isogenies (also known as Richelot isogenies) and non-split $(3, 3)$-isogenies as described by Bruin, Flynn and Testa [**5**]. We also study the multiradical nature of certain $(5, 5)$-isogenies that were described by Flynn [**21**].

REMARK 1.1. Our eventual goal is the computation of $(N^k, \ldots, N^k)$-isogenies, for arbitrary $k \geq 2$, achieved by an iterated application of our formulae. However, it is possible, and unavoidable in general, that the isogenous p.p. abelian variety A'_s marked with $\Gamma'_{s,1}, \ldots, \Gamma'_{s,g}$ does not belong to our family. For instance, if $\mathcal{S}$ parametrizes Jacobians of genus-2 curves, we may run into a product of elliptic curves. In such cases, one needs to resort to different sets of multiradical isogeny formulae in order to cover the entire isogeny chain.

We illustrate the use of multiradical isogenies in Section 6, by constructing a Charles–Goren–Lauter style hash function from $(3, 3)$-isogenies between superspecial p.p. abelian surfaces over a large quadratic finite field $\mathbb{F}_{p^2}$, similar to the $(2, 2)$-construction from our joint work with Smith [**9**]. In short, each message determines a walk in the isogeny graph (which is of size about $p^3/2880$), and the hash of the message is the end point of that walk. One should make sure that every two consecutive isogenies compose to a $(9, 9)$-isogeny, to avoid the trivial collisions described in [**22**, §2.3]. This is automatically taken care of when using multiradical isogeny formulae.

In the Richelot hash function from [**9**], a $(2, 2)$-isogeny costs about 3 square root computations, with very little overhead, and can be used to process 3 bits of the message. In our case, the cost of a $(3, 3)$-isogeny is dominated by the extraction of 3 cube roots, and now it can be used to process 3 *trits* (*i.e.*, base-3 digits) of the message. Moreover, if $p \not\equiv \pm 1 \bmod 9$ then $p^2 \not\equiv 1 \bmod 9$ and computing cube roots in $\mathbb{F}_{p^2}$ is faster than computing square roots (see Section 6.4). Altogether, this leads to an expected speed-up by a factor 9, roughly. However, a noticeable difference with [**9**] is that chaining multiradical $(3, 3)$-isogenies comes with some non-negligible overhead; our current implementation even involves three small Gröbner basis computations. Despite this overhead, the $(3, 3)$-hash function outperforms the Richelot hash function as soon as the field characteristic p is of cryptographic size (*i.e.*, 86 bits or more). The asymptotic speed-up factor ≈ 9 becomes visible when p is about 2^{1024}.

Two conventions. For any integer $N \geq 2$ we denote the ring (or the additive group) of integers modulo N by $\mathbb{Z}_N$; we thereby follow computer science customs.[2] Also, throughout this paper, we always identify a variety over a perfect field K with its set of $\overline{K}$-points equipped with the natural $\mathrm{Gal}(\overline{K}/K)$-action.

2. Background

We discuss some of the material needed for what follows, but we stress that this is not a complete overview. Our main goal is to fix notation and highlight some statements that may be known to specialists but that we did not manage to pinpoint in the existing literature, such as Lemma 2.1, Example 2.2 and Lemma 2.3. For general background on abelian varieties and isogenies we refer to [**34, 35**].

2.1. Generalized symplectic bases. We consider abelian varieties A of dimension $g \geq 1$ over a perfect field K with algebraic closure $\overline{K}$, and we always assume that A comes equipped with a principal polarization. Important examples of g-dimensional principally polarized (p.p.) abelian varieties are Jacobians of smooth projective curves C/K of genus g. Every p.p. abelian variety of dimension ≤ 3 is $\overline{K}$-isomorphic to a product of Jacobians.

For each integer $N \geq 2$ with $\operatorname{char} K \nmid N$, the N-torsion subgroup $A[N]$ can be shown to be free of rank $2g$ over $\mathbb{Z}_N$. The principal polarization induces a perfect bilinear and antisymmetric pairing

$$e_N : A[N] \times A[N] \to \mu_N \subseteq \overline{K}^*,$$

known as the Weil pairing. After fixing a primitive Nth root of unity $\zeta_N \in \mu_N$, the Weil pairing turns into a symplectic form:

$$\langle \cdot, \cdot \rangle_N : A[N] \times A[N] \to \mathbb{Z}_N : (P, Q) \mapsto \log_{\zeta_N} e_N(P, Q).$$

Thus $A[N]$ admits a symplectic basis, *i.e.*, a $\mathbb{Z}_N$-basis $P_1, \ldots, P_g, Q_1, \ldots, Q_g$ satisfying $\langle P_i, P_j \rangle_N = \langle Q_i, Q_j \rangle_N = 0$ and $\langle P_i, Q_j \rangle_N = \delta_{ij}$ for all $i, j \in \{1, \ldots, g\}$. This allows us to view $A[N]$ as $\mathbb{Z}_N^{2g}$ equipped with the standard symplectic pairing

$$\langle \cdot, \cdot \rangle : \mathbb{Z}_N^{2g} \times \mathbb{Z}_N^{2g} : (v, w) \mapsto v^T \Omega w, \qquad \Omega = \begin{pmatrix} 0 & \mathbb{I}_g \\ -\mathbb{I}_g & 0 \end{pmatrix}.$$

Changing between symplectic bases is done using matrices from the symplectic group $\mathrm{Sp}_{2g}(\mathbb{Z}_N) = \{M \in \mathrm{GL}_{2g}(\mathbb{Z}_N) \mid M^T \Omega M = \Omega\}$.

Note that the notion of a symplectic basis of $A[N]$ depends on the choice of ζ_N. If a basis is symplectic with respect to *some* choice of ζ_N, then we call it a *generalized symplectic basis*. The matrices of base change between generalized symplectic bases are now taken from the larger group

$$\mathrm{GSp}_{2g}(\mathbb{Z}_N) = \{M \in \mathrm{GL}_{2g}(\mathbb{Z}_N) \mid M^T \Omega M = d(M)\Omega \text{ for a } d(M) \in \mathbb{Z}_N^*\}, \tag{2.1}$$

which is known as the generalized symplectic group (its elements are often referred to as symplectic similitudes). An *N-level structure* on A is an isomorphism $\alpha : A[N] \to \mathbb{Z}_N^{2g}$ such that

$$\alpha^{-1}(1, 0, \ldots, 0),\ \alpha^{-1}(0, 1, \ldots, 0),\ \ldots,\ \alpha^{-1}(0, 0, \ldots, 1)$$

is a generalized symplectic basis of $A[N]$.

[2]Most pure mathematicians prefer the notation $\mathbb{Z}/N\mathbb{Z}$, especially when N is a prime number p (or a power thereof) in order to avoid confusion with p-adic rings. Our paper is free of p-adic numbers, so such confusion should not be possible.

2.2. Good chains of $(N,\ldots,N)$-isogenies. A subgroup $G \subseteq A[N]$ is called isotropic if $\langle P,Q\rangle_N = 0$ for all $P,Q \in G$. Note that this notion does not depend on the choice of ζ_N. It is called *maximal isotropic* if moreover there is no supergroup $G' \supsetneq G$ that is isotropic. This property ensures that the isogenous abelian variety $A' = A/G$ comes naturally equipped with a principal polarization. The subgroup is said to be an

$$\underbrace{(N,\ldots,N)}_{g \text{ times}}\text{-subgroup}$$

if it is a (necessarily maximal) isotropic free $\mathbb{Z}_N$-submodule of rank g, *i.e.*, an isotropic subgroup isomorphic to $\mathbb{Z}_N^g$. In that case, we say that the quotient isogeny $\varphi : A \to A'$ is an $(N,\ldots,N)$-*isogeny.*

Given an $(N,\ldots,N)$-isogeny $\varphi : A \to A'$, we say that an $(N,\ldots,N)$-isogeny $\varphi' : A' \to A''$ is a *good* extension of φ if the composition

$$A \overset{\varphi}{\to} A' \overset{\varphi'}{\to} A''$$

is an $(N^2,\ldots,N^2)$-isogeny. According to the lemma below, of which special cases can be found in [**22**, §2.2], there are $N^{g(g+1)/2}$ subgroups of $A'[N]$ that give rise to good extensions. The group $\varphi(A[N])$ is an $(N,\ldots,N)$-subgroup which is the kernel of the *dual* isogeny $\hat{\varphi} : A' \to A$. All other $(N,\ldots,N)$-subgroups of $A'[N]$ are said to give rise to *bad* extensions. These are precisely the $(N,\ldots,N)$-subgroups that differ from $\varphi(A[N])$ but that intersect it non-trivially.

LEMMA 2.1. *Consider $\mathbb{Z}_N^{2g}$ together with the standard symplectic pairing $\langle\cdot,\cdot\rangle$. Its number of $(N,\ldots,N)$-subgroups is given by*

$$N^{g(g+1)/2} \prod_{\substack{\text{primes}\\ \ell \mid N}} \prod_{i=1}^{g} \left(1 + \frac{1}{\ell^i}\right).$$

Given an $(N,\ldots,N)$-subgroup $G \subseteq \mathbb{Z}_N^{2g}$, the number of $(N,\ldots,N)$-subgroups that intersect it trivially equals $N^{g(g+1)/2}$.

PROOF. For the second count, consider generators $P_1, \ldots, P_g$ of the given subgroup G and extend to a symplectic basis $P_1, \ldots, P_g, Q_1, \ldots, Q_g$. The free rank-$g$ submodules that intersect G trivially each admit a unique basis of the form

$$\begin{aligned} P_1' &= Q_1 + a_{11}P_1 + \cdots + a_{1g}P_g, \\ &\;\;\vdots \\ P_g' &= Q_g + a_{g1}P_1 + \cdots + a_{gg}P_g, \end{aligned} \tag{2.2}$$

for certain $a_{ij} \in \mathbb{Z}_N$ and, conversely, every such basis generates a rank-g submodule intersecting G trivially. One checks that the maximal isotropy assumption $\forall i,j$: $\langle P_i', P_j'\rangle = 0$ translates into $\binom{g}{2}$ linear conditions on the a_{ij}'s. These conditions can be used to express the a_{ij}'s with $i > j$ in terms of the other a_{ij}'s. Thus we are left with $g^2 - \binom{g}{2} = g(g+1)/2$ degrees of freedom, as wanted.

As for the first count, we start with the case where $N = \ell$ is a prime number. The symplectic group $\mathrm{Sp}_{2g}(\mathbb{F}_\ell)$ acts transitively on the set of $(\ell,\ldots,\ell)$-subgroups, and our goal is to compute the size of the unique orbit. This can be done via the orbit-stabilizer theorem, which indeed yields

$$\prod_{i=1}^{g}(\ell^i + 1) = \ell^{g(g+1)/2}\prod_{i=1}^{g}\left(1 + \frac{1}{\ell^i}\right)$$

as detailed in [**27**, §1]. Next, to settle the case $N = \ell^n$ for $n > 1$, it suffices to see that the reduction-mod-ℓ map[3]

$$\{(\ell^n, \ldots, \ell^n)\text{-subgroups of } \mathbb{Z}_{\ell^n}^{2g}\} \to \{(\ell, \ldots, \ell)\text{-subgroups of } \mathbb{F}_\ell^{2g}\}$$

is $\ell^{(n-1)g(g+1)/2}$-to-1. This works as before: consider generators $Q_1, \ldots, Q_g$ of an $(\ell^n, \ldots, \ell^n)$-subgroup G, and extend to a symplectic basis $Q_1, \ldots, Q_g, P_1, \ldots, P_g$. The $(\ell^n, \ldots, \ell^n)$-subgroups having the same reduction as G admit a unique basis of the form (2.2), where each a_{ij} is now an element of $\ell\mathbb{Z}_{\ell^n}$. Again, the maximal isotropy condition translates into expressions for the a_{ij}'s with $i > j$ in terms of the other a_{ij}'s, leaving us with $\ell^{(n-1)g(g+1)/2}$ subgroups, as wanted. The count for arbitrary N then follows from the Chinese remainder theorem. □

2.3. The Tate pairing on (products of) Jacobians. We discuss the Tate pairing on Jacobians, in the sense of Frey and Rück [**24, 28**], and its natural extension to products of Jacobians. Let C/K be a curve of genus $g \geq 1$ and let $N \geq 2$ be such that $\operatorname{char} K \nmid N$. The Tate pairing is a map

$$t_N : \operatorname{Pic}_K^0(C)[N] \times \operatorname{Pic}_K^0(C)/N\operatorname{Pic}_K^0(C) \to K^*/(K^*)^N,$$

where $\operatorname{Pic}_K^0(C)$ denotes the group of K-rational degree-zero divisors on C modulo divisors of functions in $K(C)^*$, and is defined as follows. Let $\overline{D}_1 \in \operatorname{Pic}_K^0(C)[N]$ be represented by a divisor D_1 and let $\overline{D}_2 \in \operatorname{Pic}_K^0(C)/N\operatorname{Pic}_K^0(C)$ be represented by a divisor D_2 with support disjoint from that of D_1. Take a function $f_{N,D_1} \in K(C)^*$ whose divisor is ND_1. We then let

$$t_N(\overline{D}_1, \overline{D}_2) := f_{N,D_1}(D_2) \bmod (K^*)^N.$$

It can be shown that this is a well-defined bilinear pairing. In many cases of interest, the natural inclusion

$$\operatorname{Pic}_K^0(C) \hookrightarrow J_C(K)$$

into the Jacobian J_C of C is surjective, *i.e.*, it is a group isomorphism, and we obtain a pairing

$$J_C(K)[N] \times J_C(K)/NJ_C(K) \to K^*/(K^*)^N$$

that we keep denoting by t_N. Known sufficient conditions for surjectivity are that K has a trivial Brauer group (*e.g.*, this is true if K is finite) [**34**, Rmk. 1.6], that $C(K) \neq \emptyset$ [**25**, Thm. 3], or that $g = 2$ [**13**, Lem. 3.1 and Lem. 3.2].

In this paper we are mainly interested in the case where K is a certain function field over $\mathbb{Q}$, which has a non-trivial Brauer group. To avoid resulting pathologies, we only apply the Tate pairing in cases where $C(K) \neq \emptyset$ or where $g = 2$. We also consider the Tate pairing

$$t_N : A(K)[N] \times A(K)/NA(K) \to K^*/(K^*)^N$$

on abelian varieties A/K that arise as *products* of Jacobians of such curves: this is simply obtained by taking the product of the Tate pairings of the respective components.

[3]Recall from the introduction that $\mathbb{Z}_{\ell^n}$ just denotes $\mathbb{Z}/\ell^n\mathbb{Z}$, the integers modulo ℓ^n, rather than some extension of the ring of ℓ-adic integers.

EXAMPLE 2.2. For use in Section 4.2, let us consider a genus-2 curve

$$C : y^2 = G_1(x)G_2(x)G_3(x)$$

over a perfect field K of odd characteristic, where the G_i's are quadratic polynomials over K whose product is square-free. Each G_i defines an element $\overline{D}_i \in \mathrm{Pic}^0_K(C)$, namely the class of

$$D_i = (\alpha_{i1}, 0) + (\alpha_{i2}, 0) - \infty_1 - \infty_2,$$

with $\alpha_{i1}, \alpha_{i2} \in \overline{K}$ the two roots of G_i and with $\infty_1, \infty_2 \in C(\overline{K})$ the two points at infinity. An analysis of $\mathcal{L}(\infty_1 + \infty_2)$ shows that D_i is non-principal, so from $2D_i = \mathrm{div}(G_i)$ we conclude that the $\overline{D}_i$'s have order 2. Let us compute $t_2(\overline{D}_1, \overline{D}_2)$. Replace D_1 by the equivalent divisor

$$D_1' = (\alpha_{11}, 0) + (\alpha_{12}, 0) - \infty_1 - \infty_2 - \mathrm{div}(x - c)$$

for some arbitrary $c \in K$ that is not a root of G_2. Then we can take $f_{2,D_1'} = G_1/(x-c)^2$ so that

$$t_2(\overline{D}_1, \overline{D}_2) \equiv f_{2,D_1'}(D_2) \equiv \frac{G_1(\alpha_{21})G_1(\alpha_{22})}{(\alpha_{21} - c)^2(\alpha_{22} - c)^2 \,\mathrm{lc}(G_1)^2} \equiv \mathrm{res}_x(G_1, G_2)$$

modulo $(K^*)^2$. Here $\mathrm{lc}(G_1)$ denotes the leading coefficient of G_1. By symmetry, it then follows that $t_2(\overline{D}_i, \overline{D}_j) \equiv \mathrm{res}_x(G_i, G_j)$ for all pairs of distinct $i, j \in \{1, 2, 3\}$.

If K is a finite field $\mathbb{F}_q$ containing a primitive Nth root of unity, *i.e.*, $N \mid q - 1$, then the Tate pairing can be shown to be perfect. We remark that there are ways of extending Frey and Rück's definition of the Tate pairing to arbitrary abelian varieties over $\mathbb{F}_q$, where it remains perfect [**6**].

2.4. Multiradical field extensions. We say that a field extension $K \subseteq L$ is *multiradical* if there exist an integer $N \geq 1$ and elements $\alpha_1, \ldots, \alpha_r \in L$ such that $L = K(\alpha_1, \ldots, \alpha_r)$ and $\alpha_i^N \in K^*$ for all i. In this section, we discuss a sufficient Galois-theoretic condition for an extension to be multiradical. While we suspect that this is a well-known fact, we did not manage to find an exact reference, even for the case $r = 1$.

Recall that a group G is the (inner) semi-direct product $G_1 \rtimes G_2$ of a normal subgroup G_1 and a subgroup G_2 if the following three equivalent conditions hold:

- $G = G_1G_2$ and $G_1 \cap G_2 = \{e_G\}$,
- every $g \in G$ can be written as $g = g_1g_2$ for unique $g_1 \in G_1$ and $g_2 \in G_2$,
- every $g \in G$ can be written as $g = g_2g_1$ for unique $g_1 \in G_1$ and $g_2 \in G_2$.

The group structure of G is determined by that of G_1 and G_2 and by how G_2 acts on G_1 through conjugation.

The prototypical example of a multiradical extension is where $K = \mathbb{Q}$ and $L = \mathbb{Q}(\sqrt[N]{p_1}, \ldots, \sqrt[N]{p_r})$ for distinct primes p_i, which is a number field of degree N^r [**1**]. The Galois closure of L over K is $L(\zeta_N)$, with $\zeta_N \in \overline{L}$ a primitive Nth root of unity. Define

$$G_1 = \{\sigma_1^{i_1} \circ \cdots \circ \sigma_r^{i_r} \mid 0 \leq i_j < N \text{ for all } j\} \cong \mathbb{Z}_N^r,$$

where $\sigma_j : \sqrt[N]{p_j} \mapsto \zeta_N \sqrt[N]{p_j}$ for $j = 1, \ldots, r$. Letting $G_2 = \{\tau_\ell : \zeta_N \mapsto \zeta_N^\ell \mid 0 \leq \ell < N, \gcd(\ell, N) = 1\} \cong \mathbb{Z}_N^*$, one then verifies that

$$\mathrm{Gal}(L(\zeta_N)/K) = G_1 \rtimes G_2,$$

where the action is given by $\tau_\ell \circ \sigma_1^{i_1} \circ \cdots \circ \sigma_r^{i_r} \circ \tau_\ell^{-1} = \sigma_1^{i_1 \ell} \circ \cdots \circ \sigma_r^{i_r \ell}$. Of course, this example generalizes to (the Galois closures of) arbitrary multiradical extensions, as long as $\operatorname{char} K \nmid N$ and $[L : K] = N^r$.

Lemma 2.3 gives a converse statement:

LEMMA 2.3. *Let N, r be positive integers and consider a degree-N^r extension $K \subseteq L$ of fields whose characteristic does not divide N. Let $\zeta_N \in \overline{L}$ be a primitive Nth root of unity and assume that $L(\zeta_N)$ is Galois over K with Galois group*

$$\operatorname{Gal}(L(\zeta_N)/K) = \operatorname{Gal}(L(\zeta_N)/K(\zeta_N)) \rtimes \operatorname{Gal}(L(\zeta_N)/L),$$

where the first factor is isomorphic to $\mathbb{Z}_N^r$, say generated by $\sigma_1, \ldots, \sigma_r$, and where the semi-direct product is according to the rule

$$\tau_\ell \circ \sigma_1^{i_1} \circ \cdots \circ \sigma_r^{i_r} \circ \tau_\ell^{-1} = \sigma_1^{i_1 \ell} \circ \cdots \circ \sigma_r^{i_r \ell} \tag{2.3}$$

for all $i_1, \ldots, i_r \in \{0, \ldots, N-1\}$ and all $\tau_\ell : \zeta_N \mapsto \zeta_N^\ell \in \operatorname{Gal}(L(\zeta_N)/L)$. Then there exist $\alpha_1, \ldots, \alpha_r \in L$ such that $L = K(\alpha_1, \ldots, \alpha_r)$ and $\alpha_1^N, \ldots, \alpha_r^N \in K^*$.

PROOF. First assume that $r = 1$ and write σ instead of σ_1. The restricted maps $\sigma^i|_L : L \to L(\zeta_N)$ are pairwise distinct. Indeed, if $i, i' \in \{0, 1, \ldots, N-1\}$ are such that $\sigma^i|_L = \sigma^{i'}|_L$, then

$$\sigma^{i-i'} \in \operatorname{Gal}(L(\zeta_N)/K(\zeta_N)) \cap \operatorname{Gal}(L(\zeta_N)/L) = \{\mathrm{id}\},$$

which can only be true if $i = i'$. From [**41**, Lem. 0CKL] it follows that these restricted maps are linearly independent over $L(\zeta_N)$. In particular there exists some $\beta \in L$ such that

$$\alpha := \sum_{i=0}^{N-1} \zeta_N^i \sigma^i(\beta)$$

is non-zero. From

$$\tau_\ell(\alpha) = \sum_i \zeta_N^{i\ell} (\tau_\ell \circ \sigma^i)(\beta) = \sum_i \zeta_N^{i\ell} (\sigma^{i\ell} \circ \tau_\ell)(\beta) = \sum_i \zeta_N^{i\ell} \sigma^{i\ell}(\beta) = \alpha$$

it follows that $\alpha \in L$. Now observe that α was constructed in such a way that $\sigma^i(\alpha) = \zeta_N^{-i}\alpha$ for $i = 0, 1, \ldots, N-1$, which has two crucial consequences. On the one hand, it implies that $\operatorname{Gal}(L(\zeta_N)/L)$ is the exact group of automorphisms fixing $K(\alpha)$, or in other words $L = K(\alpha)$. On the other hand, it implies that $\sigma(\alpha^N) = \sigma(\alpha)^N = (\zeta_N \alpha)^N = \alpha^N$, so that α^N is fixed by the entire Galois group, i.e., $\alpha^N \in K$ as wanted.

The general case reduces to the case $r = 1$, as follows. Each element of our Galois group $\operatorname{Gal}(L(\zeta_N)/K)$ can be written as

$$\sigma_1^{i_1} \circ \cdots \circ \sigma_r^{i_r} \circ \tau_\ell$$

for unique $0 \leq i_j, \ell < N$ with $\gcd(\ell, N) = 1$. For each $j = 1, \ldots, r$, let G_j, resp. H_j, be the subgroup obtained by imposing $i_j = 0$, resp. the normal subgroup obtained by imposing $i_j = 0$ and $\ell = 1$. Defining $L_j = L(\zeta_N)^{G_j}$, it is easy to check that $L(\zeta_N)^{H_j} = L_j(\zeta_N)$ and that the chain of inclusions $K \subseteq L_j \subseteq L_j(\zeta_N)$ satisfies the hypotheses of the lemma for $r = 1$. From the first part of our proof, we conclude that there exists an $\alpha_j \in L_j$ such that $L_j = K(\alpha_j)$ and $\alpha_j^N \in K^*$. But from $\cap_j G_j = \operatorname{Gal}(L(\zeta)/L)$ one sees that L is the compositum of the L_j's, from which the lemma follows. □

Note that if $r = 1$ and L contains ζ_N then Lemma 2.3 specializes to a standard statement from Kummer theory; observe that the factor $\mathrm{Gal}(L(\zeta_N)/L)$ is trivial in this case. In fact, our proof is a tweak of that of [**41**, Lem. 09DX]. In the current paper, we are mostly interested in the other end of the spectrum, where $\langle \zeta_N \rangle \cap L$ is as small as possible, *i.e.*, contained in $\{\pm 1\}$.

2.5. Charles–Goren–Lauter style hash functions. In [**12**], Charles, Goren and Lauter introduced a hash function based on isogenies between supersingular elliptic curves. This construction was generalized to work for Richelot isogenies between superspecial p.p. abelian surfaces in [**9**], by fixing an earlier proposal due to Takashima [**44**], shown to admit trivial collisions by Flynn and Ti [**22**]. We give a rough outline of the general construction.

Fix distinct primes p and ℓ, a dimension g, and let $G_{p,\ell,g}$ be the directed multigraph with vertex set V and edge set E, which are constructed as follows. V consists of all superspecial p.p. abelian varieties over $\overline{\mathbb{F}}_p$ of dimension g up to isomorphism, which can always be defined over $\mathbb{F}_{p^2}$ [**2**, Thm. 2.13A]. The edges emanating from a vertex $v \in V$ are the $(\ell, \ldots, \ell)$-isogenies with domain v, one for each $(\ell, \ldots, \ell)$-subgroup of v. One can prove that the graph $G_{p,\ell,g}$ is connected [**31**, Thm. 43], and in the case of supersingular elliptic curves, the graph is a Ramanujan graph [**12**]. Unfortunately, this is no longer the case for dimension $g > 1$ [**31**, §10.1], but those graphs seem to exhibit strong expansion properties nonetheless; see [**20**] for an empiric analysis of the case $\ell = g = 2$. From Lemma 2.1 we see that $G_{p,\ell,g}$ is a $\prod_{i=1}^{g}(\ell^i + 1)$-regular multigraph. One can try and turn this graph into an undirected graph by considering dual isogenies, but due to p.p. abelian varieties possibly having non-trivial automorphisms, the multiplicities of the edges and their duals may not coincide. For a more in-depth discussion regarding this phenomenon, we refer to [**9**, §4].

To build a hash function from this graph, we must first fix a superspecial p.p. abelian variety and will begin a walk in the graph starting from this vertex. From this initial vertex, we label all outgoing edges in some way (*e.g.*, in lexicographical order with respect to a fixed choice of representation of $\mathbb{F}_{p^2}$). Out of these $\prod_{i=1}^{g}(\ell^i + 1)$ edges, we only consider the first $\kappa = \ell^{g(g+1)/2}$ and we walk along the edge that corresponds to the least significant digit of m when expressed in base κ.[4] We have now arrived at a new p.p. abelian variety and want to avoid any possible backtracking while walking in the graph, so for our next edge, we should not consider all possible outgoing edges. For elliptic curves, it suffices to discard the edges corresponding to the dual isogenies [**12**], but for $g > 1$ we must discard all options that have a kernel which intersects the kernel of the dual isogeny non-trivially [**9**]. In general, again in view of Lemma 2.1, this leaves us with κ possible edges to consider, which correspond to good extensions of the isogeny corresponding to the first edge we chose. Once again, we label the κ outgoing edges in some deterministic way and will walk along the one that corresponds to the second least significant digit of m in base κ. We continue this until all the digits of the message have been processed. The output of the hash function is then an invariant of the final p.p. abelian variety we encounter. In the case of elliptic curves, one can choose the j-invariant for example.

[4]There is no real reason why one cannot consider all edges in this first step. Restricting to only κ choices however streamlines the algorithm.

3. On the existence of multiradical isogeny formulae

In this section we give a group-theoretic argument in favor of the existence of multiradical isogeny formulae. The argument is motivated by Lemma 2.3.

3.1. A multiradical modular cover. For perfect field K, an integer $n \geq 2$ and a subgroup $H \subseteq \mathrm{GSp}_{2g}(\mathbb{Z}_n)$, we consider the moduli problem of parametrizing pairs (A, α) up to H-equivalence, where A is a g-dimensional p.p. abelian variety over $\overline{K}$ and α is an n-level structure on it. Two pairs (A_1, α_1) and (A_2, α_2) are called H-equivalent if there exists an isomorphism $\varphi : A_1 \to A_2$ and an element $h \in H$ such that $\alpha_1 = h \circ \alpha_2 \circ \varphi$. We write $[(A, \alpha)]_H$ for the H-equivalence class of (A, α), and denote the moduli set of such H-equivalence classes by $\mathcal{A}_g(H)$. Two extremal cases are $\mathcal{A}_g(\mathrm{GSp}_{2g}(\mathbb{Z}_n))$, which just parametrizes g-dimensional p.p. abelian varieties up to isomorphism, and $\mathcal{A}_g(\{\mathrm{id}\})$, which parametrizes g-dimensional p.p. abelian varieties A equipped with a generalized symplectic basis of $A[n]$. Note that if H' is a subgroup of H, then we have a natural map $\mathcal{A}_g(H') \to \mathcal{A}_g(H) : [(A, \alpha)]_{H'} \mapsto [(A, \alpha)]_H$.

We can construct a moduli set of g-dimensional p.p. abelian varieties A together with marked generators $P_1, \ldots, P_g$ of an $(N, \ldots, N)$-subgroup by choosing $n = N$ and letting H be

$$H_N = \left\{ \begin{pmatrix} \mathbb{I}_g & B \\ 0 & d\mathbb{I}_g \end{pmatrix} \middle| B \in \mathrm{Sym}_g(\mathbb{Z}_N), d \in \mathbb{Z}_N^* \right\} \subseteq \mathrm{GSp}_{2g}(\mathbb{Z}_N),$$

where $\mathrm{Sym}_g(\mathbb{Z}_N)$ denotes the set of symmetric $g \times g$ matrices with entries in $\mathbb{Z}_N$. Another (overcomplicated) way of arriving at a set with the same moduli interpretation is by instead letting $n = N^2$ and considering the group

$$\Gamma_{1,N} = \left\{ M \in \mathrm{GSp}_{2g}(\mathbb{Z}_{N^2}) \mid M \bmod N \in H_N \right\}.$$

This creates room for defining the subgroup

$$\Gamma'_{1,N} = \left\{ M \in \Gamma_{1,N} \subseteq \mathrm{GSp}_{2g}(\mathbb{Z}_{N^2}) \mid \text{lower-left } g \times g \text{ block of } M \text{ is zero} \right\},$$

whose associated moduli set parametrizes p.p. abelian varieties along with marked generators $Q_1, \ldots, Q_g$ of an $(N^2, \ldots, N^2)$-subgroup, considered modulo the following equivalence relation: two such sets of marked generators $Q_1, \ldots, Q_g$ and $R_1, \ldots, R_g$ are identified if and only if $R_i - Q_i \in \langle NQ_1, \ldots, NQ_g \rangle$ for $i = 1, \ldots, g$. Note that the points $P_i := NQ_i$ do not depend on the chosen representants Q_i, and neither do the cosets P'_i of Q_i modulo $\langle P_1, \ldots, P_g \rangle$.

Said differently, the set $\mathcal{A}_g(\Gamma'_{1,N})$ parametrizes g-dimensional p.p. abelian varieties A together with marked generators $P_1, \ldots, P_g$ of some $(N, \ldots, N)$-subgroup $G \subseteq A$, as well as with marked generators $P'_1, \ldots, P'_g$ of an $(N, \ldots, N)$-subgroup $G' \subseteq A/G$ which are such that the chain of quotient maps

$$A \xrightarrow{\varphi} A' = A/G \xrightarrow{\varphi'} A'/G'$$

is good, *i.e.*, $\varphi' \circ \varphi$ is an $(N^2, \ldots, N^2)$-isogeny. The natural map $\mathcal{A}_g(\Gamma'_{1,N}) \to \mathcal{A}_g(\Gamma_{1,N})$ just "forgets" about the points P'_i. Thus, the central question of our paper — given $P_1, \ldots, P_g$, how to find $P'_1, \ldots, P'_g$ — is closely related to understanding the fibers of this map.

REMARK 3.1. In the above moduli interpretation, the marked generators P_i' have the additional property that

$$\hat{\varphi}(P_i') = P_i \text{ for all } i = 1, \ldots, g, \tag{3.1}$$

where $\hat{\varphi} : A' \to A$ is the dual of φ. This feature was not explicitly asked for in the introduction. However, every subgroup $G' \subseteq A'$ for which $A' \to A'/G'$ is a good extension of φ admits a unique $\mathbb{Z}_N$-basis satisfying (3.1); we call this basis *distinguished.* It suffices to concentrate on such bases. Indeed, once we have found formulae for these distinguished generators, formulae for other sets of generators can be found by performing a base change, using arithmetic on A',[5] and this should not affect features like multiradicality, completeness and good reduction. Moreover, it seems reasonable to expect that the formulae for the distinguished generators will stand out in terms of simplicity (although we did not investigate this in detail).

The multiradical nature of the fibers of $\mathcal{A}_g(\Gamma'_{1,N}) \to \mathcal{A}_g(\Gamma_{1,N})$ is hinted at by the following lemma, which invokes the notation $d(M)$ from (2.1), in combination with Lemma 2.3. Recall that the *normal core* $\mathrm{Core}_G(H)$ of a subgroup H in a group G is the largest subgroup of H that is normal in G. For use below we remark that, under the Galois correspondence, this notion corresponds to the Galois closure of a separable field extension.

In order to state the lemma, we fix any bijection

$$k : \{1, \ldots, g(g+1)/2\} \to \{(k_1, k_2) \mid 1 \leq k_1 \leq k_2 \leq g\}$$

and for all $j = 1, \ldots, g(g+1)/2$ and $0 \leq \ell < N$, $\gcd(N, \ell) = 1$ we define the elements

$$\sigma_j = \begin{pmatrix} \mathbb{I}_g & 0 \\ NS_{k(j)} & \mathbb{I}_g \end{pmatrix}, \qquad \tau_\ell = \begin{pmatrix} \mathbb{I}_g & 0 \\ 0 & \ell\mathbb{I}_g \end{pmatrix}$$

of $\Gamma_{1,N}$, where $S_{(k_1,k_2)}$ denotes the symmetric $g \times g$ matrix having a 1 at positions (k_1, k_2) and (k_2, k_1) and 0's elsewhere.

LEMMA 3.2. *The group $\Gamma'_{1,N}$ has index $N^{g(g+1)/2}$ in $\Gamma_{1,N}$. Its normal core can be computed as*

$$\mathrm{Core}_{\Gamma_{1,N}}(\Gamma'_{1,N}) = \{M \in \Gamma'_{1,N} \mid d(M) \equiv 1 \bmod N\}$$

which has index $\varphi(N)$ in $\Gamma'_{1,N}$. Every element of $\Gamma_{1,N}/\mathrm{Core}(\Gamma'_{1,N})$ admits a unique representant of the form

$$\sigma_1^{i_1} \cdots \sigma_{g(g+1)/2}^{i_{g(g+1)/2}} \cdot \tau_\ell \tag{3.2}$$

with $0 \leq i_j < N$ for all $j = 1, \ldots, g(g+1)/2$, and $0 \leq \ell < N$, $\gcd(N, \ell) = 1$. More precisely $\Gamma_{1,N}/\mathrm{Core}_{\Gamma_{1,N}}(\Gamma'_{1,N})$ can be written as

$$\{\sigma_j^{i_j} \mid 1 \leq j \leq g(g+1)/2,\, 0 \leq i_j < N\} \rtimes \{\tau_\ell \mid 0 \leq \ell < N,\, \gcd(N, \ell) = 1\}$$
$$\cong \mathbb{Z}_N^{g(g+1)/2} \rtimes \mathbb{Z}_N^*,$$

where the semi-direct product is taken according to the rule (2.3).

[5]For example, if N is odd, then the formulae for $2P_1', \ldots, 2P_g'$ are obtained from those for $P_1', \ldots, P_g'$ by feeding the latter to a formula for doubling on A'.

PROOF. It is not hard to check that all matrices $M \in \Gamma_{1,N}$ have symmetric lower-left $g \times g$ blocks, *i.e.*, these blocks belong to $N\,\mathrm{Sym}_g(\mathbb{Z}_{N^2})$. A count shows that the resulting map

$$\Gamma_{1,N} \to N\,\mathrm{Sym}_g(\mathbb{Z}_{N^2})$$

is uniform (*i.e.*, every element in the codomain has the same number of preimages), implying that $[\Gamma_{1,N} : \Gamma'_{1,N}] = N^{g(g+1)/2}$. As for the normal core, conjugating $\Gamma'_{1,N}$ with suitable matrices (*e.g.*, one can use the matrices σ_j) reveals that

$$\mathrm{Core}_{\Gamma_{1,N}}(\Gamma'_{1,N}) \subseteq \{M \in \Gamma'_{1,N} \mid d(M) \equiv 1 \bmod N\}$$

and since the right-hand side is a normal subgroup of $\Gamma_{1,N}$, equality must hold. Finally, we have

$$[\Gamma'_{1,N} : \mathrm{Core}_{\Gamma_{1,N}}(\Gamma'_{1,N})] = \varphi(N)$$

because d defines a morphism $\Gamma'_{1,N} \to \mathbb{Z}_N^*$ which is surjective, as can be seen by evaluating it at the τ_ℓ's.

Now assume that some element of $\Gamma_{1,N}/\mathrm{Core}_{\Gamma_{1,N}}(\Gamma'_{1,N})$ admits two distinct decompositions

$$\sigma_1^{i_1} \cdots \sigma_{g(g+1)/2}^{i_{g(g+1)/2}} \cdot \tau_\ell = \sigma_1^{i'_1} \cdots \sigma_{g(g+1)/2}^{i'_{g(g+1)/2}} \cdot \tau_{\ell'}.$$

Applying d shows that $\ell \equiv \ell' \bmod N$, hence we can assume $\ell = \ell' = 1$. We then find

$$\sigma_1^{i_1 - i'_1} \cdots \sigma_{g(g+1)/2}^{i_{g(g+1)/2} - i'_{g(g+1)/2}} = \begin{pmatrix} \mathbb{I}_g & 0 \\ N \sum_{j=1}^{g(g+1)/2} (i_j - i'_j) S_{k(j)} & \mathbb{I}_g \end{pmatrix}. \tag{3.3}$$

But this is contained in $\Gamma'_{1,N}$ only if $i_j \equiv i'_j \bmod N$ for all j. In particular, the expansion (3.2) is unique. Elements of the form (3.1) are a full set of representants of $\Gamma_{1,N}/\mathrm{Core}_{\Gamma_{1,N}}(\Gamma'_{1,N})$ because there are $\varphi(N)N^{g(g+1)/2}$ such expansions.

The statement about the semi-direct product is easy to check using (3.3). □

We now give more details on how Lemma 3.2 supports the existence of multiradical isogeny formulae, although we stress that the discussion below is partly heuristic. A major ingredient is that the sets $\mathcal{A}_g(H)$ are representable by algebraic varieties over $\mathbb{Q}$.[6] Indeed, results by Artin and Faltings–Chai show that the corresponding moduli spaces exist as schemes over $\mathbb{Z}[1/N]$, see [**19**, §I.4]; it then follows from Geometric Invariant Theory that these spaces are quasi-projective [**36**, Thm. 7.9]. Consequently, the chain

$$\mathcal{A}_g(\{\mathrm{id}\}) \to \mathcal{A}_g(\Gamma'_{1,N}) \to \mathcal{A}_g(\Gamma_{1,N}) \to \mathcal{A}_g(\mathrm{GSp}_{2g}(\mathbb{Z}_{N^2}))$$

corresponds to an inclusion of function fields

$$\mathbb{Q}(\mathcal{A}_g(\mathrm{GSp}_{2g}(\mathbb{Z}_{N^2})) \subseteq \mathbb{Q}(\mathcal{A}_g(\Gamma_{1,N})) \subseteq \mathbb{Q}(\mathcal{A}_g(\Gamma'_{1,N})) \subseteq \mathbb{Q}(\mathcal{A}_g(\{\mathrm{id}\}))$$

where the outer extension is Galois, with Galois group $\mathrm{GSp}_{2g}(\mathbb{Z}_{N^2})$, and where $\mathbb{Q}(\mathcal{A}_g(\Gamma_{1,N}))$, resp. $\mathbb{Q}(\mathcal{A}_g(\Gamma'_{1,N}))$, are the subfields fixed by $\Gamma_{1,N}$, resp. $\Gamma'_{1,N}$. This extrapolates upon well-known statements from the elliptic curve case, which can be found in [**15, 37, 39**], for instance. The middle inclusion has Galois closure

$$\mathbb{Q}(\mathcal{A}_g(\{\mathrm{id}\}))^{\mathrm{Core}_{\Gamma_{1,N}}(\Gamma'_{1,N})}$$

[6] These varieties may be geometrically reducible; more precisely, for $H \subseteq \mathrm{GSp}_{2g}(\mathbb{Z}_n)$ we have that $\mathcal{A}_g(H)$ decomposes into $[\mathbb{Z}_n^* : d(H)]$ irreducible components over $\mathbb{Q}(\zeta_n)$.

which, in the same vein, is obtained from $\mathbb{Q}(\mathcal{A}_g(\Gamma'_{1,N}))$ by adding a primitive Nth root of unity ζ_N. The Galois group of this Galois closure is $\Gamma_{1,N}/\operatorname{Core}_{\Gamma_{1,N}}(\Gamma'_{1,N})$, so by Lemma 2.3 and Lemma 3.2 we have

$$\mathbb{Q}(\mathcal{A}_g(\Gamma'_{1,N})) = \mathbb{Q}(\mathcal{A}_g(\Gamma_{1,N}))(\sqrt[N]{\rho_1}, \ldots, \sqrt[N]{\rho_{g(g+1)/2}})$$

for certain functions $\rho_1, \ldots, \rho_{g(g+1)/2}$ on $\mathcal{A}_g(\Gamma_{1,N})$.

The line of thought behind multiradical isogenies is then that the coordinates of our distinguished generators $P'_1, \ldots, P'_g$ can essentially be viewed as functions on $\mathcal{A}_g(\Gamma'_{1,N})$, therefore they should be expressible in terms of the radicals $\sqrt[N]{\rho_i}$. Since we work over $\mathbb{Q}$, these expressions make sense over any perfect field K, as long as $\operatorname{char} K$ does not divide any denominators; in fact, the idea/hope behind our good reduction assumption **(3)** is that all of this can be set up over $\mathbb{Z}[1/N]$ rather than $\mathbb{Q}$.

3.2. Conjectured existence of multiradical isogeny formulae. As we have discussed in the introduction, it only makes sense to talk about multiradical isogeny formulae at the level of concrete families that come equipped with formulae of Vélu, Richelot, ... type for the codomain p.p. abelian varieties.

Let us therefore repeat, in more detail, our main surmise from Conjecture 1. For integers $r, g \geq 1, N \geq 2$, we consider a smooth family of g-dimensional p.p. abelian varieties A_s equipped with marked points $P_{s,1}, \ldots, P_{s,g}$ that generate an $(N, \ldots, N)$-subgroup $G_s \subseteq A_s$, where the parameter $s = (s_1, \ldots, s_r)$ ranges over some quasi-affine subset $\mathcal{S} \subseteq \mathbb{A}^r$. We assume that we have algebraic formulae at our disposal, explicitly describing $A'_s = A_s/G_s$ in terms of the s_i. Then we believe that there always exist accompanying multiradical formulae, producing a set of generators $P'_{s,1}, \ldots, P'_{s,g}$ of an $(N, \ldots, N)$-subgroup $G'_s \subseteq A'_s$ which is such that the extension

$$A_s \xrightarrow{\varphi} A'_s = A_s/G_s \xrightarrow{\varphi'} A'_s/G'_s$$

is good. Moreover, we believe that the formulae can be chosen such that they are complete, and such that they work over any perfect field over which the parametrization by $\mathcal{S}$ makes sense.

The radicands $\mathfrak{r}_i$ appearing in these formulae should be related to the functions ρ_i from the previous section, as follows. As before, assume we are working over $\mathbb{Q}$. By the universal property of moduli spaces, we have a natural morphism $\sigma : \mathcal{S} \to \mathcal{A}_g(\Gamma_{1,N})$, sending s to the isomorphism class of $(A_s, P_{s,1}, \ldots, P_{s,g})$. This allows us to pull back the functions $\rho_i \in \mathbb{Q}(\mathcal{A}_g(\Gamma_{1,N}))$ to $\mathbb{Q}(\mathcal{S})$; here we assume that the image of $\mathcal{S}$ is not included in the polar locus of ρ_i. These pull-backs should be our $\mathfrak{r}_i$'s. Explicitly,

$$\mathfrak{r}_1 := \rho_1 \circ \sigma, \quad \ldots, \quad \mathfrak{r}_{g(g+1)/2} := \rho_{g(g+1)/2} \circ \sigma,$$

which can indeed be viewed as algebraic expressions in the coordinates s_i.

We point out that, for the sake of flexibility, we do not require the map $\mathcal{S} \to \mathcal{A}_g(\Gamma_{1,N})$ to be injective, *i.e.*, up to isomorphism, different s may result in the same p.p. abelian variety and the same generators of an $(N, \ldots, N)$-subgroup. Our examples in Section 4 include several families featuring such a redundance.

REMARK 3.3. Our formulae should make sense at every point of $\mathcal{S}$, therefore the functions $\mathfrak{r}_1, \ldots, \mathfrak{r}_{g(g+1)/2}$ should be free of poles. In view of the completeness, they should also be free of zeroes.

REMARK 3.4. For small families, the extension

$$\mathbb{Q}(\mathcal{S}) \subseteq \mathbb{Q}(\mathcal{S})(\sqrt[N]{\mathfrak{r}_1}, \ldots, \sqrt[N]{\mathfrak{r}_{g(g+1)/2}})$$

may not be of degree $N^{g(g+1)/2}$. Indeed, when pulled back along σ, several of the radicands ρ_i may become interrelated. In such cases it is tempting to compress the formulae into versions that use fewer radicals, but then the completeness property gets lost. For instance, in the example in Section 4.3 below, as many as $g(g-1)/2$ radicands collapse to the constant 1; nevertheless one should allow the corresponding occurrences of $\sqrt[N]{1}$ to range independently over the set of Nth roots of unity if one wants to find all $N^{g(g+1)/2}$ good extensions.

If our family of p.p. abelian varieties A_s consists of (products of) Jacobians of curves C_s which, when viewed as a single curve over $\mathbb{Q}(\mathcal{S})$, is either of genus 2 or admits a rational point, then Conjecture 1 comes with the following addendum:

(4) **Tate pairings as suitable radicands.** The radicands $\mathfrak{r}_1, \ldots, \mathfrak{r}_{g(g+1)/2}$ can be taken to be representants of the Tate pairings

$$t_N(P_{s,i}, P_{s,j}) \in \mathbb{Q}(\mathcal{S})^*/(\mathbb{Q}(\mathcal{S})^*)^N$$

where $i \leq j$ range over $\{1, \ldots, g\}$.

This is motivated, again, by our examples below, and by the following observation. For each $1 \leq i \leq j \leq g$, choose a representant $\mathfrak{r}_{i,j}$ of $t_N(P_{s,i}, P_{s,j})$. Let $\mathbb{Q}(\mathcal{S})(G'_s)$ denote the field obtained from $\mathbb{Q}(\mathcal{S})$ by adjoining the coordinates of $P'_{s,1}, \ldots, P'_{s,g}$. As discussed in Remark 3.1, we can assume that $\hat{\varphi}(P'_{s,i}) = P_{s,i}$ for all i. This implies that

$$\mathfrak{r}_{i,j} = t_N(\hat{\varphi}(P'_{s,i}), \hat{\varphi}(P'_{s,j})) = t_N(P'_{s,i}, P'_{s,j})^N$$

when viewed as elements of $\mathbb{Q}(\mathcal{S})(G'_s)^*/(\mathbb{Q}(\mathcal{S})(G'_s)^*)^N$; the second equality follows from the compatibility property of the Tate pairing, see [**30**, Lem. 5]. Thus $\mathbb{Q}(\mathcal{S})(G'_s)$ contains $\mathbb{Q}(\mathcal{S})(\sqrt[N]{\mathfrak{r}_{i,j}} \mid 1 \leq i \leq j \leq g)$.

We did not manage to prove that these two fields are in fact equal, which would lend further support for our addendum.[7] While for $g = 1$ equality can be established using non-degeneracy of the Tate pairing over finite fields containing a primitive Nth root of unity [**10**, §3], for $g > 1$ non-degeneracy or even perfectness does not seem strong enough to mimic that argument.

4. Examples

In this section, we show how multiradical isogeny formulae manifest themselves for two well-known families: Richelot isogenies, and fully split isogenies from products of elliptic curves. We also show that multiradical isogeny formulae apply to a certain $(5,5)$-isogeny that was described by Flynn [**21**]. Our main example, namely non-split $(3,3)$-isogenies from Jacobians of genus-2 curves, will be discussed in Section 5. We begin by recalling an elliptic curve example from [**10**].

[7] Note however that the addendum is an even stronger statement, *e.g.*, in view of Remark 3.4.

4.1. Elliptic curves. Consider the family of elliptic curves E with a marked point $P \in E$ of order N. For $N \geq 4$ this family is conveniently parametrized by the Tate normal form:[8]

$$E : y^2 + (1-c)xy - by = x^3 - bx^2, \qquad P = (0,0).$$

Concretely, we let $\mathcal{S} \subseteq \mathbb{A}^2$ be the subset of pairs b, c for which E is non-singular and P has exact order N; we refer to [**43**] for how to obtain a concrete equation for $\mathcal{S}$, which is a model of the modular curve $Y_1(N)$ and which is naturally defined over $\mathbb{Z}[1/N]$. The existence of radical and complete isogeny formulae was discussed in [**10**], where it was argued that one can take $\mathfrak{r}_1 = f_{N,P}(-P)$, with $f_{N,P}$ the function on E with divisor $N(P) - N(\infty)$, normalized such that its expansion at ∞ with respect to the uniformizer x/y has leading coefficient 1. As mentioned there, $\mathfrak{r}_1$ is a representant of $t_N(P, -P) = t_N(P,P)^{-1}$, so in order to enforce property **(4)**, one should instead work with $\mathfrak{r}_1^{-1}$. This does not cause any issues because $\mathfrak{r}_1$ has no zeroes or poles on $\mathcal{S}$; see also Remark 3.3.

For the sake of example, let us revisit the case $N = 5$, where we have $\mathfrak{r}_1 = b$ and $\mathcal{S} = \{(b,c) \in \mathbb{A}^2 \mid b = c,\, b \neq 0, (11 \pm 5\sqrt{5})/2\}$. Vélu's formulae yield the following defining equation for $E' = E/\langle P \rangle$:

$$y^2 + (1-b)xy - by = x^3 - bx^2 - 5b(b^2 + 2b - 1)x - b(b^4 + 10b^3 - 5b^2 + 15b - 1).$$

From [**10**, §4] we see that the point

$$\begin{aligned} P' = (&5\sqrt[5]{\mathfrak{r}_1}^4 + (b-3)\sqrt[5]{\mathfrak{r}_1}^3 + (b+2)\sqrt[5]{\mathfrak{r}_1}^2 + (2b-1)\sqrt[5]{\mathfrak{r}_1} - 2b, \\ &5\sqrt[5]{\mathfrak{r}_1}^4 + (b-3)\sqrt[5]{\mathfrak{r}_1}^3 + (b^2 - 10b + 1)\sqrt[5]{\mathfrak{r}_1}^2 + (13b - b^2)\sqrt[5]{\mathfrak{r}_1} - b^2 - 11b) \end{aligned} \tag{4.1}$$

on E' is of the requested kind, *i.e.*, it is the distinguished generator of a subgroup $G' \subseteq E[5]$ such that the composed isogeny $E \to E' \to E'/G'$ is cyclic of degree 25. Varying the choice of $\sqrt[5]{\mathfrak{r}_1}$ produces the five subgroups for which this is true. The formula (4.1) satisfies the good reduction property and allows for a very fast computation of chains of 5-isogenies over finite fields; *e.g.*, over $\mathbb{F}_p$ with $p \not\equiv 1 \bmod 5$ we obtain a speed-up by roughly a factor 40 over more traditional methods [**10**, Tbl. 4]. We recall that, for general N, the good reduction property is conjectural [**10**, Conj. 1].

4.2. Richelot isogenies. A convenient reference for Richelot isogenies is [**40**, Ch. 8]. We consider genus-2 curves C equipped with two generators of a $(2,2)$-subgroup of J_C. Such marked curves can be parametrized by $\mathcal{S} = \mathbb{A}^9 \setminus \Delta$, by letting $s = (s_{ij})_{1 \leq i,j \leq 3}$ correspond to the Jacobian of

$$C : y^2 = G_1(x)G_2(x)G_3(x), \qquad G_i(x) = s_{i1}x^2 + s_{i2}x + s_{i3}$$

equipped with the divisor classes $\overline{D}_1, \overline{D}_2$ from Example 2.2. Here Δ is cut out by the discriminant of $G_1(x)G_2(x)G_3(x)$. The parametrization works over $\mathbb{Z}[1/2]$. We claim that we can take $\mathfrak{r}_1 = \mathrm{res}_x(G_2, G_3)$, $\mathfrak{r}_2 = \mathrm{res}_x(G_1, G_3)$ and $\mathfrak{r}_3 = \mathrm{res}_x(G_1, G_2)$. By Example 2.2 we know that

$$\begin{aligned} t_2(\overline{D}_1, \overline{D}_1) &\equiv \mathfrak{r}_2\mathfrak{r}_3 \\ t_2(\overline{D}_1, \overline{D}_2) &\equiv \mathfrak{r}_3 \\ t_2(\overline{D}_2, \overline{D}_2) &\equiv \mathfrak{r}_1\mathfrak{r}_2 \end{aligned}$$

[8]See [**10**, §4] for a discussion of the cases $N = 2, 3$.

modulo squares, so the validity of property **(4)** is not affected by our choice of radicands. Indeed, formulae in terms of $\sqrt{\mathfrak{r}_1}, \sqrt{\mathfrak{r}_2}, \sqrt{\mathfrak{r}_3}$ can easily be rewritten into formulae in terms of $\sqrt{\mathfrak{r}_2\mathfrak{r}_3}, \sqrt{\mathfrak{r}_3}, \sqrt{\mathfrak{r}_1\mathfrak{r}_2}$, and vice versa.

To proceed, we slightly shrink $\mathcal{S}$ by removing the zero locus of the determinant $\delta = |s_{i,j}|_{1\leq i,j\leq 3}$. This guarantees that the p.p. abelian surface $J_C/\langle \overline{D}_1, \overline{D}_2\rangle$ is again a Jacobian. More precisely, Richelot's formulae show that it is isomorphic to $J_{C'}$ with

$$C' : \delta y^2 = H_1(x) \cdot H_2(x) \cdot H_3(x),$$

where $H_1 := G_2'G_3 - G_2G_3'$, $H_2 := G_3'G_1 - G_3G_1'$ and $H_3 := G_1'G_2 - G_1G_2'$. The reader can verify that $\operatorname{disc}(H_i) = 4\mathfrak{r}_i$, so the two zeroes of H_i are algebraic expressions in $\sqrt{\mathfrak{r}_i}$ and in the s_{ij}'s, and they are obtained from one another by choosing the other square root of $\mathfrak{r}_i$; denote these two zeroes by $\alpha_{\pm i}$. Then according to [**9**, Prop. 2] the classes of

$$D_1' = (\alpha_1, 0) + (\alpha_2, 0) - \infty_1 - \infty_2, \qquad D_2' = (\alpha_{-1}, 0) + (\alpha_3, 0) - \infty_1 - \infty_2$$

generate a $(2,2)$-subgroup of $J_{C'}$ that defines a $(4,4)$-extension of the incoming isogeny $J_C \to J_{C'}$. Still according to [**9**, Prop. 2], the sign flips $\pm i$ produce the eight subgroups for which this is true. Thus we have found formulae that are multiradical and complete, and they clearly work in any characteristic different from 2.

REMARK 4.1. One could also try and study the complementary case, namely the restriction $\mathcal{S}_0$ of $\mathcal{S}$ to the zero locus of δ. In this case $J_C/\langle \overline{D}_1, \overline{D}_2\rangle$ geometrically splits as a product of two elliptic curves. Concrete equations for these elliptic curves can be found in [**40**, p. 119]. The reader can check that they are defined over the field obtained by adding a square root of $\operatorname{disc}_z(\operatorname{disc}_x(G_2+zG_3))$ which, interestingly, turns out to be $16\mathfrak{r}_1$. However, for a genuine verification of Conjecture 1, one would need a model of $J_C/\langle \overline{D}_1, \overline{D}_2\rangle$ over $\mathbb{Q}(\mathcal{S}_0)$ rather than $\mathbb{Q}(\mathcal{S}_0)(\sqrt{\mathfrak{r}_1})$. This model concerns the Weil restriction to $\mathbb{Q}(\mathcal{S}_0)$ of an elliptic curve defined over $\mathbb{Q}(\mathcal{S}_0)(\sqrt{\mathfrak{r}_1})$, which is not easy to describe explicitly; see also [**4**].

4.3. Fully split $(N,\ldots,N)$-isogenies from products of elliptic curves. In this example we consider g-fold products $E_1 \times \cdots \times E_g$ of elliptic curves, marked with generators $D_1, \ldots, D_g$ of an $(N,\ldots,N)$-subgroup that are of the following kind: each D_i is a g-tuple with ∞_{E_j} at entry j, except when $j = i$ where we then have a point $P_i \in E_i$ of order N. Assuming $N \geq 4$, such marked products are naturally parametrized by $\mathcal{S}^g \subseteq \mathbb{A}^{2g}$, with $\mathcal{S}$ the modular curve $Y_1(N)$ from Section 4.1. Note that the corresponding $(N,\ldots,N)$-isogenies split completely, *i.e.*, they are of the form

$$\Phi : E_1 \times \ldots \times E_g \to E_1' \times \ldots \times E_g',$$

decomposing as the product of cyclic N-isogenies $\phi_i : E_i \to E_i'$ with kernel $\langle P_i\rangle$. We assume that the elliptic curves E_i' are given by Vélu's formulae.

For each $i = 1, \ldots, g$, we let $\mathfrak{r}_i$ be the representant of the Tate self-pairing $t_N(P_i, P_i)$ whose inverse was described in Example 4.1. We then choose the following representants of the Tate pairings $t_N(D_i, D_j)$, $1 \leq i \leq j \leq g$: we pick 1 as soon as $i < j$, and we pick $\mathfrak{r}_i$ if $i = j$.

We are interested in identifying all $(N,\ldots,N)$-subgroups of $(E_1' \times \cdots \times E_g')[N]$ that have trivial intersection with the kernel of the dual of Φ. Indeed, these are precisely the subgroups that can occur as $\ker \Psi$ for a good extension Ψ of Φ. To get a handle on the kernel of $\hat{\Phi}$, which is just the product of the $\hat{\phi}_i$'s, we rely on Lemma 4.2 below. When applied over $\mathbb{Q}(\mathcal{S})$, it implies that for each $i = 1, \ldots, g$ we can find a formula $P_i'(\sqrt[N]{1})$ which, when reading $\sqrt[N]{1}$ as ζ_N, produces a generator P_i' of $\ker \hat{\phi}_i$ and which, when reading $\sqrt[N]{1}$ as ζ_N^k, produces the point kP_i' for $0 \leq k \leq N-1$. Then $\ker \hat{\Phi}$ can be written as $\langle C_1', \ldots, C_g' \rangle$, where each C_i' is a g-tuple with $\infty_{E_j'}$ at each entry, except at $j = i$ where we have P_i'.

LEMMA 4.2. *Let E be an elliptic curve over a perfect field K with* $\operatorname{char} K \nmid N$ *and let $P \in E(K)$ be a point of order N. Let $\phi : E \to E' = E/\langle P \rangle$ be the corresponding quotient isogeny, where E' is given by Vélu's formulae. Let P' be a generator of the dual isogeny. Then there exist polynomials $F, G, H \in K[z]$ such that*

$$[F(\zeta_N^k) : G(\zeta_N^k) : H(\zeta_N^k)] = kP'$$

for all $0 \leq k \leq N-1$.

PROOF. The Weil pairing gives a group isomorphism between $\ker \hat{\phi}$ and μ_N that is compatible with the action of $\operatorname{Gal}(\overline{K}/K)$. In particular P' has coordinates in $K(\zeta_N)$. Define $F(z)$ to be the classical Lagrange polynomial that interpolates the x-coordinates of kP' for $0 \leq k \leq N-1$. More precisely,

$$F(z) = \sum_{k=0}^{N-1} x(kP')\ell_k(z), \text{ with } \ell_k(z) = \prod_{\substack{0 \leq m \leq N-1 \\ m \neq k}} \frac{z - \zeta_N^m}{\zeta_N^k - \zeta_N^m}.$$

Then it suffices to show that for any $\sigma \in \operatorname{Gal}(K(\zeta_N)/K)$ it holds that $F(z) = F^\sigma(z)$. Note that $\sigma : \zeta_N \mapsto \zeta_N^a$ for some a coprime to N. One verifies that

$$\ell_k^\sigma(z) = \prod_{\substack{0 \leq m \leq N-1 \\ m \neq k}} \frac{z - \zeta_N^{am}}{\zeta_N^{ak} - \zeta_N^{am}} = \prod_{\substack{0 \leq m \leq N-1 \\ m \neq ak}} \frac{z - \zeta_N^{m}}{\zeta_N^{ak} - \zeta_N^{m}} = \ell_{ak}(z).$$

Furthermore, we can assume that the x-coordinates of the points of $\ker \hat{\phi}$ within the same Galois orbit were chosen compatibly, i.e. $\sigma(x(kP')) = x(\sigma(kP'))$ for all $\sigma \in \operatorname{Gal}(K(\zeta_N)/K)$ and for all $0 \leq k \leq N-1$. Then because of the aforementioned isomorphism we must have $\sigma(x(kP')) = x(akP')$, such that indeed $F(z) = F^\sigma(z)$ as wanted. An analogous argument applies to the polynomials G and H. □

We also know that, for each $i = 1, \ldots, g$, there exists a formula $Q_i'(\sqrt[N]{\mathfrak{r}_i})$ producing a point Q_i' that extends P_i' to a basis of $E_i'[N]$. Furthermore, we know that by scaling $\sqrt[N]{\mathfrak{r}_i}$ with ζ_N^k for $0 \leq k \leq N-1$, we cycle through all elements $Q_i' + kP_i'$.

We are ready to give multiradical and complete formulae that produce g-tuples $D_1', \ldots, D_g' \in E_1' \times \cdots \times E_g'$ generating the kernel of a good extension Ψ of Φ. Fix

$$D_1' = (Q_1'(\sqrt[N]{\mathfrak{r}_1}), P_2'(\sqrt[N]{1}), \ldots, P_g'(\sqrt[N]{1})),$$

which has g degrees of freedom. Next, choose

$$D_2' = (\infty_{E_1'}, Q_2'(\sqrt[N]{\mathfrak{r}_2}), P_3'(\sqrt[N]{1}), \ldots, P_g'(\sqrt[N]{1})),$$

where we fixed the first coordinate at $\infty_{E_1'}$ in order to avoid repetitions in the subgroups generated by D_1' and D_2'. This results in $g-1$ degrees of freedom. Continuing this inductively, we end up with

$$D_g' = (\infty_{E_1'}, \ldots, \infty_{E_{g-1}'}, Q_g'(\sqrt[N]{\mathfrak{r}_g}))$$

with only 1 degree of freedom left. In total, we have $\sum_{j=1}^{g} j = g(g+1)/2$ degrees of freedom as wanted, and running through all possible interpretations of the radicals (including the $g(g-1)/2$ occurrences of $\sqrt[N]{1}$) provides the kernels of all possible good extensions.

4.4. Flynn's family of $(5,5)$-isogenies from genus-2 curve Jacobians. Consider the family of genus-2 curves with given $(5,5)$-subgroup from [**21**], involving a single parameter r. In this section, we illustrate that multiradical isogeny formulae apply to this family. We do not aim at a full analysis including completeness, etc; in fact, for simplicity we will restrict to the curve at $r = 1$. We remark that the absolute Igusa invariants of Flynn's family are in fact parameterless, so up to isomorphism this is the only curve in the family.

In order for the generators of the $(5,5)$-subgroup to be rational (and not just the subgroup), we will fix the base field as $\mathbb{Q}(\zeta_5)$, where ζ_5 is a fifth root of unity.[9] Writing $\gamma_1 = \sqrt{5} = 2\zeta_5^3 + 2\zeta_5^2 + 1 \in \mathbb{Q}(\zeta_5)$, we have

$$C : y^2 = x^5 + 25x^4 - 200x^3 + 560x^2 - 640x + 256,$$
$$T_1 = (4, 16\gamma_1) - \infty, \qquad T_2 = (0, 16) - \infty,$$

where $T_1, T_2 \in J_C[5]$. Writing $\gamma_2 = \sqrt{2(1/\gamma_1 - 1)} = \frac{2\zeta_5^3 - 6\zeta_5^2 - 4\zeta_5 - 2}{5} \in \mathbb{Q}(\zeta_5)$, the genus-2 curve associated with the isogenous abelian surface obtained by quotienting out $\langle T_1, T_2 \rangle$ can be written as

$$\widetilde{C} : y^2 = x^5 - 125x^4 + 5000x^3 - 175000x^2 + 1250000x - 81250000,$$
$$\widetilde{T}_1 = (10\gamma_1, 10000\gamma_2) - \infty, \quad \widetilde{T}_2 = (-10\gamma_1, 5000\gamma_2(\gamma_1 + 1)) - \infty,$$

where $\langle \widetilde{T}_1, \widetilde{T}_2 \rangle$ is the kernel of the dual isogeny (in particular, $\widetilde{T}_1, \widetilde{T}_2 \in J_{\widetilde{C}}[5]$). In order to extend $\langle \widetilde{T}_1, \widetilde{T}_2 \rangle$ to a basis for the 5-torsion of the Jacobian of $\widetilde{C}$, with conjectured property **(4)** in mind we compute the following Tate pairings:

$$t_5(T_1, T_1) \equiv \gamma_1, \quad t_5(T_1, T_2) \equiv (\gamma_1 - 1)/2, \quad t_5(T_2, T_2) \equiv 1.$$

Defining $\mathfrak{r}_1 = \gamma_1$ and $\mathfrak{r}_2 = (\gamma_1 - 1)/2$, Conjecture 1 predicts that we can expect to find the 5-torsion of $J_{\widetilde{C}}$ in $\mathbb{Q}(\zeta_5, \sqrt[5]{\mathfrak{r}_1}, \sqrt[5]{\mathfrak{r}_2})$. In order to compute this 5-torsion, we use techniques from [**26**] that build upon the work of [**8**].

Concretely, a typical 5-torsion point is expected to be represented by a divisor $D = P_1 + P_2 - 2\infty = (x_1, y_1) + (x_2, y_2) - 2\infty$, for two affine points $(x_1, y_1), (x_2, y_2)$ on $\widetilde{C}$. We read the condition $5D \equiv 0$ as $5(P_1 - \infty) \equiv -5(P_2 - \infty)$. In [**8**], recursive formulae are derived to express $5((x_1, y_1) - \infty)$ in function of x_1, y_1 and the coefficients of our genus-2 curve $\widetilde{C}$. The same can be done for $-5((x_2, y_2) - \infty)$ and the aforementioned equality results in a system of equations that can be solved by a Gröbner basis computation. Note that for D to be rational over a certain field, x_1, y_1, x_2, y_2 need not necessarily be defined over that same field. In Mumford coordinates, we can write $D = \left(x^2 - (x_1 + x_2)x + x_1x_2, y_1 + (y_2 - y_1)\frac{x - x_1}{x_2 - x_1}\right)$ and it suffices for the coefficients of these polynomials to be defined over the field. In practice, it is most convenient to simply add an extra variable and corresponding equation to the Gröbner basis computation from before, such as $X - (x_1 + x_2)$, and then compute the minimal polynomial of X (*i.e.*, put it last in a lexicographic

[9] Remark that the quadratic extension $\mathbb{Q}(\sqrt{5})$ would suffice, but adding ζ_5 makes for easier notation up ahead.

monomial ordering for the Gröbner basis computation). The roots of this polynomial will then correspond to all possible $x_1 + x_2$ such that the class of D is 5-torsion.

There are $5^4 - 1 = 624$ nontrivial elements in $J_{\widetilde{C}}[5]$, but since D and $-D$ correspond to the same $x_1 + x_2$, we expect the minimal polynomial of X to be of degree 312 generically. In this specific case though, we have multiple 5-torsion divisors of the form $(x_1, y_1) - \infty$ rather than $(x_1, y_1) + (x_2, y_2) - 2\infty$ (*e.g.*, this is the case for $\widetilde{T}_1$ and $\widetilde{T}_2$). The techniques of [**26**] do not capture such points. Nonetheless, all other 5-torsion divisors can be found this way and the minimal polynomial of X turns out to be of degree 305. Factoring this polynomial over $\mathbb{Q}(\zeta_5, \sqrt[5]{\mathfrak{r}_1}, \sqrt[5]{\mathfrak{r}_2})[X]$ we see that it splits completely as expected, thereby lending support to Conjecture 1.

A similar computation can be done for the other coefficients of the Mumford coordinates, which allows us to define

$$\begin{aligned}
\widetilde{T}_3 = \Bigg(& x^2 + 100\left(\frac{\alpha_2^4 - (\zeta_5+1)^2\alpha_2^3 - (\zeta_5^4+1)\alpha_2^2 + (\zeta_5^3 - 2\zeta_5 - 2)\alpha_2}{\gamma_1\zeta_5^3(\zeta_5+1)^2} + 1\right)x \\
&+500\left(\frac{10\alpha_2^4 - 2(\zeta_5-1)^2\alpha_2^3 - 2(7\zeta_5^3 + 11\zeta_5^2 + 7\zeta_5)\alpha_2^2 + 10(\zeta_5^3 - 2\zeta_5 - 2)\alpha_2}{\gamma_1\zeta_5^3(\zeta_5+1)^2} + 1\right), \\
&100\Big((7\zeta_5^2 - \zeta_5 + 7)\alpha_2^4 - (2\zeta_5^3 + 5\zeta_5^2 + 2\zeta_5)\alpha_2^3 + (7\zeta_5^3 + 5\zeta_5 + 5)\alpha_2^2 \\
&-(6\zeta_5^3 + 7\zeta_5^2 + 7\zeta_5 + 6)\alpha_2 - 7\Big)x + 5000\Big(- (3\zeta_5^2 + 3\zeta_5 + 3)\alpha_2^4 \\
&-(2\zeta_5^3 - \zeta_5^2 + 2\zeta_5)\alpha_2^3 + (\zeta_5^3 - \zeta_5 - 1)\alpha_2^2 + (6\zeta_5^3 + 3\zeta_5^2 + 3\zeta_5 + 6)\alpha_2 - 5\Big)\Bigg),
\end{aligned}$$

where $\alpha_2 = \sqrt[5]{\mathfrak{r}_2}$. One can easily verify that $\widetilde{T}_3 \in J_{\widetilde{C}}[5] \setminus \langle \widetilde{T}_1, \widetilde{T}_2 \rangle$. The expression for a fourth element $\widetilde{T}_4$ that completes a basis for $\mathrm{Jac}(\widetilde{C})[5]$ is too voluminous to reproduce here, but can be found online in our repository at `https://github.com/KULeuven-COSIC/Multiradical-Isogenies`. From this basis, the 125 maximal isotropic $(5,5)$-subgroups that determine a kernel which intersects the kernel of the dual isogeny trivially can easily be computed.

5. Multiradical $(3,3)$-isogenies

5.1. The parametrization by Bruin, Flynn and Testa. Over any perfect field K with $\operatorname{char} K \nmid 6$, we consider $\mathbb{A}^3$ with coordinates r, s, t, and we let $\mathcal{S} \subseteq \mathbb{A}^3$ be the joint complement of the zero loci of[10]

$$\begin{aligned}
\delta_1 &= t, \\
\delta_2 &= s, \\
\delta_3 &= st + 1, \\
\delta_4 &= r^3 - 3rt + t^2 + t, \\
\delta_5 &= r^3 s - 3rst + st^2 + st + t, \\
\delta_6 &= r^3 s^2 - 3rs^2 t - 3rs + s^2 t^2 + s^2 t + 2st + s + 1,
\end{aligned}$$

[10]Note that [**5**] define $\delta_1 = s$ and $\delta_2 = t$, so some care is needed when comparing our formulae with the ones from this reference.

$$\begin{aligned}\delta_7 &= r^3s^2t + r^3s - 3rs^2t^2 - 3rst + s^2t^3 + s^2t^2 + 2st^2 + t,\\ \Delta &= r^6s^2 - 6r^4s^2t - 3r^4s + 2r^3s^2t^2 + 2r^3s^2t + 3r^3st + r^3s + r^3\\ &\quad + 9r^2s^2t^2 + 6r^2st - 6rs^2t^3 - 6rs^2t^2 - 9rst^2 - 3rst - 3rt + s^2t^4\\ &\quad + 2s^2t^3 + s^2t^2 + 2st^3 + 3st^2 + t^2 + t\end{aligned}$$

and also of $r-1$, r^2-t and $rs-st-1$ (we don't give a name to these last three polynomials since their role is less essential, see Remark 5.2 below). Following Bruin, Flynn and Testa [5], to r,s,t we then attach the genus-2 curve $C_{rst} : y^2 = F_{rst}(x)$, where

$$F_{rst}(x) = G_1(x)^2 + \lambda_1 H_1(x)^3 = G_2(x)^2 + \lambda_2 H_2(x)^3$$

and

$$\begin{aligned}H_1(x) &= x^2 + rx + t,\\ \lambda_1 &= 4s,\\ G_1(x) &= (s - st - 1)x^3 + 3s(r-t)x^2 + 3sr(r-t)x - st^2 + sr^3 + t,\\ H_2(x) &= x^2 + x + r,\\ \lambda_2 &= 4st,\\ G_2(x) &= (s - st + 1)x^3 + 3s(r-t)x^2 + 3sr(r-t)x - st^2 + sr^3 - t.\end{aligned}$$

One can calculate that $\operatorname{disc}(F_{rst}) = -2^{12}3^6\delta_1^3\delta_2^3\delta_3\delta_4^3\delta_5\delta_6^3\delta_7^3 \neq 0$, so C_{rst} is a genus-2 curve. We write J_{rst} for the Jacobian of C_{rst}.

PROPOSITION 5.1. *For $i = 1,2$, write $T_i \in J_{rst}(K)$ for the divisor class of*

$$(H_i, G_i) := (\alpha_{i1}, G_i(\alpha_{i1})) + (\alpha_{i2}, G_i(\alpha_{i2})) - \infty_1 - \infty_2,$$

where $\alpha_{i1}, \alpha_{i2} \in \overline{K}$ denote the zeroes of $H_i(x)$. Then $\langle T_1, T_2\rangle$ is a maximal isotropic subgroup of J_{rst}, and the quotient $J_{rst}/\langle T_1, T_2\rangle$ is isomorphic over K to the Jacobian $J^{(-3)}_{r's't'}$ of the genus-2 curve

$$C^{(-3)}_{r's't'} : -3y^2 = F_{r's't'}(x)$$

where $(r', s', t') = \psi_0(r,s,t) :=$

$$\left(\frac{-s(r-1)(r^2-t)(\delta_5 - r)}{(rs-st-1)^2\delta_4}, \frac{(rs-st-1)^3\delta_4^2}{st(r-1)^3\Delta}, \frac{s^2(r-1)^3(r^2-t)^3}{(rs-st-1)^3\delta_4^2}\right).$$

Writing $F_{r',s',t'}(x) = G_1'(x)^2 + \lambda_1' H_1'(x)^3 = G_2'(x)^2 + \lambda_2' H_2'(x)^3$ as above, the kernel of the dual isogeny is generated by the corresponding points T_i', by which we mean the divisor classes of

$$(H_i', G_i'/\sqrt{-3}) = (\alpha_{i1}', G_i'(\alpha_{i1}')/\sqrt{-3}) + (\alpha_{i2}', G_i'(\alpha_{i2}')/\sqrt{-3}) - \infty_1 - \infty_2,$$

with $\alpha_{i1}', \alpha_{i2}' \in \overline{K}$ the zeroes of $H_i'(x)$, for $i = 1,2$.

PROOF. This follows from [**5**, Thm. 6 & Lem. 10]. □

We call (H_i, G_i) the *Mumford coordinates* of T_i, because of the clear analogy with the Mumford coordinates in the case of hyperelliptic curves with an imaginary Weierstrass model, *i.e.*, with a unique place at infinity.[11]

[11] For an even better analogy, one should reduce the degree of the second component by writing $(H_i, G_i \bmod H_i)$.

All sufficiently general triples (C, T_1, T_2) with C a genus-2 curve and T_1, T_2 generating a $(3,3)$-subgroup of J_C are reached by the above parametrization. One exception is where the effective parts of (the natural representants of) the divisor classes corresponding to the generators T_1, T_2 have non-disjoint supports. This is how one should understand the role of $r-1$, r^2-t, $rs-st-1$: if any one of these expressions is zero, then one can still consider C_{rst}, T_1, T_2 as above,[12] but the formulae of [**5**] will produce generators of the kernel of the dual isogeny that have non-disjoint supports.

REMARK 5.2. While for certain curves the parametrization misses certain pairs T_1, T_2 generating a $(3,3)$-subgroup, every $(3,3)$-subgroup is reached. Indeed, by [**5**, Lem. 3] in combination with the paragraph preceding [**5**, Thm. 6], at least one choice of basis with generators from $\{T_1, T_2, T_1+T_2, T_1-T_2\}$ will be in sufficiently general form.

The role of Δ is more fundamental: it should not vanish because otherwise the quotient $J_{rst}/\langle T_1, T_2\rangle$ is $\overline{K}$-isomorphic to a product of elliptic curves.

We discuss the multiradical isogeny formulae corresponding to the family $\mathcal{S}$ in Section 5.2. First, as an intermezzo, let us elaborate and discuss how to handle the case $\Delta = 0$, as well as how to walk away from products of elliptic curves. None of the material below is new, however, to the best of our knowledge, there is no article containing all these formulae, so we felt it was worth gathering them.

From Jacobians to products If $\Delta = 0$, then any algebraic software package can easily verify that the polynomial $F_{rst}(x)$ factors in two cubic polynomials over the ring $\mathbb{Q}(r,s,t,\zeta_3)[x]/(\Delta)$, where ζ_3 is a primitive cubic root of unity. This factorization induces an isogeny to a product of elliptic curves, and we refer to [**33**] for the general construction for (ℓ,ℓ)-split Jacobians. In the specific case of a (3,3)-split Jacobian, we mention the complete characterization by [**3**, Prop. A.2].

PROPOSITION 5.3. *Let C be a genus-2 curve over a perfect field K with* $\operatorname{char} K \nmid 6$*, and J the Jacobian of C. If J is $(3,3)$-isogenous to a product of elliptic curves $E_1 \times E_2$, then there exist elements $a,b,c,d,t \in K$ with*

$$12ac + 16bd = 1, \qquad \Delta_1 = a^3 + b^2 \neq 0, \qquad \Delta_2 = c^3 + d^2 \neq 0, \qquad t \neq 0,$$

such that C is isomorphic to $C_{abcdt} : ty^2 = f(x)$ and E_i is isomorphic to $E_{i,abcdt} : ty^2 = f_i(x)$ for $i \in \{1,2\}$, with

$$\begin{aligned} f(x) &= (x^3 + 3ax + 2b)(2dx^3 + 3cx^2 + 1),\\ f_1(x) &= x^3 + 12(2a^2d - bc)x^2 + 12(16ad^2 + 3c^2)\Delta_1 x + 512\Delta_1^2 d^3,\\ f_2(x) &= x^3 + 12(2bc^2 - ad)x^2 + 12(16b^2c + 3a^2)\Delta_2 x + 512\Delta_2^2 b^3. \end{aligned}$$

The corresponding morphisms $\varphi_i : C_{abcdt} \to E_{i,abcdt}$ are given by

$$\begin{aligned} \varphi_1(x,y) &\mapsto \left(12\Delta_1 \frac{-2dx + c}{x^3 + 3ax + 2b}, y\Delta_1 \frac{16dx^3 - 12cx^2 - 1}{(x^3 + 3ax + 2b)^2}\right),\\ \varphi_2(x,y) &\mapsto \left(12\Delta_2 \frac{x^2(ax - 2b)}{2dx^3 + 3cx^2 + 1}, y\Delta_2 \frac{x^3 + 12ax - 16b}{(2dx^3 + 3cx^2 + 1)^2}\right). \end{aligned}$$

[12] As long as no δ_i vanishes.

As mentioned, the Jacobian of a genus-2 curve is generically not $(3,3)$-split. If it is, however, the curves $E_{1,abcdt}$ and $E_{2,abcdt}$ will typically be unique up to isomorphism, *i.e.*, the Jacobian should not be expected to split in more than one way. Up to isomorphism, there are only two genus-2 curves which are $(3,3)$-isogenous to distinct products of elliptic curves [**38**].

Ideally, we would like more uniform formulae to identify the curves C_{rst} and C_{abcdt} with one another in the case Δ equals zero. Unfortunately, these formulae would be extremely lengthy and finding an isomorphism from one to the other in practice can be done relatively easily by a Gröbner basis computation since isomorphisms between genus-2 curves are well-understood.

Isogenies from products Let $E_1 \times E_2$ be a product of elliptic curves, both defined over a perfect field K with $\operatorname{char} K \nmid 6$, and $T_1, T_2 \in (E_1 \times E_2)(K)[3]$ such that $\langle T_1, T_2 \rangle$ is maximal isotropic with respect to the 3-Weil pairing. Then $(E_1 \times E_2)/\langle T_1, T_2 \rangle$ is again a product of elliptic curves in two scenarios. The first scenario is the most common one, where T_1, T_2 correspond to 3-torsion points on the separate elliptic curves E_1, E_2. The codomain of the isogeny can be computed using Vélu's formulae.

PROPOSITION 5.4. *Consider elliptic curves E_1, E_2 over a perfect field K with $\operatorname{char} K \nmid 6$, with non-trivial $T_1 \in E_1[3], T_2 \in E_2[3]$. Then E_i can be written as $E_i : y^2 + a_i xy + b_i y = x^3$ for $i \in \{1,2\}$, where the T_i have been translated to $(0,0)$ on the respective curves. Write $G = \langle (T_1, \infty_{E_2}), (\infty_{E_1}, T_2) \rangle$. Then the codomain of the isogeny with kernel G is again a product of elliptic curves $E_1' \times E_2'$, where for $i \in \{1,2\}$ we can write*

$$E_i' : y^2 + a_i xy + b_i y = x^3 - 5a_i b_i x - a_i^3 b_i - 7b_i^2.$$

The second situation where the codomain of a $(3,3)$-isogeny with domain $E_1 \times E_2$ is again a product of elliptic curves, is the relatively rare occurrence when there exists a 2-isogeny $\theta : E_1 \to E_2$. In this case, the isogeny is the endomorphism

$$\begin{aligned} \phi : E_1 \times E_2 &\to E_1 \times E_2 \\ (P, Q) &\mapsto (P + \hat{\theta}(Q), -Q + \theta(P)), \end{aligned}$$

with kernel the graph of the 2-isogeny $\theta|_{E_1[3]}$, see for example [**23**, §1].

In all other scenarios, $(E_1 \times E_2)/\langle T_1, T_2 \rangle$ is the Jacobian of a genus-2 curve, where the kernel is the graph of an anti-isometry with respect to the 3-Weil pairing (see for example [**16**, Prop. 5.6] or [**32**, Thm. 3]). By this we mean that there exists an isomorphism $\psi : E_1[3] \to E_2[3]$ such that $e_3(\psi(P), \psi(Q)) = e_3(P,Q)^{-1}$ for all $P, Q \in E_1[3]$. The formulae in this case are simply the dual isogenies of the split Jacobians in Proposition 5.3.

Of the 40 $(3,3)$-isogenies with domain $E_1 \times E_2$, generically there are 16 with codomain a product of elliptic curves, and 24 with codomain the Jacobian of a genus-2 curve. The only exception to this is by means of an aforementioned 2-isogeny $\theta : E_1 \to E_2$.

5.2. Multiradical formulae. We are interested in finding good extensions of our $(3,3)$-isogeny

$$J_{rst} \longrightarrow J^{(-3)}_{r's't'} = J_{rst}/\langle T_1, T_2 \rangle. \tag{5.1}$$

In view of the conjectured property **(4)**, let us compute the relevant Tate pairings. The reader might want to compare the following lemma with the Weil pairing computation from [**5**, Lem. 4].

LEMMA 5.5. *Let*

$$C : y^2 = G_1^2 + \lambda_1 H_1^3 = G_2^2 + \lambda_2 H_2^3$$

be a genus-2 curve over K *with* $G_1, G_2, H_1, H_2 \in K[x]$ *and* H_1, H_2 *quadratic, and consider the corresponding points* $T_1 = (H_1, G_1), T_2 = (H_2, G_2) \in J_C[3]$. *Then* $t_3(T_1, T_2) \equiv \operatorname{res}_x(G_1 - G_2, H_2)/\lambda_1$.

PROOF. Write α_{11}, α_{12}, resp., α_{21}, α_{22}, for the roots of $H_1(x)$, resp., $H_2(x)$. It is easy to check that $G_1(x) - y$ has divisor $3(H_1, G_1)$; however, in order to move away from infinity, as we did in Example 2.2, we instead work with $(G_1(x) - y)/(x - c)^3$ for some $c \in K$ that is different from α_{21}, α_{22}. Evaluating this function in (H_2, G_2) yields

$$t_3(T_1, T_2) \equiv -\frac{(G_1(\alpha_{21}) - G_2(\alpha_{21})(G_1(\alpha_{22}) - G_2(\alpha_{22}))}{(\alpha_{21} - c)^3(\alpha_{22} - c)^3 \lambda_1 \operatorname{lc}(H_1)^3} \equiv \operatorname{res}_x(G_1 - G_2, H_2)/\lambda_1$$

modulo $(K^*)^3$. □

Applying this to our instances of T_1, T_2, one checks that $\operatorname{res}_x(G_1 - G_2, H_2)/\lambda_1$ equals δ_4/δ_2. As for the other pairings: Bruin, Flynn and Testa have also provided an explicit Mumford representation (H_3, G_3) for $T_3 := T_1 + T_2$, see [**5**, Thm. 6], and the analogous computations yield $t_3(T_1, T_3) \equiv \delta_7^2$ and $t_3(T_3, T_2) \equiv \delta_1 \delta_6^2$. From these outcomes it follows that

$$t_3(T_1, T_1) \equiv \delta_2 \delta_4^2 \delta_7^2, \qquad t_3(T_1, T_2) \equiv \delta_2^2 \delta_4, \qquad t_3(T_2, T_2) \equiv \delta_1 \delta_2 \delta_4^2 \delta_6^2.$$

We will instead work with the radicands

$$\begin{aligned} \mathfrak{r}_1 &= \delta_7 \equiv t_3(T_1, T_1) t_3(T_1, T_2), \\ \mathfrak{r}_2 &= \delta_2 \delta_4^2 \equiv t_3(T_1, T_2)^{-1}, \\ \mathfrak{r}_3 &= \delta_1 \delta_6^2 \equiv t_3(T_1, T_2) \cdot t_3(T_2, T_2), \end{aligned}$$

which does not affect the validity of property **(4)**. Indeed, formulae in terms of $\sqrt[3]{\mathfrak{r}_1}, \sqrt[3]{\mathfrak{r}_2}, \sqrt[3]{\mathfrak{r}_3}$ can easily be rewritten into formulae in terms of $\sqrt[3]{\mathfrak{r}_1 \mathfrak{r}_2} = \sqrt[3]{t_3(T_1, T_1)}$, $\sqrt[3]{1/\mathfrak{r}_2} = \sqrt[3]{t_3(T_1, T_2)}$, $\sqrt[3]{\mathfrak{r}_2 \mathfrak{r}_3} = \sqrt[3]{t_3(T_2, T_2)}$, and vice versa.

The good extensions of (5.1) are characterized by the fact that their kernel intersects the kernel $\langle T_1', T_2' \rangle$ of the dual isogeny trivially. In order to find such kernels, we are first and foremost interested in extending T_1', T_2' to a basis of the 3-torsion. To this end, we try to find all $b_1, \ldots, b_7$ such that

$$F_{r's't'}(x) = (b_4 x^3 + b_3 x^2 + b_2 x + b_1)^2 + b_7 (x^2 + b_5 x + b_6)^3. \tag{5.2}$$

Indeed, every such tuple produces a divisor D with Mumford coordinates

$$(x^2 + b_5 x + b_6, (b_4 x^3 + b_3 x^2 + b_2 x + b_1)/\sqrt{-3})$$

satisfying $3D = (b_4 x^3 + b_3 x^2 + b_2 x + b_1 - \sqrt{-3} y)$, hence $\overline{D} \in J_{r's't'}^{(-3)}[3]$. Conversely, every 3-torsion point arises in this way, see for example [**7**, §3.1]. Over an algebraic closure of the base field, 80 nontrivial 3-torsion elements exist and hence 80 tuples $(b_1, \ldots, b_7)$ satisfy the above equation. We remark that for every solution $(b_1, b_2, b_3, b_4, b_5, b_6, b_7)$ corresponding to a divisor D, there exists another solution

$(-b_1, -b_2, -b_3, -b_4, b_5, b_6, b_7)$ corresponding to the opposite divisor, whose class is $-\overline{D}$.

The parametrization from Section 5.1 already gives rise to eight solution tuples $(b_1, \ldots, b_7)$ corresponding to the elements in $\{iT_1' + jT_2' : 0 \leq i, j \leq 2\} \setminus \{0\}$. To find the rest of the tuples, one can write out the equation of $F_{r's't'}(x)$ as well as the right-hand side of (5.2), and equate coefficients of the degree-six polynomials found. One can then compute a reduced Gröbner basis of these seven expressions with respect to the lexicographic monomial order.[13]

Assuming we put b_4 last in the monomial ordering, the last polynomial of the Gröbner basis will be a degree-80 polynomial in just b_4, whose roots correspond to possible solutions for b_4 in (5.2). Up to some constant factor, this minimal polynomial of b_4 is of the form

$$M(b_4) = \prod_{i=1}^{4} (b_4^2 - \beta_i^2) \prod_{k=1}^{4} f_k(b_4),$$

where the $f_k(b_4)$ are polynomials of degree 18, and the β_i are the (necessarily rational) solutions corresponding to $\{iT_1' + jT_2' : 0 \leq i, j \leq 2\} \setminus \{0\}$. These β_i appear in pairs, which on the level of divisors coincides with the correspondence between D and $-D$, and for the same reason one can see that the polynomials f_k ought to be even. We will write $f_k'(b_4)$ for the polynomial obtained by halving the exponents of the monomials of $f_k(b_4)$.

One can verify that the polynomials $f_k'(b_4) \in \mathbb{Q}(r, s, t)[b_4]$ all have Galois group $(\mathbb{Z}_3 \times \mathbb{Z}_3) \rtimes \mathbb{Z}_3^*$, but the action of $\mathbb{Z}_3^*$ originates from a cubic root of unity, and their Galois groups over $\mathbb{Q}(r, s, t, \zeta_3)$ are thus $\mathbb{Z}_3 \times \mathbb{Z}_3$. Writing $\alpha_1 = \sqrt[3]{\mathfrak{r}_1}$, $\alpha_2 = \sqrt[3]{\mathfrak{r}_2}$, $\alpha_3 = \sqrt[3]{\mathfrak{r}_3}$, it turns out that they split completely when extending the field $\mathbb{Q}(r, s, t, \zeta_3)$ with $\{\alpha_1, \alpha_2\}$, $\{\alpha_1, \alpha_3\}$, $\{\alpha_2, \alpha_3\}$ or $\{\alpha_1\alpha_2, \alpha_1\alpha_3\}$. All roots of one specific $f_k'(b_4)$ can be obtained from a single given root, by scaling the cubic roots with powers of ζ_3. On the level of divisors, these associated roots correspond to adding a linear combination of T_1' and T_2'. More precisely, if x_k denotes a root of $f_k'(b_4)$, we can make the following identification:

$$\begin{aligned}
x_1(\zeta_3^i\alpha_1, \zeta_3^j\alpha_2) &\longleftrightarrow T_3' + iT_1' + jT_2' \text{ for } 0 \leq i, j \leq 2,\\
x_2(\zeta_3^i\alpha_1, \zeta_3^j\alpha_3) &\longleftrightarrow T_4' + iT_1' + jT_2' \text{ for } 0 \leq i, j \leq 2,\\
x_3(\zeta_3^i\alpha_2, \zeta_3^j\alpha_2) &\longleftrightarrow T_3' + T_4' + iT_1' + jT_2' \text{ for } 0 \leq i, j \leq 2,\\
x_4(\zeta_3^i\alpha_1\alpha_2, \zeta_3^j\alpha_1\alpha_3) &\longleftrightarrow T_3' - T_4' + iT_1' + jT_2' \text{ for } 0 \leq i, j \leq 2,
\end{aligned}$$

for any T_3', T_4' that extend $\langle T_1', T_2' \rangle$ to a basis of $J_{r's't'}^{(-3)}$ [3]. This correspondence can be seen from the fact that all $f_k'(b_4)$ split over different fields, yet T_1' and T_2' are rational over the ground field. Furthermore, for any fixed choice of $i, j, k \in \{0, 1, 2\}$, any two distinct divisors from this correspondence coinciding with the choice of $\zeta_3^i\alpha_1, \zeta_3^j\alpha_2, \zeta_3^k\alpha_3$ generate a $(3,3)$-subgroup that intersects $\langle T_1', T_2' \rangle$ trivially. Hence, to find the 27 (up to sign) distinct b_4 that correspond to a $(3,3)$-subgroup which

[13] Performing a straightforward Gröbner basis computation in $\mathbb{Q}[r, s, t, b_1, \ldots, b_7]$ will quickly result in memory issues. Instead, one can first transform $F_{r's't'}$ to the more generic form $x^6 + ax^4 + bx^3 + cx^2 + dx + e$ to suppress the high degrees of r', s', t'. Next, one can compute the Gröbner basis over $\mathbb{F}_p[a, b, c, d, e, b_1, \ldots, b_7]$ for many p, then lift the solution to $\mathbb{Q}[a, b, c, d, e, b_1, \ldots, b_7]$ with the Chinese remainder theorem.

is the kernel of a good extension relative to the original isogeny, it suffices to scale the radicands with cubic roots of unity.

In the appendix, we have included two expressions for b_4 which we believe are the easiest amongst the b_4 in terms of arithmetic. Alternatively, the formulae can also be extracted from the code of our hash function from Section 6, which can be found in our online repository at `https://github.com/KULeuven-COSIC/Multiradical-Isogenies`. One can derive closed algebraic expressions for b_i in function of b_4 for $i \in \{1, 2, 3, 5, 6, 7\}$. However, in practice, it is more efficient to only partially do this for the easier expressions, and the remainder by means of a small Gröbner basis computation. Finding the 27 distinct pairs of tuples $(b_1, \ldots, b_7)$ corresponding to good extensions is done by simply scaling the radicands in the expressions of the b_4 with cubic roots of unity before computing the rest of the b_i.

REMARK 5.6. Observe that our formulae involve a factor $\sqrt{-3}$ (called `twist`), but this factor disappears when considering the corresponding Mumford coordinates.

Iterated application. Using this new (3,3)-subgroup $\langle T_3', T_4' \rangle$ as kernel for a new isogeny is easiest if we first transform $C_{r's't'}$ into an isomorphic curve C_{RST}, where T_3' and T_4' have now taken the role of the T_1 and T_2 from Section 5.1 again. This isomorphism allows us to only need to perform the rational transformation $\psi_0(R, S, T)$ from Proposition 5.1 to compute the next isogenous curve. To find this isomorphism, one can use the construction of [**5**] that has been implemented in Magma in [**22**]. This construction makes use of somewhat expensive field extensions though, and in practice, a Gröbner basis computation is more efficient.

6. Hash function from (3,3)-isogenies

We can use the (3,3)-isogenies from the previous section to construct a hash function similar to the hash function from [**9**]. We start by describing a general outline, then present a more in-depth discussion regarding choices that must be made.

6.1. The graph $\mathcal{G}_p$. For a large prime p, we denote the (directed multi-)graph $\mathcal{G}_{p,3,2}$ from Section 2.5 as $\mathcal{G}_p$ and recall its construction. The vertices are all the $\overline{\mathbb{F}}_{p^2}$-isomorphism classes of superspecial p.p. abelian surfaces, which can always be defined over $\mathbb{F}_{p^2}$. In practice we assume $p \equiv 2 \bmod 3$ and work with representants $A/\mathbb{F}_{p^2}$ on which Frobenius acts as multiplication with $-p$; see [**2**]. A consequence of this choice is that $A[3] \subseteq A(\mathbb{F}_{p^2})$; indeed, on 3-torsion points Frobenius acts as multiplication by $-p \equiv 1 \bmod 3$. The edges are all possible (3,3)-isogenies between these p.p. abelian surfaces (in the sense of Section 2.5), where multiplicities need to be taken into account. Given that only the superspecial surfaces are considered, the graph $\mathcal{G}_p$ is a directed 40-regular finite multigraph. In order to hash a given message in this graph, we first choose an arbitrary — yet fixed — starting vertex. Next, we order the 40 outgoing edges from this vertex according to some fixed order (*e.g.*, lexicographic), and choose the first 27 to continue with. The message that needs to be hashed is then converted into a base-3 number, of which the digits are called trits. We choose to walk along the edge that corresponds to the three least significant trits of the message towards the next vertex. At this vertex, we consider the 27 outgoing edges that correspond to (3,3)-isogenies whose kernel intersect the kernel of the dual of the previous isogeny trivially. Now we follow the edge that corresponds to the next three trits of the message. By excluding the other 13 $(3,3)$-isogenies, we avoid trivial cycles in our path by not (partially or fully) backtracking. This process is repeated until the entire message has been hashed. As output, an invariant of the resulting p.p. abelian surface is then returned.

Given that we will have to compute cubic roots in the computations, p should ideally be chosen such that the valuation of $p^2 - 1$ at 3 is 1 in order to speed up the computations. In combination with our assumption $p \equiv 2 \bmod 3$, this means we want $p \equiv 2, 5 \bmod 9$. Of course, we want p large enough to provide ample security. The graph $\mathcal{G}_p$ was proven to be connected, see for example [**31**, Thm. 43]. Even though the graph is not Ramanujan, in the $(2,2)$-case it still exhibits strong expander properties so we assume this to be the case for $(3,3)$-isogenies as well. The set of edges of the graph is of size $\mathcal{O}(p^3)$, of which the majority consists of p.p. abelian surfaces corresponding to Jacobians of genus-2 curves, and only $\mathcal{O}(p^2)$ corresponding to products of elliptic curves.

REMARK 6.1. Since $p^2 \equiv 1 \bmod 3$ we have $\sqrt{-3} \in \mathbb{F}_{p^2}$. Consequently, we can ignore the twisting factor -3 from Proposition 5.1 and identify $J_{rst}/\langle T_1, T_2\rangle$ with $J_{r's't'}$. This comes at the (negligible) expense of carrying an extra factor $\sqrt{-3}$ in our multiradical isogeny formulae (called `twist` in our code); see Remark 5.6.

6.2. Starting p.p. abelian surface. It is still an open problem whether one can generate a supersingular elliptic curve over a large prime field in reasonable time without knowing its endomorphism ring. This knowledge can in fact compromise the security of the associated cryptographic protocols, see for example [**18**]. Even though this has not been explicitly written down yet for superspecial p.p. abelian surfaces, it is not too far-fetched to assume the knowledge of its endomorphism ring can pose similar security risks. On the same note, it is not known how to generate a genus-2 curve over a large prime field whose Jacobian is superspecial in reasonable time without knowing its endomorphism ring. Some exceptional curves are known, see for example [**29**, §1]. Note that all of these are curves with many automorphisms, possibly leading to small collisions at the start of the hash function. Therefore, a better starting vertex in our graph should be obtained by taking a long enough random walk in the graph starting from one of these exceptional cases. Given that the isomorphism classes corresponding to products of elliptic curves represent a negligible proportion of vertices in $\mathcal{G}_p$ for cryptographically large p, we can assume our starting vertex to be the Jacobian of a genus-2 curve. Furthermore, we are interested in only 27 of the 40 $(3,3)$-subgroups of this Jacobian. Hence our starting point can be chosen as an (r,s,t)-parametrization from Section 5.1, where the 27 $(3,3)$-subgroups correspond precisely to those that intersect the $(3,3)$-subgroup determined by the (r,s,t)-parametrization trivially. Making this choice can be seen as having performed a step 0 in the hash function, where the kernel of the dual isogeny corresponding to this step is determined by this (r,s,t)-parametrization.

6.3. Genus-2 curves versus products of elliptic curves. Vertices corresponding to the Jacobians of genus-2 curves or the product of two elliptic curves will of course need to be handled differently with regard to computing the next edge in our walk. Apart from this internal code distinction, it is more user-friendly for a hash function to have a fixed size as output. The isomorphism class of the Jacobians of genus-2 curves can be classified by their absolute Igusa invariants, which are ordered triplets of elements in $\mathbb{F}_{p^2}$, whereas products of elliptic curves are completely determined by an unordered pair of j-invariants in $\mathbb{F}_{p^2}$. In order to unify these two types of invariants in one output, we first note that the number of possible output values is only $3\log p$, and not $6\log p$ as the absolute Igusa invariants may suggest. If the application for the hash function is not impeded by taking values in a set that is sparse in a much larger set, one can apply the following method during the hashing. Whenever we arrive at a vertex corresponding to a product of elliptic curves, we (deterministically) take one more step in the graph without processing information, to a vertex corresponding to the Jacobian of a genus-2 curve again. Alternatively, if one only wants an output of the same length as there is entropy, one needs to choose a function to reduce both the absolute Igusa invariants as well as the pair of j-invariants to something of size $3\log p$.

6.4. Implementation. We implemented our $(3,3)$-hash function in Magma (version 2.26-1) and ran it on an Intel(R) Xeon(R) CPU E5-2630 v2 @ 2.60GHz with 128 GB of memory. For every prime size considered we averaged the computation times over 100 random inputs of 1000 bits. A summary of our timed results can be found in the following table, where we included the timings of the $(2,2)$-hash function from [**9**] for comparison. The security claims in the table are the same as in [**9**, §7.4] and to the best of our knowledge, no advancements have been made in that area. In particular, the best known classical attack is based on the general Pollard-ρ attack, whereas the best known quantum attack is based on Grover's claw-finding algorithm.

	$p \approx 2^{86}$	$p \approx 2^{128}$	$p \approx 2^{171}$	$p \approx 2^{256}$
bits of classical security	128	192	256	384
bits of quantum security	86	128	170	256
output bits	516	768	1026	1536
time per bit processed $(2,2)$	5.01ms	6.52ms	9.33ms	15.70ms
time per bit processed $(3,3)$ (this work)	4.70ms	4.87ms	5.54ms	6.36ms

To understand why the $(3,3)$-hash function scales much better than the $(2,2)$-hash function, we take a look at the decomposition of the computation cost in the following table.

	$p \approx 2^{86}$	$p \approx 2^{128}$	$p \approx 2^{171}$	$p \approx 2^{256}$
1) Tate pairings (cubic roots)	7.0%	8.5%	11.2%	14.3%
2) Compute b_4's (arithmetic)	20.5%	18.9%	18.9%	17.0%
3) Find other b_i's (two GCD's)	16.4%	15.9%	15.8%	15.2%
4) Reparametrize r, s, t (Gröbner basis)	54.6%	55.3%	52.7%	52.2%
5) Isogenous curve (arithmetic)	1.5%	1.4%	1.4%	1.3%

As p grows, the degrees of the polynomials involved in steps 3 and 4 in this table don't change, hence the complexity of these steps depends only on the arithmetic of the field $\mathbb{F}_{p^2}$. Asymptotically, root finding over finite fields $\mathbb{F}_{p^2}$ for large p, *e.g.*, with the Tonelli–Shanks algorithm, scales a lot worse than addition and multiplication. Therefore, in the $(3,3)$-hash function step 1 in the table takes up a larger relative amount of work as p grows. For p large enough, this part of the computation will dominate the total cost. In the $(2,2)$-hash function on the other hand, the computation is already heavily dominated by the three (square) roots for small p, with only a handful of basic arithmetic operations.

Furthermore, the valuation of p^2-1 at N determines the complexity of finding an Nth root of an element in $\mathbb{F}_{p^2}$, see for instance [**17**, Thm. 1]. One can choose p such that $9 \nmid p^2-1$ but at the very least we always have $8 \mid p^2-1$, which means cubic roots can be computed significantly faster than square roots. In practice, Magma can compute cubic roots over $\mathbb{F}_{p^2}$ faster than square roots with a factor of about 2.7 for large enough p.

Additionally, for every three computed roots, the $(3,3)$-hash function can process 3 trits, whereas the $(2,2)$-hash function can only process 3 bits. Asymptotically we can thus expect the $(3,3)$-hash function to outperform the $(2,2)$-hash function by a total factor of $2.7\cdot(3/2)^3\approx 9$. For $\mathbb{F}_{p^2}$ with $p=2^{1024}+643$ for example, we see that $(2,2)$-hashing a 100-bit message takes about 20.4 seconds, whereas $(3,3)$-hashing a 100-bit message takes about 2.26 seconds.

Appendix: code for 3-torsion

The following is the Magma code that accompanies Section 5.2. The formulae can be extracted as part of the hash function code found in our online repository at `https://github.com/KULeuven-COSIC/Multiradical-Isogenies`, but we deem the formulae important enough to be displayed in the appendix as well.

The variables `r,s,t` in the code represent the domain of the $(3,3)$-isogeny, whereas `R,S,T` represent the codomain.[14] The variables `a,b,c` represent cubic roots of factors of the Tate pairings. The variables `b4ab` and `b4bc` represent solutions for b_4 in (5.2). Note that we work with b_4 instead of b_5 since in practice we want to be able to distinguish between a divisor and its opposite. From these two solutions for b_4, we compute a Gröbner basis to find solutions for the other coefficients b_i. Note that the formulae are general, but Magma struggles to work over a degree-54 extension of a function field in 3 variables. Hence, to make the code work standalone, we opted to work with a concrete example where $(R,S,T)=(2,5,-3)$. To verify the formulae in general, one works over $\mathbb{Q}(R,S,T)$ and adjoin only the cubic roots `a,b`, for example. Then, one checks that one of the degree-18 factors from the minimal polynomial of b_4 coincides with the product $\prod_{i=1}^3\prod_{j=1}^3(x^2-b_4(\zeta_3^i a,\zeta_3^j b))$, where the product ranges over all possible cubic roots `a,b`.

```
clear;
Q := Rationals();
R := 2; S := 5; T := -3;
Qx<x> := PolynomialRing(Q);
Q<twist> := ext<Q | x^2 + 3>;
Qx<x> := PolynomialRing(Q);

D1 := T;
D2 := S;
D3 := S*T + 1;
D4 := R^3 - 3*R*T + T^2 + T;
D5 := R^3*S - 3*R*S*T + S*T^2 + S*T + T;
D8 := R^2 - T;
D9 := R - 1;
D10 := R*S - S*T - 1;
D11 := S*T - S + 1;
DELTA := R^6*S^2 - 6*R^4*S^2*T - 3*R^4*S + 2*R^3*S^2*T^2 + 2*R^3*S^2*T
+ 3*R^3*S*T + R^3*S + R^3 + 9*R^2*S^2*T^2 + 6*R^2*S*T
- 6*R*S^2*T^3 - 6*R*S^2*T^2 - 9*R*S*T^2 - 3*R*S*T - 3*R*T + S^2*T^4
+ 2*S^2*T^3 + S^2*T^2 + 2*S*T^3 + 3*S*T^2 + T^2 + T;
```

[14] Remark that we want the codomain curve to have small integer parameters, so in the code these are defined first, after which we use the dual isogeny to compute the more elaborate rational parameters of the domain curve.

```
r := -D2*D9*D8*(D5-R)/(D10^2*D4);
s := D10^3*D4^2/(D1*D2*D9^3*DELTA);
t := D2^2*D9^3*D8^3/(D10^3*D4^2);

d1 := t;
d2 := s;
d4 := r^3 - 3*r*t + t^2 + t;
d6 := r^3*s^2 - 3*r*s^2*t - 3*r*s + s^2*t^2 + s^2*t + 2*s*t + s + 1;
d7 := r^3*s^2*t + r^3*s - 3*r*s^2*t^2 - 3*r*s*t + s^2*t^3 + s^2*t^2 + 2*s*t^2 + t;

Q<a> := ext<Q | x^3 - d7>;
Q<b> := ext<Q | x^3 - d2*d4^2>;
Q<c> := ext<Q | x^3 - d1*d6^2>;

cofab1 := D1^2     *D4^4     *D10^8 /(D2^3*D8^6*D9^2*DELTA^2);
cofab2 := D1^2     *D4^4*D8  *D10^7      *D11 /(D2^2*D8^6*D9^2*DELTA^2);
cofab3 := D1       *D4^4*D8^2*D10^6 /(D2  *D8^6*D9^2*DELTA^2);
cofab4 := D1^2     *D4^2*     D10^5      *D11 /(D2^2*D8^4*D9  *DELTA);
cofab5 := D1       *D4^2*D8*  D10^4 /(D2  *D8^4*D9  *DELTA);
cofab6 := D1  *D3  *D4^2*D8^2*D10^3 /(     D8^4*D9  *DELTA);
cofab7 := D1^2                  *D10^2 /      D8^2;
cofab8 := D1                    *D10 /      D8;
cofab9 :=                              D11;

cofab1 *:= -6*S*T-2;
cofab2 *:= -2;
cofab3 *:= 6*S*T+4;
cofab4 *:= 2;
cofab5 *:= -6*S*T-2;
cofab6 *:= -6;
cofab7 *:= 6;
cofab8 *:= 6*S*T+4;
cofab9 *:= 2*S*T+1;

b4ab := twist* ((cofab9 + cofab8*a + cofab7*a^2) + (cofab6 + cofab5*a + cofab4*a^2)*b
+ (cofab3 + cofab2*a + cofab1*a^2)*b^2);

cofbc1 := 1 /(D2  *D4^3);
cofbc2 := D1^2   *D9  *D10 /(D2  *D4^3     *D8);
cofbc3 := D1^3   *D9^2*D10^2 /(D2  *D4^3*D5  *D8^2);
cofbc4 := D1          *D10^3 /(D2^2*D4       *D8^2*D9  *DELTA);
cofbc5 := D1^2        *D10^4 /(D2^2*D4       *D8^3     *DELTA);
cofbc6 := D1^3   *D9  *D10^5 /(D2^2*D4  *D5  *D8^4     *DELTA);
cofbc7 := D1  *D4     *D10^6 /(D2^3          *D8^4*D9^2*DELTA^2);
cofbc8 := D1^2*D4     *D10^7 /(D2^3          *D8^5*D9  *DELTA^2);
cofbc9 := D1^4*D4     *D10^8 /(D2^3     *D5  *D8^6     *DELTA^2);

cofbc1 *:= R^9*S^2*T + R^9*S^2 - R^9*S - 6*R^8*S^2*T - 3*R^7*S^2*T^2 - 3*R^7*S^2*T
- 5*R^7*S*T + R^6*S^2*T^3 + 40*R^6*S^2*T^2 + R^6*S^2*T + 13*R^6*S*T^2 + 13*R^6*S*T
- 2*R^6*T - 21*R^5*S^2*T^3 - 21*R^5*S^2*T^2 + 3*R^5*S*T^2 + 6*R^4*S^2*T^4 - 54*R^4*S^2*T^3
+ 6*R^4*S^2*T^2 - 52*R^4*S*T^3 - 52*R^4*S*T^2 - 6*R^4*T^2 - R^3*S^2*T^5 + 64*R^3*S^2*T^4
+ 64*R^3*S^2*T^3 - R^3*S^2*T^2 + 11*R^3*S*T^4 + 103*R^3*S*T^3 + 11*R^3*S*T^2 + 14*R^3*T^3
+ 14*R^3*T^2 - 33*R^2*S^2*T^5 - 48*R^2*S^2*T^4 - 33*R^2*S^2*T^3 - 15*R^2*S*T^4 - 15*R^2*S*T^3
- 18*R^2*T^3 + 9*R*S^2*T^6 + 15*R*S^2*T^5 + 15*R*S^2*T^4 + 9*R*S^2*T^3 + 7*R*S*T^5
- 40*R*S*T^4 + 7*R*S*T^3 - 6*R*T^4 - 6*R*T^3 - S^2*T^7 - 2*S^2*T^6 - 2*S^2*T^5
- 2*S^2*T^4 - S^2*T^3 - 3*S*T^6 + 9*S*T^5 + 9*S*T^4 - 3*S*T^3 - 2*T^5 + 14*T^4 - 2*T^3;
cofbc2 *:= -2*R^7*S + 8*R^6*S - 6*R^5*S + 6*R^5 + 2*R^4*S*T^2 - 22*R^4*S*T - 12*R^4*T
+ 22*R^3*S*T^2 + 28*R^3*S*T + 6*R^3*T - 18*R^2*S*T^3 - 24*R^2*S*T^2 - 6*R^2*S*T
+ 6*R^2*T^2 - 12*R^2*T + 4*R*S*T^4 + 20*R*S*T^3 - 2*R*S*T^2 + 6*R*T^3 + 6*R*T^2 -
4*S*T^4 - 2*S*T^3 + 2*S*T^2 - 12*T^3 + 6*T^2;
cofbc3 *:= 2*R^8*S - 6*R^7*S + 4*R^6*S*T - 8*R^5*S*T^2 + 16*R^5*S*T + 6*R^5*T
- 30*R^4*S*T^2 - 12*R^4*T + 44*R^3*S*T^3 + 14*R^3*S*T^2 + 6*R^3*T^2 - 10*R^2*S*T^4
- 32*R^2*S*T^3 - 22*R^2*S*T^2 - 12*R^2*T^3 + 6*R^2*T^2 - 6*R*S*T^4 + 36*R*S*T^3
+ 6*R*S*T^2 + 6*R*T^3 + 6*R*T^2 + 4*S*T^5 - 4*S*T^4 - 8*S*T^3 + 6*T^4 - 12*T^3;
cofbc4 *:= 2*R^9*S^2 - 2*R^8*S^2*T - 4*R^8*S^2 + 4*R^8*S - 8*R^7*S^2*T + 2*R^7*S^2
- 10*R^7*S + 16*R^6*S^2*T^2 + 26*R^6*S^2*T - 4*R^5*S^2*T^3 - 18*R^5*S^2*T^2 - 20*R^5*S^2*T
- 10*R^5*S*T^2 + 32*R^5*S*T + 6*R^5*T - 22*R^4*S^2*T^3 - 24*R^4*S^2*T^2 + 4*R^4*S^2*T
- 38*R^4*S*T^2 - 2*R^4*S*T - 12*R^4*T + 14*R^3*S^2*T^4 + 72*R^3*S^2*T^3 + 40*R^3*S^2*T^2
+ 60*R^3*S*T^3 + 6*R^3*S*T^2 + 6*R^3*T^2 - 2*R^2*S^2*T^5 - 32*R^2*S^2*T^4 - 64*R^2*S^2*T^3
- 16*R^2*S^2*T^2 - 14*R^2*S*T^4 - 40*R^2*S*T^3 - 26*R^2*S*T^2 - 12*R^2*T^3 + 6*R^2*T^2
+ 4*R*S^2*T^5 + 22*R*S^2*T^4 + 20*R*S^2*T^3 + 2*R*S^2*T^2 - 10*R*S*T^4 + 52*R*S*T^3
+ 8*R*S*T^2 + 6*R*T^3 + 6*R*T^2 - 2*S^2*T^5 - 4*S^2*T^4 - 2*S^2*T^3 + 6*S*T^5
```

```
- 6*S*T^4 - 12*S*T^3 + 6*T^4 - 12*T^3;
cofbc5 *:= -2*R^7*S + 4*R^6*S*T + 4*R^6*S - 2*R^6 - 6*R^5*S*T - 10*R^4*S*T^2 - 10*R^4*S*T
- 6*R^4*T + 2*R^3*S*T^3 + 46*R^3*S*T^2 + 2*R^3*S*T + 14*R^3*T^2 + 14*R^3*T - 24*R^2*S*T^3
- 24*R^2*S*T^2 - 18*R^2*T^2 + 10*R*S*T^4 + 2*R*S*T^3 + 10*R*S*T^2 - 6*R*T^3 - 6*R*T^2
- 2*S*T^5 - 2*S*T^2 - 2*T^4 + 14*T^3 - 2*T^2;
cofbc6 *:= 2*R^8*S - 6*R^7*S*T + 4*R^6*S*T + 16*R^5*S*T^2 - 8*R^5*S*T + 6*R^5*T
- 30*R^4*S*T^2 - 12*R^4*T^2 + 14*R^3*S*T^3 + 44*R^3*S*T^2 + 6*R^3*T^2 - 22*R^2*S*T^4
- 32*R^2*S*T^3 - 10*R^2*S*T^2 + 6*R^2*T^3 - 12*R^2*T^2 + 6*R*S*T^5 + 36*R*S*T^4
- 6*R*S*T^3 + 6*R*T^4 + 6*R*T^3 - 8*S*T^5 - 4*S*T^4 + 4*S*T^3 - 12*T^4 + 6*T^3;
cofbc7 *:= 2*R^9*S^2 - 4*R^8*S^2*T - 2*R^8*S^2 + 4*R^8*S + 2*R^7*S^2*T^2 - 8*R^7*S^2*T
- 10*R^7*S*T + 26*R^6*S^2*T^2 + 16*R^6*S^2*T - 20*R^5*S^2*T^3 - 18*R^5*S^2*T^2 - 4*R^5*S^2*T
+ 32*R^5*S*T^2 - 10*R^5*S*T + 6*R^5*T + 4*R^4*S^2*T^4 - 24*R^4*S^2*T^3 - 22*R^4*S^2*T^2
- 2*R^4*S*T^3 - 38*R^4*S*T^2 - 12*R^4*T^2 + 40*R^3*S^2*T^4 + 72*R^3*S^2*T^3 + 14*R^3*S^2*T^2
+ 6*R^3*S*T^3 + 60*R^3*S*T^2 + 6*R^3*T^2 - 16*R^2*S^2*T^5 - 64*R^2*S^2*T^4 - 32*R^2*S^2*T^3
- 2*R^2*S^2*T^2 - 26*R^2*S*T^4 - 40*R^2*S*T^3 - 14*R^2*S*T^2 + 6*R^2*T^3 - 12*R^2*T^2
+ 2*R*S^2*T^6 + 20*R*S^2*T^5 + 22*R*S^2*T^4 + 4*R*S^2*T^3 + 8*R*S*T^5 + 52*R*S*T^4
- 10*R*S*T^3 + 6*R*T^4 + 6*R*T^3 - 2*S^2*T^6 - 4*S^2*T^5 - 2*S^2*T^4 - 12*S*T^5
- 6*S*T^4 + 6*S*T^3 - 12*T^4 + 6*T^3;
cofbc8 *:= -2*R^7*S + 8*R^6*S*T - 6*R^5*S*T^2 + 6*R^5*T - 22*R^4*S*T^2 + 2*R^4*S*T
- 12*R^4*T + 28*R^3*S*T^3 + 22*R^3*S*T^2 + 6*R^3*T^2 - 6*R^2*S*T^4 - 24*R^2*S*T^3
- 18*R^2*S*T^2 - 12*R^2*T^3 + 6*R^2*T^2 - 2*R*S*T^4 + 20*R*S*T^3 + 4*R*S*T^2 + 6*R*T^3
+ 6*R*T^2 + 2*S*T^5 - 2*S*T^4 - 4*S*T^3 + 6*T^4 - 12*T^3;
cofbc9 *:= -8*R^7*S + 10*R^6*S*T + 10*R^6*S - 2*R^6 + 12*R^5*S*T - 40*R^4*S*T^2 - 40*R^4*S*T
- 6*R^4*T + 8*R^3*S*T^3 + 64*R^3*S*T^2 + 8*R^3*S*T + 14*R^3*T^2 + 14*R^3*T - 6*R^2*S*T^3
- 6*R^2*S*T^2 - 18*R^2*T^2 + 4*R*S*T^4 - 28*R*S*T^3 + 4*R*S*T^2 - 6*R*T^3 - 6*R*T^2
- 2*S*T^5 + 6*S*T^4 + 6*S*T^3 - 2*S*T^2 - 2*T^4 + 14*T^3 - 2*T^2;

b4bc := twist* ((cofbc1 + cofbc2*c + cofbc3*c^2) + (cofbc4 + cofbc5*c + cofbc6*c^2)*b
+ (cofbc7 + cofbc8*c + cofbc9*c^2)*b^2);

Qbi<b1,b2,b3,b5,b6,b7> := PolynomialRing(Q,6);
Qx<x> := PolynomialRing(Qbi);

H1 := x^2 + R*x + T;
lambda1 := 4*S;
G1 := (S - S*T - 1)*x^3 + 3*S*(R - T)*x^2 + 3*S*R*(R - T)*x - S*T^2 + S*R^3 + T;
F := G1^2 + lambda1*H1^3;

bis := [];
for b4 in [b4ab,b4bc] do
  Fbi := (b4*x^3 + b3*x^2 + b2*x + b1)^2 + b7*(x^2 + b6*x + b5)^3;
  I := {Eltseq(F)[i] - Eltseq(Fbi)[i] : i in [1..7]};
  GB := GroebnerBasis(I);
  roots := [Roots(UnivariatePolynomial(GB[i]))[1][1] : i in [1..6]];
  bi := roots[1..3] cat [b4] cat roots[4..6];
  Append(~bis, bi);
end for;

C := HyperellipticCurve(F);
J := Jacobian(C);

for bi in bis do
  T := J ! [Qx ! (bi[5..6] cat [1]), Qx ! bi[1..4]];
  assert 3*T eq J ! 0;
end for;
```

Acknowledgments

We thank Marc Houben, Frederik Vercauteren and the anonymous referees for several helpful remarks.

References

[1] Iurie Boreico, *My favorite problem – linear independence of radicals*, The Harvard College Mathematics Review, vol. 2, 2008, pp. 87–92.

[2] Bradley Wayne Brock, *Superspecial curves of genera two and three*, ProQuest LLC, Ann Arbor, MI, 1993. Thesis (Ph.D.)–Princeton University. MR2689446

[3] Reinier Bröker, Everett W. Howe, Kristin E. Lauter, and Peter Stevenhagen, *Genus-2 curves and Jacobians with a given number of points*, LMS J. Comput. Math. **18** (2015), no. 1, 170–197, DOI 10.1112/S1461157014000461. MR3349314

[4] Nils Bruin and Kevin Doerksen, *The arithmetic of genus two curves with* $(4,4)$*-split Jacobians*, Canad. J. Math. **63** (2011), no. 5, 992–1024, DOI 10.4153/CJM-2011-039-3. MR2866068

[5] Nils Bruin, E. Victor Flynn, and Damiano Testa, *Descent via* $(3,3)$*-isogeny on Jacobians of genus 2 curves*, Acta Arith. **165** (2014), no. 3, 201–223, DOI 10.4064/aa165-3-1. MR3263947

[6] Peter Bruin, *The Tate pairing for Abelian varieties over finite fields* (English, with English and French summaries), J. Théor. Nombres Bordeaux **23** (2011), no. 2, 323–328. MR2817932

[7] Frank Calegari, Shiva Chidambaram, and David P. Roberts, *Abelian surfaces with fixed 3-torsion*, ANTS XIV—Proceedings of the Fourteenth Algorithmic Number Theory Symposium, Open Book Ser., vol. 4, Math. Sci. Publ., Berkeley, CA, 2020, pp. 91–108, DOI 10.2140/obs.2020.4.91. MR4235108

[8] David G. Cantor, *On the analogue of the division polynomials for hyperelliptic curves*, J. Reine Angew. Math. **447** (1994), 91–145, DOI 10.1515/crll.1994.447.91. MR1263171

[9] Wouter Castryck, Thomas Decru, and Benjamin Smith, *Hash functions from superspecial genus-2 curves using Richelot isogenies*, J. Math. Cryptol. **14** (2020), no. 1, 268–292, DOI 10.1515/jmc-2019-0021. MR4134760

[10] Wouter Castryck, Thomas Decru, and Frederik Vercauteren, *Radical isogenies*, Advances in cryptology—ASIACRYPT 2020. Part II, Lecture Notes in Comput. Sci., vol. 12492, Springer, Cham, [2020] ©2020, pp. 493–519, DOI 10.1007/978-3-030-64834-3_17. MR4210348

[11] Wouter Castryck, Tanja Lange, Chloe Martindale, Lorenz Panny, and Joost Renes, *CSIDH: an efficient post-quantum commutative group action*, Advances in cryptology—ASIACRYPT 2018. Part III, Lecture Notes in Comput. Sci., vol. 11274, Springer, Cham, 2018, pp. 395–427, DOI 10.1007/978-3-030-03332-3_15. MR3897883

[12] Denis X. Charles, Kristin E. Lauter, and Eyal Z. Goren, *Cryptographic hash functions from expander graphs*, J. Cryptology **22** (2009), no. 1, 93–113, DOI 10.1007/s00145-007-9002-x. MR2496385

[13] Daniel Coray and Constantin Manoil, *On large Picard groups and the Hasse principle for curves and* $K3$ *surfaces*, Acta Arith. **76** (1996), no. 2, 165–189, DOI 10.4064/aa-76-2-165-189. MR1393513

[14] Jean-Marc Couveignes, *Hard homogeneous spaces*, Cryptology ePrint Archive, available at `https://eprint.iacr.org/2006/291`, 2006.

[15] P. Deligne and M. Rapoport, *Les schémas de modules de courbes elliptiques* (French), Modular functions of one variable, II (Proc. Internat. Summer School, Univ. Antwerp, Antwerp, 1972), Springer, Berlin, 1973, pp. 143–316. Lecture Notes in Math., Vol. 349. MR0337993

[16] Martin Djukanović, *Families of (3, 3)-split jacobians*, Cornell University arXiv, available at `arXiv:1811.10075`, 2018.

[17] Javad Doliskani and Éric Schost, *Taking roots over high extensions of finite fields*, Math. Comp. **83** (2014), no. 285, 435–446, DOI 10.1090/S0025-5718-2013-02715-9. MR3120598

[18] Kirsten Eisenträger, Sean Hallgren, Kristin Lauter, Travis Morrison, and Christophe Petit, *Supersingular isogeny graphs and endomorphism rings: reductions and solutions*, Advances in cryptology—EUROCRYPT 2018. Part III, Lecture Notes in Comput. Sci., vol. 10822, Springer, Cham, 2018, pp. 329–368, DOI 10.1007/978-3-319-78372-7_11. MR3794837

[19] Gerd Faltings and Ching-Li Chai, *Degeneration of abelian varieties*, Ergebnisse der Mathematik und ihrer Grenzgebiete (3) [Results in Mathematics and Related Areas (3)], vol. 22, Springer-Verlag, Berlin, 1990. With an appendix by David Mumford, DOI 10.1007/978-3-662-02632-8. MR1083353

[20] Enric Florit and Benjamin Smith, *Automorphisms and isogeny graphs of abelian varieties, with applications to the superspecial Richelot isogeny graph*, Cornell University arXiv, available at `arXiv:2101.00919`, 2020.

[21] E. V. Flynn, *Descent via* $(5,5)$*-isogeny on Jacobians of genus 2 curves*, J. Number Theory **153** (2015), 270–282, DOI 10.1016/j.jnt.2015.01.018. MR3327574

[22] E. V. Flynn and Yan Bo Ti, *Genus two isogeny cryptography*, Post-quantum cryptography, Lecture Notes in Comput. Sci., vol. 11505, Springer, Cham, 2019, pp. 286–306, DOI 10.1007/978-3-030-25510-7_16. MR3989010

[23] Gerhard Frey and Ernst Kani, *Curves of genus 2 covering elliptic curves and an arithmetical application*, Arithmetic algebraic geometry (Texel, 1989), Progr. Math., vol. 89, Birkhäuser Boston, Boston, MA, 1991, pp. 153–176. MR1085258

[24] Gerhard Frey and Hans-Georg Rück, *A remark concerning m-divisibility and the discrete logarithm in the divisor class group of curves*, Math. Comp. **62** (1994), no. 206, 865–874, DOI 10.2307/2153546. MR1218343

[25] Steven D. Galbraith, Sachar Paulus, and Nigel P. Smart, *Arithmetic on superelliptic curves*, Mathematics of Computation **71** (2002), no. 237, 393–405, (The cited theorem refers to a preliminary version of this paper, published as Hewlett-Packard Labs technical report HPL-98-179, available at `https://www.hpl.hp.com/techreports/98/HPL-98-179.pdf`).

[26] P. Gaudry and É. Schost, *Modular equations for hyperelliptic curves*, Math. Comp. **74** (2005), no. 249, 429–454, DOI 10.1090/S0025-5718-04-01682-5. MR2085901

[27] Genevieve Hanlon, *Counting points in* $\mathrm{Sp}(2n, \mathbb{F}_q)$*/maximal parabolic subgroup*, Course notes available at `http://www-math.mit.edu/~dav/symplectic_parabolic.pdf`, 2005.

[28] F. Hess, *A note on the Tate pairing of curves over finite fields*, Arch. Math. (Basel) **82** (2004), no. 1, 28–32, DOI 10.1007/s00013-003-4773-2. MR2034467

[29] Tomoyoshi Ibukiyama, Toshiyuki Katsura, and Frans Oort, *Supersingular curves of genus two and class numbers*, Compositio Math. **57** (1986), no. 2, 127–152. MR827350

[30] Sorina Ionica, *Pairing-based algorithms for Jacobians of genus 2 curves with maximal endomorphism ring*, J. Number Theory **133** (2013), no. 11, 3755–3770, DOI 10.1016/j.jnt.2013.04.023. MR3084299

[31] Bruce W. Jordan and Yevgeni Zaytman, *Isogeny graphs of superspecial abelian varieties and Brandt matrices*, Cornell University arXiv, available at `arXiv:2005.09031`, 2021.

[32] Ernst Kani, *The number of curves of genus two with elliptic differentials*, J. Reine Angew. Math. **485** (1997), 93–121, DOI 10.1515/crll.1997.485.93. MR1442190

[33] Robert M. Kuhn, *Curves of genus 2 with split Jacobian*, Trans. Amer. Math. Soc. **307** (1988), no. 1, 41–49, DOI 10.2307/2000749. MR936803

[34] J. S. Milne, *Abelian varieties*, Arithmetic geometry (Storrs, Conn., 1984), Springer, New York, 1986, pp. 103–150. MR861974

[35] David Mumford, *Abelian varieties*, Tata Institute of Fundamental Research Studies in Mathematics, vol. 5, Published for the Tata Institute of Fundamental Research, Bombay; by Hindustan Book Agency, New Delhi, 2008. With appendices by C. P. Ramanujam and Yuri Manin; Corrected reprint of the second (1974) edition. MR2514037

[36] D. Mumford, J. Fogarty, and F. Kirwan, *Geometric invariant theory*, 3rd ed., Ergebnisse der Mathematik und ihrer Grenzgebiete (2) [Results in Mathematics and Related Areas (2)], vol. 34, Springer-Verlag, Berlin, 1994, DOI 10.1007/978-3-642-57916-5. MR1304906

[37] David E. Rohrlich, *Modular curves, Hecke correspondence, and L-functions*, Modular forms and Fermat's last theorem (Boston, MA, 1995), Springer, New York, 1997, pp. 41–100. MR1638476

[38] T. Shaska, *Genus 2 fields with degree 3 elliptic subfields*, Forum Math. **16** (2004), no. 2, 263–280, DOI 10.1515/form.2004.013. MR2039100

[39] Samir Siksek, *Explicit arithmetic of modular curves*, Summer school notes, available at `https://homepages.warwick.ac.uk/staff/S.Siksek/teaching/modcurves/lecturenotes.pdf`, 2019.

[40] Benjamin Smith, *Explicit endomorphisms and correspondences*, Ph.D. thesis, University of Sydney, 2005.

[41] The Stacks project authors, *The Stacks project*, Available at `https://stacks.math.columbia.edu`, 2021.

[42] Anton Stolbunov, *Public-key encryption based on cycles of isogenous elliptic curves*, Master's thesis, Saint-Petersburg State Polytechnical University, 2004, In Russian.

[43] Marco Streng, *Generators of the group of modular units for* $\Gamma_1(N)$ *over the rationals*, Cornell University arXiv, available at `arXiv:1503.08127v2`, 2015.

[44] Katsuyuki Takashima, *Efficient algorithms for isogeny sequences and their cryptographic applications*, Mathematical modelling for next-generation cryptography, Math. Ind. (Tokyo), vol. 29, Springer, Singapore, 2018, pp. 97–114. MR3586863

[45] Jacques Vélu, *Isogénies entre courbes elliptiques* (French), C. R. Acad. Sci. Paris Sér. A-B **273** (1971), A238–A241. MR294345

COSIC, RESEARCH GROUP AT IMEC AND KU LEUVEN, KASTEELPARK ARENBERG 10/2452, 3001 LEUVEN (HEVERLEE), BELGIUM; AND DEPARTMENT OF MATHEMATICS: ALGEBRA AND GEOMETRY, GHENT UNIVERSITY, KRIJGSLAAN 281 – S25, 9000 GENT, BELGIUM

Email address: wouter.castryck@esat.kuleuven.be

COSIC, RESEARCH GROUP AT IMEC AND KU LEUVEN, KASTEELPARK ARENBERG 10/2452, 3001 LEUVEN (HEVERLEE), BELGIUM

Email address: thomas.decru@esat.kuleuven.be

Contemporary Mathematics
Volume **779**, 2022
https://doi.org/10.1090/conm/779/15677

Arithmetic monodromy groups of dynamical Belyi maps

Özlem Ejder

Abstract. We consider a large family of dynamical Belyi maps of arbitrary degree and study the arithmetic monodromy groups attached to the iterates of such maps. Building on the results of Bouw-Ejder-Karemaker on the geometric monodromy groups of these maps, we show that the quotient of the arithmetic monodromy group by the geometric monodromy group has order either 1 or 2. Prior to this article, a result of this kind was only known for quadratic maps (Pink) and a few examples in degree 3.

1. Introduction

Let $f : \mathbb{P}^1_k \to \mathbb{P}^1_k$ be a rational map of degree d defined over a number field k. For each $n \geq 1$, define the n-th iterate of f by $f^n = f \circ \ldots \circ f$. It was Odoni [**Odo85**] who first studied the Galois theory of the iterates of f mainly for its applications in dynamical systems. Assume f is postcritically finite (PCF), i.e., the orbit of each critical point is finite. Let $P = \{f^n(x) \in \mathbb{P}^1(k) : x \text{ is a critical point of } f \text{ and } n \geq 1\}$. Then the iterates f^n are unbranched outside the finite set P. For a point $x_0 \in \mathbb{P}^1_k \backslash P$, one can construct a tree T whose leaves are the points in $f^{-n}(x_0)$ for $n \geq 1$. We obtain a representation of the étale fundamental group of $\mathbb{P}^1_k \backslash P$ (resp., $\mathbb{P}^1_{\bar{k}} \backslash P$) inside the automorphism group of the tree T. We call the image of such map the arithmetic (resp., geometric) monodromy group G^{arith} (resp., G^{geom}) of f. See Section 4 for details. For PCF maps, the geometric fundamental group is topologically finitely generated.

Pink [**Pin13b**] [**Pin13a**] has studied the case $d = 2$ extensively. He showed that the arithmetic and geometric monodromy groups of the quadratic PCF maps are determined only by the combinatorial data of the postcritical orbit P. Moreover he described the quotient group $G^{\text{arith}}/G^{\text{geom}}$ for quadratic polynomials and quadratic morphisms with infinite postcritical orbit P.

Similarly the article [**BEK21**] determines the Galois groups attached to a large class of Belyi maps of any degree d. A rational map $f : \mathbb{P}^1_k \to \mathbb{P}^1_k$ is called a *Belyi map* if it is branched exactly over $\{0, 1, \infty\}$. The authors of [**BEK21**] show that the geometric monodromy group is again only determined by their combinatorial type for Belyi maps $f : \mathbb{P}^1_k \to \mathbb{P}^1_k$ with exactly three ramification points; $0, 1, \infty$ which are fixed by f. We call these maps normalized, single cycle genus zero Belyi maps. A dynamical Belyi map is a Belyi map $f : \mathbb{P}^1_k \to \mathbb{P}^1_k$ where $f(\{0, 1, \infty\}) \subset \{0, 1, \infty\}$.

2020 *Mathematics Subject Classification.* Primary 11G32, 12F10; Secondary 37P05, 37P15.

All maps considered in [**BEK21**] are all normalized dynamical Belyi maps which are all PCF.

This article is essentially a sequel to [**BEK21**] where a condition is described for the quotient of the arithmetic and the geometric monodromy group of normalized single cycle dynamical Belyi maps to be trivial. This is done by introducing a product discriminant. In this article, we use group theoretical tools to describe the normalizer of the geometric monodromy group inside the automorphism group of the tree T. This helps us bound the size of the quotient group $G^{\text{arith}}/G^{\text{geom}}$ for general single cycle dynamical Belyi maps. In particular we prove Theorem 6.1 which states that the quotient $G^{\text{arith}}/G^{\text{geom}}$has order at most 2 for all but finitely many of these maps.

See also the article [**JM14**] for the Galois groups of iterates of quadratic maps, [**ABC$^+$21**] for a similar result for $f(x) = x^2 - 1$, [**BFH$^+$17**] for an example in degree 3, and [**Jon13**] for a fantastic survey on the subject.

2. Automorphism group of T

Let $d \geq 3$. Let T be the infinite regular d-ary tree whose vertices are the finite words over the alphabet $\{0, \ldots, d-1\}$. For any integer $n \geq 1$, we let T_n denote the finite rooted subtree whose vertices are the words of length at most n. We will call the set of words of length n the level n of T. We will use the notation $W := \text{Aut}(T)$ and $W_n := \text{Aut}(T_n)$.

We embed $W^d := W \times \ldots \times W$ into W by identifying the complete subtrees rooted at level one of the tree T with T itself. The image of the embedding $W^d \hookrightarrow W$ is given by the set of automorphisms acting trivially on the first level. The exact sequence $1 \to W^d \to W \to W_1 \to 1$ splits and gives the semi direct product

$$W \simeq W^d \rtimes S_d \text{ and } W_n \simeq W_{n-1}^d \rtimes S_d.$$

In other words, W and W_n have a wreath product structure:

$$W \simeq W \wr S_d \text{ and } W_n \simeq W_{n-1} \wr S_d$$

for $n \geq 2$.

We denote an element of W (resp. W_n) by $(x_1, \ldots x_d)\tau$ where x_i are in W (resp. W_{n-1}) and $\tau \in S_d$. We have the following relations in W:

$$(2.1) \qquad \begin{aligned} (x_1, \ldots, x_d)(y_1, \ldots, y_d) &= (x_1 y_1, \ldots, x_d y_d) \\ \tau(x_1, \ldots, x_d) &= (x_{\tau^{-1}(1)}, \ldots, x_{\tau^{-1}(d)})\tau \end{aligned}$$

We embed S_d into W by the map $\tau \mapsto (-, \ldots, -)\tau$. We denote the automorphism $(-, \ldots, -)\tau$ simply by τ in W. Here $-$ denotes the identity. For every $m \leq n$, we write π_m for the natural projection

$$\pi_m : W_n \to W_m,$$

which corresponds to restricting the action of an element of W_n to the subtree T_m consisting of the levels 0, 1, ..., m. Abusing the notation let $\pi_m : W \to W_m$ also denote the natural projection onto the finite level m. We denote the image of an element w under π_m as $w_{|T_m}$.

Let G be a subgroup of W. For each $n \geq 1$, we define $G_n := \pi_n(G) \subset W_n$. Next we define some subgroups of W which will be essential in Theorem 4.2 where we describe the geometric monodromy groups of dynamical Belyi maps. To define these groups, we first need to define a *product sign* map on W.

DEFINITION 2.1. Define $\mathrm{sgn}_2 : W_2 \to \{\pm 1\}$ by setting

$$\mathrm{sgn}_2((x_1, \dots, x_d)\tau) = \mathrm{sgn}(\tau) \prod_{i=1}^{d} \mathrm{sgn}(x_i). \tag{2.2}$$

Here sgn is the usual sign on W_1 via the identification $W_1 \simeq S_d$ induced by the choice of labeling of the vertices. We define

$$\mathrm{sgn}_2 := \mathrm{sgn}_2 \circ \pi_2 : W \to \{\pm 1\} \tag{2.3}$$

Note that we abuse the notation and denote the maps $W \to W_2$ and $W_n \to W_2$ both by π_2.

DEFINITION 2.2.

(1) Define the subgroup $E_n \subseteq W_n$ by

$$E_n = \begin{cases} W_1 & \text{if } n = 1, \\ (E_{n-1} \wr E_1) \cap \ker(\mathrm{sgn}_2) \subseteq W_n & \text{otherwise} \end{cases}$$

and

$$E := \varprojlim_n E_n \subseteq W$$

(2) Define the subgroup $U_n \subseteq W_n$ as the n-fold iterated wreath product of A_d and

$$U := \varprojlim_n U_n \subseteq W$$

REMARK 2.3. Note that it follows from Definition 2.2 that $E \simeq (E \wr S_d) \cap \ker(\mathrm{sgn}_2)$. Let $w = (x_1, \dots, x_d)\tau \in E$. By definition $w_{|T_n}$ is in E_n for any $n \geq 1$. Hence $w_{|T_2} \in \ker(\mathrm{sgn}_2)$ and $x_{i|T_{n-1}}$ is in E_{n-1} for all i which implies that $x_i \in E$ for all $i \geq 1$. Conversely, let $w = (x_1, \dots, x_d)\tau \in W$. If $x_i \in E$ for all $i \geq 1$ and $w_{|T_2} \in \ker(\mathrm{sgn}_2)$, then $w_{|T_n} \in E_n$ and hence w is in E.

Similarly, $w \in U$ if and only if $x_i \in U$ for each i and $\tau \in A_d$.

3. Belyi Maps

A (*genus zero*) *Belyi map* is a rational map $f : \mathbb{P}^1_{\mathbb{C}} \to \mathbb{P}^1_{\mathbb{C}}$ such that f is branched exactly over $x_1 = 0$, $x_2 = 1$, and $x_3 = \infty$. It is called a *dynamical Belyi* map if $f(\{0, 1, \infty\}) \subset \{0, 1, \infty\}$. Hence the iterates of a dynamical Belyi map are also dynamical Belyi maps.

A Belyi map is called *single cycle* if there is a unique ramification point over each of the three branch points. It is called *normalized* if $f(0) = 0$, $f(1) = 1$, and $f(\infty) = \infty$. Hence a normalized Belyi map is dynamical.

The *combinatorial type* of a single cycle Belyi map is the tuple $(d; e_1, e_2, e_3)$ where d denotes the degree of f and e_i denotes the ramification index of the unique ramification point above each x_i. An abstract type is a tuple $(d; e_1, e_2, e_3)$ such that $2 \leq e_1 \leq e_2 \leq e_3 \leq d$ and $e_1 + e_2 + e_3 = 2d + 1$. For each abstract type $\underline{C}$, there exists a unique normalized Belyi map of type $\underline{C}$ which can be defined over $\mathbb{Q}$. See [**ABE$^+$18**, Proposition 1] for a proof.

Notice that a normalized dynamical Belyi map is postcritically finite (PCF), i.e., the orbit of each critical point is finite. In this paper, we will focus on the genus zero, single cycle normalized Belyi maps which form a large class of dynamical Belyi maps.

4. Monodromy groups of dynamical Belyi maps

Let k be a number field. Fix an algebraic closure $\bar{k}$ of k. Let $P = \{0, 1, \infty\}$ and let f be a dynamical Belyi map of combinatorial type $\underline{C}$ defined over k.

Let $x_0 \in \mathbb{P}^1_k(k)\backslash P$. Then each f^n is a connected unramified covering of $\mathbb{P}^1_k\backslash P$, hence it is determined by the monodromy action of $\pi_1^{\acute{e}t}(\mathbb{P}^1_k\backslash P, x_0)$ on $f^{-n}(x_0))$ up to isomorphism. Let T_{x_0} be the tree defined as follows: it is rooted at x_0, the leaves of T_{x_0} are the points of $f^{-n}(x_0)$ for all $n \geq 1$, and the two leaves p, q are connected if $f(p) = q$. Varying n, associated monodromy defines a representation

$$\rho : \pi_1^{\acute{e}t}(\mathbb{P}^1_k\backslash P, x_0) \to \mathrm{Aut}(T_{x_0}) \tag{4.1}$$

whose image we call the *arithmetic monodromy group* $G^{\mathrm{arith}}(f)$ of f. One can also study this representation over $\bar{k}$ and obtain

$$\pi_1^{\acute{e}t}(\mathbb{P}^1_{\bar{k}}\backslash P, x_0) \to \mathrm{Aut}(T_{x_0}).$$

We call the image of the map in this case the *geometric monodromy group* $G^{\mathrm{geom}}(f)$. We note that these monodromy groups are unique up to conjugation by the elements of $\mathrm{Aut}(T_{x_0})$. The arithmetic and the geometric monodromy groups of f fit into an exact sequence as follows:

$$\begin{array}{ccccccccc} 1 & \longrightarrow & \pi_1^{\acute{e}t}(\mathbb{P}^1_{\bar{k}}\backslash P, x_0) & \longrightarrow & \pi_1^{\acute{e}t}(\mathbb{P}^1_k\backslash P, x_0) & \longrightarrow & \mathrm{Gal}(\bar{k}/k) & \longrightarrow & 1 \\ & & \downarrow & & \downarrow & & \downarrow & & \\ 1 & \longrightarrow & G^{\mathrm{geom}}(f) & \longrightarrow & G^{\mathrm{arith}}(f) & \longrightarrow & \mathrm{Gal}(L/k) & \longrightarrow & 1 \end{array} \tag{4.2}$$

for some field extension L of k. Determining this field L or its degree is a fundamental problem. The groups G^{geom} and G^{arith} are profinite groups and they are embedded into $\mathrm{Aut}(T_{x_0})$ by construction.

Let $x_1, \ldots, x_d$ denote the points in $f^{-1}(x_0)$. Then $f^{-(n+1)}(x_0) = f^{-n}(x_1) \cup \ldots \cup f^{-n}(x_d)$. Hence after deleting the root x_0, the tree T_{x_0} decomposes into the d regular trees $T_{x_1}, \ldots, T_{x_d}$. Let $\tilde{P}$ denote the set $f^{-1}(P)$. Then by functoriality, $f : \mathbb{P}^1_k\backslash\tilde{P} \to \mathbb{P}^1_k\backslash P$ induces a map on the fundamental groups:

$$f_*^i : \pi_1^{\acute{e}t}(\mathbb{P}^1_k\backslash\tilde{P}, x_i) \to \pi_1^{\acute{e}t}(\mathbb{P}^1_k\backslash P, x_0). \tag{4.3}$$

for any $i \in \{1, \ldots, d\}$. Here we use the notation f_*^i to specify the base point x_i of $\pi_1^{\acute{e}t}(\mathbb{P}^1_k\backslash\tilde{P}, x_i)$. This should not be confused with the i'th iteration of f. Similarly, by functoriality, the inclusion map $\mathbb{P}^1_k\backslash\tilde{P} \hookrightarrow \mathbb{P}^1_k\backslash P$ induces the surjective map

$$id_* : \pi_1^{\acute{e}t}(\mathbb{P}^1_k\backslash\tilde{P}, x_i) \to \pi_1^{\acute{e}t}(\mathbb{P}^1_k\backslash P, x_i). \tag{4.4}$$

for any $i \in \{1, \ldots, d\}$. The action of $\pi_1^{\acute{e}t}(\mathbb{P}^1_k\backslash\tilde{P}, x_i)$ on T_{x_1} through f_*^i coincides with its natural action on T_{x_i}. That is, the image of the composition of f_*^i with (4.1) is exactly the automorphisms in the image of ρ that fix x_i. However these automorphisms do not have to fix any of the other x_j for $j \neq i$. Since we would like to describe the action on the subtrees T_{x_i} for all i, we construct the following subgroup:

$$F_k := \{(\gamma_1, \ldots, \gamma_d) \in \prod_{i=1}^{d} \pi_1^{\acute{e}t}(\mathbb{P}^1_k\backslash\tilde{P}, x_i) : \rho \circ f_*^i(\gamma_i) = \rho \circ f_*^j(\gamma_j)$$
$$\text{for any } i, j \in \{1, \ldots, d\}\}.$$

Now an automorphism in the image of the composition $F_k \to \pi_1^{\acute{e}t}(\mathbb{P}^1_k\backslash\tilde{P}, x_i)$ (for any i) and $\rho \circ {f^i}_*$ fixes x_i for all i. This action coincides with the natural action of F_k composed with (4.4) on $\sqcup_{i=1}^d T_{x_i}$. This is given in the left upper square of Diagram (4.5). The first vertical isomorphism on the left is obtained by a change of base point from x_1 to x_0. We identify the trees $T_{x_0}, T_{x_1}, \ldots, T_{x_d}$ with the regular d-ary tree T introduced earlier. We can do this exactly because $x_0 \notin P$.

$$\begin{array}{ccccc}
\pi_1^{\acute{e}t}(\mathbb{P}^1_k\backslash P, x_0) & \xrightarrow{\rho} & \mathrm{Aut}(T_{x_0}) & \xrightarrow{\simeq} & W \\
{\scriptstyle f_*}\uparrow & & \uparrow & & \uparrow \\
F_k & \longrightarrow & \mathrm{Aut}(T_{x_1}) \times \ldots \times \mathrm{Aut}(T_{x_d}) & \xrightarrow{\simeq} & W^d \\
\downarrow{\scriptstyle pr_1} & & \Big\downarrow{\scriptstyle pr_1} & & \Big\downarrow \\
\pi_1^{\acute{e}t}(\mathbb{P}^1_k\backslash\tilde{P}, x_1) & \searrow & & & \\
\downarrow{\scriptstyle \simeq} & & & & \\
\pi_1^{\acute{e}t}(\mathbb{P}^1_k\backslash\tilde{P}, x_0) & & \mathrm{Aut}(T_{x_1}) & \xrightarrow{\simeq} & W \\
\downarrow{\scriptstyle id_*} & \searrow & \downarrow{\scriptstyle \simeq} & & \| \\
\pi_1^{\acute{e}t}(\mathbb{P}^1_k\backslash P, x_0) & \longrightarrow & \mathrm{Aut}(T_{x_0}) & \xrightarrow{\simeq} & W
\end{array} \tag{4.5}$$

We also note here that a discussion of this kind is given in [**Pin13b**, pg 20] for rational maps of degree 2. We generalize Pink's argument to any d here. In the case $d = 2$, if an automorphism fixes x_1, then it also has to fix x_2. Hence there is no need to define a subgroup F_k and hence Pink only uses $\pi_1^{\acute{e}t}(\mathbb{P}^1_k\backslash\tilde{P}, x_1)$.

By the construction of G^{arith}, (4.5) induces a commutative diagram

$$\begin{array}{ccccc}
\pi_1^{\acute{e}t}(\mathbb{P}^1_k\backslash P, x_0) & \xrightarrow{\rho} & G^{\mathrm{arith}} & \hookrightarrow & W \\
{\scriptstyle f_*}\uparrow & & \uparrow & & \uparrow \\
F_k & \longrightarrow & G^{\mathrm{arith}} \cap W^d & \longrightarrow & W^d \\
\downarrow{\scriptstyle pr_1} & & \downarrow{\scriptstyle pr_1} & & \downarrow \\
\pi_1^{\acute{e}t}(\mathbb{P}^1_k\backslash P, x_0) & \twoheadrightarrow & G^{\mathrm{arith}} & \hookrightarrow & W
\end{array} \tag{4.6}$$

We obtain a similar diagram for G^{geom} when we replace k by $\bar{k}$.

$$\begin{array}{ccccc}
\pi_1^{\acute{e}t}(\mathbb{P}^1_{\bar{k}}\backslash P, x_0) & \xrightarrow{\rho} & G^{\mathrm{geom}} & \hookrightarrow & W \\
{\scriptstyle f_*}\uparrow & & \uparrow & & \uparrow \\
F_{\bar{k}} & \longrightarrow & G^{\mathrm{geom}} \cap W^d & \longrightarrow & W^d \\
\downarrow{\scriptstyle pr_1} & & \downarrow{\scriptstyle pr_1} & & \downarrow \\
\pi_1^{\acute{e}t}(\mathbb{P}^1_{\bar{k}}\backslash P, x_0) & \twoheadrightarrow & G^{\mathrm{geom}} & \hookrightarrow & W
\end{array} \tag{4.7}$$

Diagrams (4.6) and (4.7) will be used in the proof of the main theorem in Section 6. For single cycle normalized Belyi maps, the group G^{geom} is determined in [**BEK21**]. Remember the groups E and U are defined in Definition 2.2. We first give an existing result on the geometric monodromy groups of Belyi maps at level one.

THEOREM 4.1 ([**LO08**, Theorem 5.3]). *Let f be a normalized Belyi map of type $\underline{C} = (d; e_1, e_2, e_3) \notin \{(4;3,3,3), (6;4,4,5)\}$.*

(1) *If at least one of the e_j is even, then $G_1^{\mathrm{geom}}(f) \simeq S_d$.*
(2) *If all e_j are odd, then $G_1^{\mathrm{geom}}(f) \simeq A_d$.*

The next result describes the geometric monodromy group G^{geom} for single cycle, normalized dynamical Belyi maps.

THEOREM 4.2 ([**BEK21**, Theorem 2.3.1]). *Let f be a normalized Belyi map of type $\underline{C} = (d; e_1, e_2, e_3) \notin \{(4;3,3,3), (6;4,4,5)\}$.*

(1) *Assume that at least one of the e_j is even. Then $G^{\mathrm{geom}}(f) \simeq E$.*
(2) *Assume that e_i are all odd. Then $G^{\mathrm{geom}}(f) \simeq U$.*

In the same paper, a criteria for the triviality of the quotient $G^{\mathrm{arith}}/G^{\mathrm{geom}}$ is given. See [**BEK21**, Corollary 2.4.6]. In this article, we prove that the order of the quotient group $G^{\mathrm{arith}}/G^{\mathrm{geom}}$ is either 1 or 2. As seen in (4.2), G^{arith} can be seen as a subgroup of the normalizer of G^{geom} in W. We will study the normalizer of the subgroups E and U in the next section.

5. Normalizer of E and U inside W

Let G be either E or U. Let $N(G)$ denote the normalizer of G in W. Given an element $x \in N(G)$, we denote its image in the quotient N/G by $x \pmod{G}$.

PROPOSITION 5.1. *Let G be one of the groups E or U. Let $x = (x_1, \dots, x_d)\tau \in N(G)$. Then*

(1) *x_i is in $N(G)$ for all $1 \leq i \leq d$.*
(2) *$x_i x_j^{-1}$ is in G for all $1 \leq i, j \leq d$.*

PROOF. Let $x = (x_1, \dots, x_d)\tau \in N(G)$ and let $g \in G$. We will show that $x_i g x_i^{-1}$ is in G for all $1 \leq i \leq d$. We will assume $i = 1$ for simplicity and let $j = \tau^{-1}(1)$. Let $y = (y_1, \dots, y_d)$ be an element of W such that $y_j = g$ and $y_i = \mathrm{id}$ for $i \neq j$. Similarly, let $y' = (z_1, \dots, z_d) \in W$ with $z_1 = z_j = g$ and $z_i = \mathrm{id}$ for $i \notin \{1, j\}$. Since g is in G, either y or y' is in G. This only depends on the sign of $g_{|T_1}$ in S_d. Assume $y \in G$. We compute

$$
\begin{aligned}
xyx^{-1} &= (x_1, \dots, x_d)\tau(y_1, \dots, y_d)\tau^{-1}(x_1^{-1}, \dots x_d^{-1}) \\
&= (x_1 y_{\tau^{-1}(1)} x_1^{-1}, \dots, x_d y_{\tau^{-1}(d)} x_d^{-1}).
\end{aligned}
$$

Since x is in $N(G)$, $xyx^{-1} \in G$, and by the construction of y,

$$
x_1 y_{\tau^{-1}(1)} x_1^{-1} = x_1 y_j x_1^{-1} = x_1 g x_1^{-1}
$$

is also in G and this finishes the proof of the first claim. Notice that if $y \notin G$, then one can use y' instead.

For the second claim, fix i and j in $\{1, \dots, d\}$. We take an element $y = \sigma \in G$ where σ is an even cycle such that $\tau\sigma\tau^{-1}(j) = i$. Then

$$
\begin{aligned}
xyx^{-1} &= (x_1, \dots, x_d)\tau\sigma\tau^{-1}(x_1^{-1}, \dots, x_d^{-1}) \\
&= (x_1 x_{\tau'^{-1}(1)}^{-1}, \dots, x_d x_{\tau'^{-1}(d)}^{-1})
\end{aligned}
$$

where $\tau' = \tau\sigma\tau^{-1}$. Since $\tau'^{-1}(i) = j$ and $xyx^{-1} \in G$, we have $x_i x_j^{-1}$ is in G. $\square$

REMARK 5.2. Note that one can replace G and $N(G)$ by G_n and $N(G)_n$ in Proposition 5.1 and its proof.

Let $\sigma = (12) \in S_d$. Remember that we embed S_d into W by $\tau \mapsto (-,\dots,-)\tau$ for any $\tau \in S_d$.

DEFINITION 5.3. For each $i \geq 1$, define an element of W as follows:

$$w_i := \begin{cases} \sigma & \text{for } i = 1 \\ (w_{i-1}, \dots, w_{i-1}) & \text{for } i \geq 2. \end{cases}$$

Let $N(E)$ and $N(U)$ denote the normalizer of E and U in W respectively. Remember that we denote the restriction of $N(E)$ and $N(U)$ to level n by $N(E)_n$ and $N(U)_n$.

LEMMA 5.4.

(1) *For $k \geq 1$, the automorphism w_k has order* 2.
(2) *For $k \geq 1$, the automorphism w_k is in $N(E)\backslash E$ (resp., $N(U)\backslash U$).*

PROOF. The first part follows by induction on k using the equality $w_k^2 = (w_{k-1}^2, \dots, w_{k-1}^2)$. We prove the second part by induction on k as well. Let $k = 1$ and let $g = (g_1, \dots, g_d)\tau \in W$. We compute

$$w_1 g w_1^{-1} = \sigma(g_1, \dots, g_d)\tau\sigma^{-1} = (g_2, g_1, g_3, \dots, g_d)\sigma\tau\sigma^{-1}.$$

Assume g is in E (resp., in U), then each g_i are in E (resp., in U). Since the sign of a permutation is invariant under conjugation and $g_i \in E$ (rep., $\in U$), we have $w_1 g w_1^{-1} \in E$ (resp., $\in U$). This proves that w_1 is in $N(E)$ (resp., in $N(U)$). Assume w_{k-1} is in $N(E)$ (resp., in $N(U)$). Then

$$\begin{aligned} w_k g w_k^{-1} &= (w_{k-1}, \dots, w_{k-1})(g_1, \dots, g_d)\tau(w_{k-1}^{-1}, \dots w_{k-1}^{-1}) \\ &= (w_{k-1} g_1 w_{k-1}^{-1}, \dots, w_{k-1} g_d w_{k-1}^{-1})\tau. \end{aligned}$$

By the induction hypothesis $w_{k-1} g_i w_{k-1}^{-1}$ is in E (resp., in U) for all i and $\operatorname{sgn}_2(w_k g w_k^{-1}) = \operatorname{sgn}_2(g) = 1$, hence w_k is in $N(E)$. Since $\operatorname{sgn}_2(w_1) = -1$ (resp., $\operatorname{sgn}(\sigma) \neq 1$), w_1 is not in E (resp., not in U) and similarly $w_k \notin E$ (resp., $\notin U$) since $w_1 \notin E$ (resp., $\notin U$). □

LEMMA 5.5. *We have $w_i w_j = w_j w_i$ for all $i, j \geq 1$.*

PROOF. We first observe that w_1 commutes with w_i for any $i \geq 1$. This follows from the relations given in Equation (2.1). Assume w_i and w_j commute for all $i, j \leq n$. Let $i > 1$. Then

$$w_i w_{n+1} = (w_{i-1}, \dots, w_{i-1})(w_n, \dots, w_n) = (w_{i-1}w_n, \dots, w_{i-1}w_n).$$

By our assumption $w_{i-1}w_n = w_n w_{i-1}$ and hence $w_i w_{n+1} = w_{n+1} w_i$. □

DEFINITION 5.6.

(1) Let $n \geq 2$. Define $\varphi_n : \prod_{i=1}^{n-1} \mathbb{F}_2 \to N(E)_n/E_n$ such that

$$\varphi((k_1, \dots, k_{n-1})) = \prod_{i=1}^{n-1} {w_{i|T_n}}^{k_i} \pmod{E_n}$$

(2) Let $n \geq 1$. Define $\phi_n : \prod_{i=1}^n \mathbb{F}_2 \to N(U)_n/U_n$ such that

$$\phi((k_1, \dots, k_n)) = \prod_{i=1}^n w_{i|T_n}{}^{k_i} \pmod{U_n}.$$

LEMMA 5.7. *The map φ_n (resp., ϕ_n) is a well defined homomorphism for any $n \geq 2$ (resp., for any $n \geq 1$).*

PROOF. By Lemma 5.4(2), the order of w_i is 2 for any $i \geq 1$, hence the maps φ_n and ϕ are well-defined. By Lemma 5.5, they are both homomorphisms. □

PROPOSITION 5.8. *The homomorphism φ_n (resp., ϕ_n) is injective for any $n \geq 2$ (resp., for any $n \geq 1$).*

PROOF. We will do induction on n to show that φ_n is injective. If $w_{1|T_2}^{k_1} = \mathrm{id} \pmod{E_2}$, then $w_{1|T_2}^{k_1}$ is in E_2. Since $\mathrm{sgn}_2(w_1) = -1$, we have $k_1 = 0$ and hence φ_2 is injective. Similarly ϕ_1 is injective since $w_{1|T_2}$ is not in $U_1 = A_d$.

Assume that φ_n (resp., ϕ_n) is injective for some $n \geq 2$ (resp., $n \geq 1$). Let $\varphi_{n+1}(k_1, \dots, k_n) = \prod_{i=1}^n w_{i|T_{n+1}}^{k_i} = \mathrm{id} \pmod{E_{n+1}}$. We compute that

$$\begin{aligned}\prod_{i=1}^n w_{i|T_{n+1}}^{k_i} &= w_{1|T_{n+1}}^{k_1} \prod_{i=2}^n (w_{i-1|T_n}^{k_i}, \dots, w_{i-1|T_n}^{k_i}) \\ &= w_{1|T_{n+1}}^{k_1} (\prod_{i=1}^{n-1} w_{i|T_n}^{k_{i+1}}, \dots, \prod_{i=1}^{n-1} w_{i|T_n}^{k_{i+1}}) \in E_{n+1}\end{aligned}$$

By Proposition 5.1, $\prod_{i=1}^{n-1} w_{i|T_n}^{k_{i+1}}$ is in E_n. By the induction hypothesis $k_{i+1} = 0$ for all $i = 1, \dots, n-2$ and $\varphi_{n+1}(k_1, \dots, k_n) = w_{1|T_{n+1}}^{k_1}$. We are left to show that $k_1 = 0$. This follows since $w_{1|T_n}$ is not in E_n. Since Proposition 5.1 holds for both E and U, the proof for the injectivity of ϕ_n follows the same argument. □

COROLLARY 5.9. *Let $k_j \in \mathbb{F}_2$ for $1 \leq j \leq n$. If the product $\prod_{j=1}^{n-1} w_j{}^{k_j}$ (resp., $\prod_{j=1}^n w_j{}^{k_j}$) is in E_n (resp., in U_n), then $k_j = 0$ for all j.*

PROPOSITION 5.10. *Let $n \geq 2$. The homomorphism φ_n (resp., ϕ_n) is surjective for any $n \geq 2$ (resp., for any $n \geq 1$).*

PROOF. The surjectivity of φ_2 follows from the fact that $[W_2 : E_2] = 2$ and that φ_2 is injective. Similarly $[W_1 : U_1] = 2$ and ϕ_1 is injective implies that ϕ_1 is surjective.

Assume φ_n is surjective. Let $x = (x_1, \dots, x_d)\tau$ be in $N(E)$ (resp. $N(U)$). If $\tau_{|T_1}$ is an even permutation, then τ is in E. Otherwise $\tau w_{|T_1}^{-1}$ is an even permutation and again τw^{-1} is in E. A similar argument works for U. Hence $\tau = w_1{}^k \pmod{E}$ (resp., $\pmod{U}$) for some $k \in \mathbb{F}_2$. So we may assume $x = (x_1, \dots, x_d)$. By Proposition 5.1, $x_{i|T_n}$ is in $N(E)_n$ for all $1 \leq i \leq d$. Since φ_n is surjective, we have $x_{i|T_n} = \prod_{j=1}^{n-1} w_{j|T_n}^{k_{i,j}} g_i$ for some $k_{i,j} \in \mathbb{F}_2$ and $g_i \in E_n$ for all $1 \leq i \leq d$. Hence

$x_{|T_{n+1}}$ is

$$\begin{aligned} x_{|T_{n+1}} &= (\prod_{j=1}^{n-1} w_{j\,|T_n}^{k_{1,j}}, \dots, \prod_{j=1}^{n-1} w_{j\,|T_n}^{k_{d,j}})(g_1, \dots, g_d) \\ &= (\prod_{j=1}^{n-1} w_{j\,|T_n}^{k_{1,j}}, \dots, \prod_{j=1}^{n-1} w_{j\,|T_n}^{k_{d,j}}) w_{1\,|T_{n+1}}^{k} \pmod{E_{n+1}} \end{aligned}$$

for some $k \in \mathbb{F}_2$. By Lemma 5.4, $w_{1\,|T_{n+1}}^{k} \in N(E)_{n+1}$ and hence

$$(\prod_{j=1}^{n-1} w_{j\,|T_n}^{k_{1,j}}, \dots, \prod_{j=1}^{n-1} w_{j\,|T_n}^{k_{d,j}}) \in N(E)_{n+1}$$

and by Proposition 5.1(2), we find that

$$\prod_{j=1}^{n-1} w_{j\,|T_n}^{k_{1,j}-k_{i,j}}$$

is in E_n for all i. Hence by Corollary 5.9, $k_{1,j} = k_{2,j} = \ldots = k_{d,j}$ for all $1 \leq j \leq n-1$. Let this number be k_j for each $j = 1, \ldots, n-1$. Now we have

$$\begin{aligned} x_{|T_{n+1}} &= (\prod_{j=1}^{n-1} w_{j\,|T_n}^{k_j}, \dots, \prod_{j=1}^{n-1} w_{j\,|T_n}^{k_j}) w_{1\,|T_{n+1}}^{k} && \pmod{E_{n+1}} \\ &= \prod_{j=1}^{n-1} (w_{j\,|T_n}^{k_j}, \dots, w_{j\,|T_n}^{k_j}) w_{1\,|T_{n+1}}^{k} && \pmod{E_{n+1}} \\ &= w_{1\,|T_{n+1}}^{k} \prod_{j=1}^{n-1} w_{j+1\,|T_{n+1}}^{k_j} && \pmod{E_{n+1}} \end{aligned}$$

Hence φ_n is surjective. The proof is same for ϕ_n since Proposition 5.1 and Lemma 5.4 holds for U as well. □

COROLLARY 5.11. *Let $N(E)$ denote the normalizer of E in W and let $N(U)$ denote the normalizer of U in W. Then*

$$\prod_{i=1}^{\infty} \mathbb{F}_2 \to N(E)/E \text{ and } \prod_{i}^{\infty} \mathbb{F}_2 \to N(U)/U$$

given by $(k_1, \dots, k_i, \dots) \mapsto w_1^{k_1} w_2^{k_2} \dots w_n^{k_n} \dots$ is an isomorphism.

6. Arithmetic monodromy groups of dynamical Belyi maps

In this section f denotes a dynamical Belyi map of combinatorial type $\underline{C} = (d; e_1, e_2, e_3) \notin \{(4; 3, 3, 3), (6; 4, 4, 5)\}$. To ease the notation we will drop f from the notation and denote $G^{\text{geom}}(f)$ by G. Remember that G is either E or U (Definition 2.2) by Theorem 4.2. Now we are ready to prove the main theorem.

THEOREM 6.1. *Let f be a dynamical Belyi map of combinatorial type $\underline{C} = (d; e_1, e_2, e_3) \notin \{(4; 3, 3, 3), (6; 4, 4, 5)\}$. The order of the quotient $G^{\text{arith}}(f)/G^{\text{geom}}(f)$ divides 2.*

Proof. The idea of the proof comes from [**Pin13a**, Lemma 4.8.4]. Using (4.2), we have

$$\pi_1^{\acute{e}t}(\mathbb{P}^1_k\backslash P, x_0)/\pi_1^{\acute{e}t}(\mathbb{P}^1_{\bar{k}}\backslash P, x_0) \to G^{\mathrm{arith}}/G.$$

and that $\pi_1^{\acute{e}t}(\mathbb{P}^1_k\backslash P, x_0)/\pi_1^{\acute{e}t}(\mathbb{P}^1_{\bar{k}}\backslash P, x_0)$ is isomorphic to $\mathrm{Gal}(\bar{k}/k)$. Using (4.6) and (4.7), we obtain the following diagram.

$$\begin{array}{ccc} \pi_1^{\acute{e}t}(\mathbb{P}^1_k\backslash P, x_0)/\pi_1^{\acute{e}t}(\mathbb{P}^1_{\bar{k}}\backslash P, x_0) & \longrightarrow & G^{\mathrm{arith}}/G \\ \uparrow & & \uparrow \\ F_k/F_{\bar{k}} & \longrightarrow & G^{\mathrm{arith}} \cap W^d/G \cap W^d \\ \downarrow & & \downarrow \\ \pi_1^{\acute{e}t}(\mathbb{P}^1_k\backslash P, x_0)/\pi_1^{\acute{e}t}(\mathbb{P}^1_{\bar{k}}\backslash P, x_0) & \longrightarrow & G^{\mathrm{arith}}/G \end{array} \tag{6.1}$$

The quotient $F_k/F_{\bar{k}}$ is $\{(\gamma_1, \dots, \gamma_d) \in \prod_{i=1}^d \mathrm{Gal}(\bar{k}/k) : \gamma_1 = \dots = \gamma_d\}$. Putting all of these discussions together, we obtain the diagram below.

$$\begin{array}{ccc} \mathrm{Gal}(\bar{k}/k) & \longrightarrow & G^{\mathrm{arith}}/G \\ \| & & \uparrow \\ \mathrm{Gal}(\bar{k}/k) & \longrightarrow & G^{\mathrm{arith}} \cap W^d/G \cap W^d \\ \| & & \downarrow \\ \mathrm{Gal}(\bar{k}/k) & \longrightarrow & G^{\mathrm{arith}}/G \end{array} \tag{6.2}$$

Since G is a normal subgroup of G^{arith}, the quotient G^{arith}/G is a subgroup of $N(G)/G$ which we studied in detail in Corollary 5.11. We first note that by Corollary 5.11, $N(G)/G$ is isomorphic to a direct product of copies of $\mathbb{F}_2$ and hence the exact sequence

$$1 \to N(G) \cap W^d/G \cap W^d \to N(G)/G \to \langle w_{1|T_1} \rangle \to 1.$$

splits with trivial action. Hence

$$N(G)/G \simeq (N(G) \cap W^d/G \cap W^d) \times \langle w_{1|T_1} \rangle.$$

We note that w_i is in $N(G) \cap W^d$ for all $i \geq 2$. Therefore we have a projection map

$$N(G)/G \to N(G) \cap W^d/G \cap W^d$$

sending the automorphism $\prod_{i=1}^{\infty} w_i^{k_i} \pmod G$ to $\prod_{i=2}^{\infty} w_i^{k_i} \pmod G$. Furthermore we can compose it with the projection of W^d onto the first component. The composition of these two maps gives a homomorphism $N(G)/G \to N(G)/G$ that maps

$$w_1^{k_1} w_2^{k_2} \dots = w_1^{k_1}\Big(\prod_{i=1}^{\infty} w_i^{k_{i+1}}, \dots, \prod_{i=1}^{\infty} w_i^{k_{i+1}}\Big) \mapsto \prod_{i=1}^{\infty} w_i^{k_{i+1}} \tag{6.3}$$

since $w_i = (w_{i-1}, \dots, w_{i-1})$ for $i \geq 2$.

We can see these maps in the following diagram.

(6.4)

$$\begin{array}{ccccccc}
\mathrm{Gal}(\bar{k}/k) & \longrightarrow & G^{\mathrm{arith}}/G & \longrightarrow & N(G)/G & \xrightarrow{\simeq} & \prod_i^\infty \mathbb{F}_2 \\
\| & & \downarrow & & \downarrow & & \downarrow \\
\| & & G^{\mathrm{arith}} \cap W^d/G \cap W^d & \longrightarrow & N(G) \cap W^d/G \cap W^d & & \downarrow \\
\| & & \downarrow & & \downarrow & & \downarrow \\
\mathrm{Gal}(\bar{k}/k) & \longrightarrow & G^{\mathrm{arith}}/G & \longrightarrow & N(G)/G & \xrightarrow{\simeq} & \prod_i^\infty \mathbb{F}_2
\end{array}$$

We obtain the left part of the diagram from (6.2). We define the vertical map on the right hand side as $(k_1, k_2, \ldots) \to (k_2, k_3, \ldots)$ so that the diagram commutes.

Let $w = \prod_{i=1}^\infty w_i^{k_i} \pmod{G}$ be an element of G^{arith}/G. Since $\mathrm{Gal}(\bar{k}/k) \to G^{\mathrm{arith}}/G$ is surjective, there is a $\tau \in \mathrm{Gal}(\bar{k}/k)$ that maps to w. The image of τ in the top row is $(k_1, k_2, \ldots)$ and in the bottom row $(k_2, k_3, \ldots)$. Since the diagram commutes, they are equal and $k_1 = k_2 = \ldots = k_n$ for all n which shows that G^{arith}/G is a subset of $\{(1,1,1,\ldots),(0,0,0,\ldots)\}$. □

REMARK 6.2. In [**BEK21**, Lemma 2.4.3], a product discriminant D is defined. Moreover, it is shown that $G^{\mathrm{arith}} = G^{\mathrm{geom}}$ if and only if D is a square in $\mathbb{Q}(t)$. One can generalize this to $k(t)$ where k is a number field. By Theorem 6.1, we know that $\mathrm{Gal}(\bar{k}/k) \to G^{\mathrm{arith}}/G$ factors through a quadratic extension L of k. Furthermore, in [**BEK21**, Proposition 2.4.5], this discriminant D is explicitly calculated. From this we see that the quadratic extension L is given by $k(\sqrt{u})$ where $D = u(1-t)^{2(e_2-1)}t^{2(e_1-1)}$ for a dynamical Belyi map of combinatorial type $(d; e_1, e_2, e_3)$.

References

[ABC+21] Faseeh Ahmad, Robert L. Benedetto, Jennifer Cain, Gregory Carroll, and Lily Fang, *The arithmetic basilica: A quadratic pcf arboreal galois group*, Journal of Number Theory (2021).

[ABE+18] Jacqueline Anderson, Irene I. Bouw, Ozlem Ejder, Neslihan Girgin, Valentijn Karemaker, and Michelle Manes, *Dynamical Belyi maps*, Women in numbers Europe II, Assoc. Women Math. Ser., vol. 11, Springer, Cham, 2018, pp. 57–82, DOI 10.1007/978-3-319-74998-3_5. MR3882706

[BEK21] Irene I. Bouw, Özlem Ejder, and Valentijn Karemaker, *Dynamical Belyi maps and arboreal Galois groups*, Manuscripta Math. **165** (2021), no. 1-2, 1–34, DOI 10.1007/s00229-020-01204-3. MR4242559

[BFH+17] Robert L. Benedetto, Xander Faber, Benjamin Hutz, Jamie Juul, and Yu Yasufuku, *A large arboreal Galois representation for a cubic postcritically finite polynomial*, Res. Number Theory **3** (2017), Paper No. 29, 21, DOI 10.1007/s40993-017-0092-8. MR3736808

[JM14] Rafe Jones and Michelle Manes, *Galois theory of quadratic rational functions*, Comment. Math. Helv. **89** (2014), no. 1, 173–213, DOI 10.4171/CMH/316. MR3177912

[Jon13] Rafe Jones, *Galois representations from pre-image trees: an arboreal survey* (English, with English and French summaries), Actes de la Conférence "Théorie des Nombres et Applications", Publ. Math. Besançon Algèbre Théorie Nr., vol. 2013, Presses Univ. Franche-Comté, Besançon, 2013, pp. 107–136. MR3220023

[LO08] Fu Liu and Brian Osserman, *The irreducibility of certain pure-cycle Hurwitz spaces*, Amer. J. Math. **130** (2008), no. 6, 1687–1708, DOI 10.1353/ajm.0.0031. MR2464030

[Odo85] R. W. K. Odoni, *The Galois theory of iterates and composites of polynomials*, Proc. London Math. Soc. (3) **51** (1985), no. 3, 385–414, DOI 10.1112/plms/s3-51.3.385. MR805714

[Pin13a] Richard Pink, *Profinite iterated monodromy groups arising from quadratic morphisms with infinite postcritical orbits*, Preprint, arXiv:1309.5804, 2013.

[Pin13b] Richard Pink, *Profinite iterated monodromy groups arising from quadratic polynomials*, Preprint, arXiv:1307.5678, 2013.

Department of Mathematics, Boğaziçi University, Istanbul
Email address: ozlem.ejder@boun.edu.tr

Contemporary Mathematics
Volume **779**, 2022
https://doi.org/10.1090/conm/779/15672

Automorphisms and isogeny graphs of abelian varieties, with applications to the superspecial Richelot isogeny graph

Enric Florit and Benjamin Smith

Abstract. We investigate special structures due to automorphisms in isogeny graphs of principally polarized abelian varieties, and abelian surfaces in particular. We give theoretical and experimental results on the spectral and statistical properties of $(2,2)$-isogeny graphs of superspecial abelian surfaces, including stationary distributions for random walks, bounds on eigenvalues and diameters, and a proof of the connectivity of the Jacobian subgraph of the $(2,2)$-isogeny graph. Our results improve our understanding of the performance and security of some recently-proposed cryptosystems, and are also a concrete step towards a better understanding of general superspecial isogeny graphs in arbitrary dimension.

1. Introduction

When studying the internal structure of isogeny classes of abelian varieties from an algorithmic point of view, we work with *isogeny graphs*: the vertices are isomorphism classes of abelian varieties, and the edges are isomorphism classes of isogenies, often of some fixed degree. For elliptic curves, these graphs have already had a wealth of applications. Mestre [**32**] used his *méthode des graphes* to compute a basis of the space $S_2(N)$ of modular forms of weight 2, level N, and trivial character. Kohel [**27**] used isogeny graphs to compute endomorphism rings of elliptic curves over finite fields, and Fouquet and Morain turned this around to improve point-counting algorithms for elliptic curves [**17**]. Bröker, Lauter, and Sutherland [**8**] developed an algorithm for computing modular polynomials using isogeny graph structures; Sutherland [**41**] has used the difference between the structures of ordinary and supersingular isogeny graphs to give a remarkable and efficient deterministic supersingularity test for elliptic curves.

More recently, isogeny graphs have become a setting for post-quantum cryptographic algorithms, especially in the supersingular case. Charles, Goren, and Lauter proposed a cryptographic hash function with provable security properties based on

2020 *Mathematics Subject Classification.* Primary 14K02; Secondary 14G50, 14Q05, 11T99, 05C81.

Key words and phrases. Superspecial abelian varieties, isogeny graphs, isogeny-based cryptography.

The second author was supported in part by l'Agence nationale de la recherche (ANR) program CIAO ANR-19-CE48-0008.

combinatorial properties of the supersingular elliptic 2-isogeny graph [**12**]. Rostovtsev and Stolbunov proposed a key exchange scheme based on ordinary isogeny graphs [**38,40**]; this was vastly accelerated by Castryck, Lange, Martindale, Panny, and Renes by transposing it to a subgraph of the supersingular isogeny graph, where it is known as CSIDH [**10**]. Jao and De Feo's SIDH key exchange algorithm [**14,24**], the basis of SIKE [**2**] (a third-round alternate candidate in the NIST post-quantum cryptography standardization process), is based on the difficulty of finding paths in the elliptic supersingular 2- and 3-isogeny graphs. These applications all depend, both in their constructions and in their security arguments, on a precise understanding of the combinatorial properties of supersingular isogeny graphs.

It is natural to try to extend these applications to the setting of isogeny graphs of higher-dimensional principally polarized abelian varieties (PPAVs). First steps in this direction have been made by Charles, Goren, and Lauter [**11**], Takashima [**42**], Flynn and Ti [**16**], and Castryck, Decru, and Smith [**9**]. Costello and Smith have proposed an attack on cryptosystems based on the difficulty of computing isogenies between higher-dimensional superspecial abelian varieties [**13**].

But so far, the efficiency and security of these algorithms is conjectural—even speculative—because of a lack of information on combinatorial properties of supersingular isogeny graphs in higher dimension, such as their connectedness, their diameter, and their expansion constants. For example, the hash functions typically depend on the rapid convergence of random walks to the uniform distribution on the isogeny graph; but while this is well-known for the elliptic case, it is not yet well-understood even in $g = 2$. Indeed, even the connectedness of the superspecial graph for $g = 2$ has only recently been proven by Jordan and Zaytman [**25**].

Our ultimate aim is a deeper understanding of the combinatorial and spectral properties of the superspecial graph, such as its diameter and the limit distribution of random walks. In this article we give some theoretical results on general superspecial graphs, and experimental results focused on the *Richelot isogeny graph*: that is, the graph formed by $(2,2)$-isogenies of 2-dimensional PPAVs. Richelot isogeny graphs are the most amenable to explicit computation (apart from elliptic graphs), and already exhibit a particularly rich structure.

After recalling basic results in §2, we explore the impact of automorphisms of g-dimensional PPAVs on edge weights in the $(\ell,\ldots,\ell)$-isogeny graph for general g and ℓ in §3. Automorphisms are a complicating factor that can almost be ignored in elliptic isogeny graphs, since only two vertices (corresponding to j-invariants 0 and 1728) have automorphisms other than ± 1. In higher dimensions, however, extra automorphisms are much more than an isolated corner-case: every general product PPAV $\mathcal{A} \times \mathcal{B}$ has an involution $[1]_{\mathcal{A}} \times [-1]_{\mathcal{B}}$ which may induce nontrivial weights in the isogeny graph, and entire families of simple PPAVs can come equipped with extra automorphisms, as we will see in §5 for dimension $g = 2$. The *ratio principle* proven in Lemma 3.2, which relates automorphism groups of $(\ell,\ldots,\ell)$-isogenous PPAVs with the weights of the directed edges between them in the isogeny graph, is an essential tool for our later investigations.

We consider the spectral and statistical properties of isogeny graphs, still in the most general setting, in §4. Here we prove results which, combined with an understanding of the automorphism groups of vertices, allow us to state general theoretical bounds on eigenvalues, and compute stationary distributions for random walks in the superspecial isogeny graph—and also in interesting subgraphs of the superspecial graph, such as the Jacobian subgraph.

We then narrow our focus to the Richelot isogeny graph: that is, the case $g = 2$ and $\ell = 2$. We recall Bolza's classification of automorphism groups of genus-2 Jacobians in §5, and apply it in the context of Richelot isogeny graphs (extending the results of Katsura and Takashima [**26**]). In §6 we specialize our general results to $g = 2$ and $\ell = 2$, and give experimental data for diameters and second eigenvalues of superspecial Richelot isogeny graphs (and Jacobian subgraphs) for $17 \leq p \leq 601$. This allows us to prove that the Jacobian subgraph of the Richelot isogeny graph is connected and aperiodic, and to bound its diameter relative to the diameter of the entire superspecial graph in §7.

Our results have consequences for the security and efficiency arguments of the cryptographic algorithms described in [**42**], [**16**], [**9**], and [**13**]. For example, we can estimate the frequency with which elliptic products are encountered during random walks in the superspecial graph, which is essential for understanding the true efficiency of the attack in [**13**]; and we can understand the stationary distribution for random walks restricted to the Jacobian subgraph (which were used in [**9**]). These cryptographic implications are further discussed in §6. Our results also offer a concrete step towards a better understanding of the situation for general superspecial isogeny graphs—that is, in arbitrary dimension g, and with $(\ell, \ldots, \ell)$-isogenies for arbitrary primes ℓ.

2. Isogeny graphs

DEFINITION 2.1. Let $\mathcal{A}/\Bbbk$ be a principally polarized abelian variety (PPAV) and ℓ a prime, not equal to the characteristic of $\Bbbk$. A subgroup of $\mathcal{A}[\ell]$ is **Lagrangian** if it is maximally isotropic with respect to the ℓ-Weil pairing. An **$(\ell, \ldots, \ell)$-isogeny** is an isogeny $\mathcal{A} \to \mathcal{A}'$ of PPAVs whose kernel is a Lagrangian subgroup of $\mathcal{A}[\ell]$.

If $\mathcal{A}$ is a g-dimensional PPAV, then every Lagrangian subgroup of $\mathcal{A}[\ell]$ is necessarily isomorphic to $(\mathbb{Z}/\ell\mathbb{Z})^g$, though the converse does not hold. Since its kernel is Lagrangian, an $(\ell, \ldots, \ell)$-isogeny $\phi : \mathcal{A} \to \mathcal{A}'$ respects the principal polarizations: if λ and λ' are the principal polarizations on $\mathcal{A}$ and $\mathcal{A}'$, respectively, then the pullback $\phi^*(\lambda')$ is equal to $\ell\lambda$.

Given another g-dimensional PPAV $\mathcal{A}'$, we say two Lagrangian subgroups K of $\mathcal{A}[\ell]$ and K' of $\mathcal{A}'[\ell]$ yield **isomorphic isogenies** ϕ and ϕ', if there are isomorphisms $\alpha \colon \mathcal{A} \to \mathcal{A}'$ and $\beta \colon \mathcal{A}/K \to \mathcal{A}/K'$ respecting the principal polarizations, such that the following diagram commutes:

$$\begin{array}{ccc} \mathcal{A} & \xrightarrow{\ \alpha\ } & \mathcal{A}' \\ {\scriptstyle\phi}\downarrow & & \downarrow{\scriptstyle\phi'} \\ \mathcal{A}/K & \xrightarrow[\ \beta\ ]{} & \mathcal{A}'/K' \end{array}$$

In this case, the dual isogenies $\phi^\dagger$ and $\phi'^\dagger$ are also isomorphic.

DEFINITION 2.2. Fix a positive integer g and a prime p. The **$(\ell, \ldots, \ell)$-isogeny graph**, denoted $\Gamma_g(\ell; p)$, is the directed weighted multigraph defined as follows.

- The **vertices** are isomorphism classes of PPAVs defined over $\bar{\mathbb{F}}_p$. If $\mathcal{A}$ is a PPAV, then $[\mathcal{A}]$ denotes the corresponding vertex.

- The **edges** are isomorphism classes of $(\ell, \ldots, \ell)$-isogenies, **weighted** by the number of distinct kernels yielding isogenies in the class. The weight of an edge $[\phi]$ is denoted by $w([\phi])$.

If $[\phi] : [\mathcal{A}] \to [\mathcal{A}']$ is an edge, then $w([\phi]) = n$ if and only if there are n Lagrangian subgroups $K \subset \mathcal{A}[\ell]$ such that $\mathcal{A}' \cong \mathcal{A}/K$ (this definition is independent of the choice of representative isogeny ϕ). Equivalently, if there is an $(\ell, \ldots, \ell)$-isogeny $\phi : \mathcal{A} \to \mathcal{A}'$, then $w([\phi])$ is equal to the size of the orbit of $\ker \phi$ under the action of $\mathrm{Aut}(\mathcal{A})$ on the set of Lagrangian subgroups of $\mathcal{A}[\ell]$.

The isogeny graph breaks up into components; there are at least as many connected components as there are isogeny classes over $\Bbbk$. We are particularly interested in the superspecial isogeny class.

DEFINITION 2.3. A PPAV $\mathcal{A}/\overline{\mathbb{F}}_p$ of dimension g is **superspecial** if its Hasse–Witt matrix vanishes identically. Equivalently, $\mathcal{A}$ is superspecial if it is isomorphic *as an unpolarized abelian variety* to a product of supersingular elliptic curves.

For general facts and background on superspecial and supersingular abelian varieties, we refer to Li and Oort [**29**], and Brock's thesis [**6**] (especially for $g \leq 3$).

DEFINITION 2.4. The $(\ell, \ldots, \ell)$-isogeny graph of g-dimensional superspecial PPAVs over $\overline{\mathbb{F}}_p$ is denoted by $\Gamma_g^{SS}(\ell; p)$. We often refer to $\Gamma_g^{SS}(\ell; p)$ as the *superspecial graph*, with g, ℓ, and p implicit.

The graph $\Gamma_g^{SS}(\ell; p)$ is regular (every vertex has the same weighted out-degree), and Jordan and Zaytman recently proved that $\Gamma_g^{SS}(\ell; p)$ is connected (see [**25**]; though this result was already implicit, in a different language, in [**34**, Lemma 7.9]). If an elliptic curve is supersingular, then it is isomorphic to a curve defined over $\mathbb{F}_{p^2}$. Similarly, if $\mathcal{A}/\overline{\mathbb{F}}_p$ is superspecial, then $\mathcal{A}$ is isomorphic to a PPAV defined over $\mathbb{F}_{p^2}$, so in our experiments involving superspecial graphs, we work over $\mathbb{F}_{p^2}$ for various p.

3. Isogenies and automorphisms

Isogeny graphs are weighted directed graphs, and before going any further, we should pause to understand the weights. The weights of the edges are closely related to the automorphism groups of the vertices that they connect, as we shall see.

Let $\mathcal{A}$ be a PPAV, let K be a Lagrangian subgroup of $\mathcal{A}[\ell]$ for some ℓ, and let α be an automorphism of $\mathcal{A}$. We write K_α for $\alpha(K)$.

If $K_\alpha = K$, then α induces an automorphism of $\mathcal{A}/K$. Going further, if S is the stabiliser of K in $\mathrm{Aut}(\mathcal{A})$, then S induces an isomorphic subgroup S' of $\mathrm{Aut}(\mathcal{A}/K)$.

Now suppose that $K_\alpha \neq K$. If $\phi : \mathcal{A} \to \mathcal{A}/K$ and $\phi_\alpha : \mathcal{A} \to \mathcal{A}/K_\alpha$ are the quotient isogenies, then α induces an isomorphism $\alpha_* : \mathcal{A}/K \to \mathcal{A}/K_\alpha$ such that $\alpha_* \circ \phi = \phi' \circ \alpha$. (Note that ϕ and ϕ_α are only defined up to isomorphism, but if we fix a choice of ϕ and ϕ_α, then α_* is unique.) Let $\phi_\alpha = \alpha_*^{-1} \circ \phi'$. The isogenies ϕ and ϕ_α have identical domains and codomains, but distinct kernels; thus, they both represent the same edge in the isogeny graph, and $w([\phi]) > 1$. Going further, if O_K is the orbit of K under $\mathrm{Aut}(\mathcal{A})$, then there are $\#O_K$ distinct kernels of isogenies representing $[\phi]$: that is, $w([\phi]) = \#O_K$.

Looking at the dual isogenies, we see that $\alpha^{-1} \circ (\phi_\alpha)^\dagger \circ \phi = [\ell]_\mathcal{A}$, so $\phi^\dagger$ and $\phi_\alpha^\dagger$ have the same kernel. Hence, while automorphisms of $\mathcal{A}$ may lead to increased weight on the edge $[\phi]$, they have no effect on the weight of the dual edge $[\phi^\dagger]$.

Every PPAV has a nontrivial involution $[-1]$, but $[-1]$ fixes every kernel and commutes with every isogeny. It therefore has no impact on edges or weights in the isogeny graph, so can simplify our analysis by quotienting it away. Indeed, since $\langle[-1]\rangle$ is contained in the centre of $\mathrm{Aut}(\mathcal{A})$, the quotient $\mathrm{Aut}(\mathcal{A})/\langle[-1]\rangle$ acts on the set of Lagrangian subgroups of $\mathcal{A}[\ell]$. This is crucial in what follows.

DEFINITION 3.1. If $\mathcal{A}$ is a PPAV, then its **reduced automorphism group**[1] is

$$\mathrm{RA}(\mathcal{A}) := \mathrm{Aut}(\mathcal{A})/\langle[-1]\rangle.$$

LEMMA 3.2. *Let $\phi : \mathcal{A} \to \mathcal{A}'$ be an $(\ell, \ldots, \ell)$-isogeny, and let S be the stabiliser of $\ker(\phi)$ in $\mathrm{RA}(\mathcal{A})$.*

(1) *The isogeny ϕ induces a subgroup S' of $\mathrm{RA}(\mathcal{A}')$ isomorphic to S, and S' is the stabiliser of $\ker\phi^\dagger$ in $\mathrm{RA}(\mathcal{A}')$.*
(2) *If $s := \#S$ (so $s = \#S'$), then in the $(\ell, \ldots, \ell)$-isogeny graph we have*

$$w([\phi]) = \#\mathrm{RA}(\mathcal{A})/s \quad \textit{and} \quad w([\phi^\dagger]) = \#\mathrm{RA}(\mathcal{A}')/s.$$

In particular,

$$\#\mathrm{RA}(\mathcal{A}) \cdot w([\phi^\dagger]) = \#\mathrm{RA}(\mathcal{A}') \cdot w([\phi]). \tag{3.1}$$

PROOF. Let $K := \ker(\phi)$ be the kernel of ϕ. As discussed above, each α in $\mathrm{Aut}(\mathcal{A})$ induces an isomorphism $\alpha_* : \mathcal{A}' \to \mathcal{A}/\alpha(K)$, and if α stabilises K, then α_* is an automorphism of $\mathcal{A}'$. As α stabilises $\mathcal{A}[\ell]$, this gives an inclusion of S into the stabiliser of $\ker\phi^\dagger$. The reverse inclusion comes from the symmetric argument on the dual. The second statement follows from the orbit-stabiliser theorem. Note we only need to consider the action by reduced automorphisms, as $[-1]$ acts trivially on all subgroups of $\mathcal{A}$. □

To understand the isogeny graph, then, we need to understand the reduced automorphism groups of its vertices. A generic PPAV $\mathcal{A}$ has $\mathrm{Aut}(\mathcal{A}) = \langle[-1]\rangle$, so $\mathrm{RA}(\mathcal{A}) = 1$. The simplest examples of nontrivial reduced automorphism groups are the elliptic curves with j-invariants 0 and 1728. Moving into higher dimensions, nontrivial reduced automorphism groups are much more common: for example, if $\mathcal{A} = \mathcal{E} \times \mathcal{E}'$ is a product of elliptic curves, then $[1]_\mathcal{E} \times [-1]_{\mathcal{E}'}$ is a nontrivial involution in $\mathrm{RA}(\mathcal{E} \times \mathcal{E}')$. We will see many more examples of nontrivial reduced automorphism groups below.

EXAMPLE 3.3. Consider the graph $\Gamma_2^{SS}(2; 11)$, shown in Figure 1. It has five vertices:

- $[\mathcal{A}_1] = [\mathcal{J}(\mathcal{C}_1)]$, for $\mathcal{C}_1 \colon y^2 = x^6 - 1$, with $\mathrm{RA}(\mathcal{A}_1) = D_{2\times 6}$.
- $[\mathcal{A}_2] = [\mathcal{J}(\mathcal{C}_2)]$, for $\mathcal{C}_2 \colon y^2 = (x^3 - 1)(x^3 - 3)$, with $\mathrm{RA}(\mathcal{A}_2) = S_3$.
- $[\mathcal{E}_{1728}^2]$, where $\mathcal{E}_{1728} \colon y^2 = x^3 - x$, and $\#\mathrm{RA}(\mathcal{E}_{1728}^2) = 16$.
- $[\mathcal{E}_0^2]$, where $\mathcal{E}_0 \colon y^2 = x^3 - 1$, and $\#\mathrm{RA}(\mathcal{E}_0^2) = 36$.
- $[\Pi] = [\mathcal{E}_0 \times \mathcal{E}_{1728}]$, with $\#\mathrm{RA}(\Pi) = 12$.

[1] Reduced automorphism groups are usually defined for hyperelliptic curves, not abelian varieties, but if $\mathcal{A} = \mathcal{J}(\mathcal{C})$ is the Jacobian of a hyperelliptic curve and ι is the hyperelliptic involution, then $\mathrm{RA}(\mathcal{J}(\mathcal{C}))$ is canonically isomorphic to $\mathrm{RA}(\mathcal{C}) = \mathrm{Aut}(\mathcal{C})/\langle\iota\rangle$; so our definition is consistent for hyperelliptic Jacobians.

The weights indicated in the figure indeed satisfy Equation (3.1). For instance, there is a unique $(2,2)$-isogeny $\phi\colon \mathcal{E}_{1728}^2 \to \mathcal{E}_0^2$ (up to isomorphism), and

$$\frac{w([\phi])}{w([\phi^\dagger])} = \frac{4}{9} = \frac{16}{36} = \frac{\#\mathrm{RA}(\mathcal{E}_{1728}^2)}{\#\mathrm{RA}(\mathcal{E}_0^2)}.$$

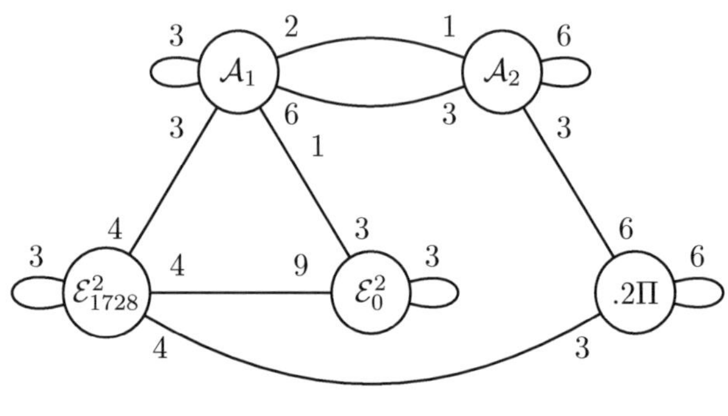

FIGURE 1. The graph $\Gamma_2^{SS}(2;11)$, with isogeny weights.

4. Random walks

Let $G = (V, E, w)$ be a directed weighted multigraph with finite vertex set V. The weight of an edge e is denoted by $w(e) > 0$. Given subsets $S, T \subset V$, we denote the multiset of edges from S to T by $E(S,T)$, omitting the curly braces when S or T is a singleton $\{u\}$. For each pair of vertices $u, v \in V$ we write $w_{uv} = \sum_{e\in E(u,v)} w(e)$, and for each vertex $u \in V$ we have $\deg u = \sum_{e\in E(u,V)} w(e)$. The set of neighbors of a vertex $u \in V$ (that is, the set of vertices v such that $E(u,v) \neq \emptyset$) is denoted $N(u)$.

We define a **random walk** on G with starting vertex $v_0 \in V$ in the usual way: for each natural $t \geq 0$ and pair of vertices $u, v \in V$, we have

$$P(v_{t+1} = v \mid v_t = u) = \frac{w_{uv}}{\deg u},$$

with the remark that this probability is zero whenever $E(u,v) = \emptyset$. The random walk transition matrix is the matrix M given by $M_{v,u} = \frac{w_{uv}}{\deg u}$.

If G is a strongly connected aperiodic graph, then the Perron–Frobenius Theorem tells us there is a unique positive vector $\varphi = (\varphi(u))_{u\in V}$ with $\|\varphi\|_1 = 1$ such that $M\varphi = \varphi$ (see [**28**, Proposition 1.14 and Theorem 4.9]). This vector φ is called the **stationary distribution** of G. Moreover, for any starting distribution ψ on the vertices of G, we have $\lim_{n\to\infty} M^n\psi = \varphi$.[2]

When G is an *undirected* graph, the stationary distribution is the vector φ where

$$\varphi(u) = \frac{\deg u}{2|E|} \quad \text{for} \quad u \in V;$$

we see immediately that this is indeed the stationary distribution, because

$$\varphi(u) = \frac{\deg u}{2|E|} = \sum_{v\in N(u)} \frac{1}{\deg v}\frac{\deg v}{2|E|}.$$

[2]If we drop the connectivity hypothesis, then φ is neither positive nor unique. Meanwhile, a periodic graph will still have a stationary distribution, but convergence to it is not granted.

However, when G is a *directed* graph, there is no closed-form formula for the stationary distribution of the random walk. Even the principal ratio $\frac{\max_{u\in V}\varphi(u)}{\min_{u\in V}\varphi(u)}$ of the distribution can be difficult to bound, and it can be exponentially large even when degree bounds such as $\delta \le \deg u \le \Delta$, for all $u \in V$, are known [**1**].

4.1. Directed graphs and linear imbalance. The following definition tries to restrict the amount of allowed "directedness" in a graph, so that we are able to find closed-form stationary distributions for isogeny graphs. It applies directly to the graph $\Gamma_2^{SS}(2;11)$ displayed in Figure 1.

DEFINITION 4.1. Let $G = (V, E, w)$ be a directed weighted graph. We say G has **linear imbalance** if there exists a vertex partition $V = A_1 \sqcup \cdots \sqcup A_n$ and a bijection

$$E(u,v) \overset{(\cdot)^\dagger}{\to} E(v,u)$$

for each pair of adjacent vertices $u, v \in V$, such that

(1) If $u, v \in A_i$, then for each $e \in E(u,v)$, $w(e) = w(e^\dagger)$.
(2) For each $i \neq j$ there exists a rational number m_{ij}, such that if $u \in A_i$, $v \in A_j$, and $e \in E(u,v)$, then $w(e) = m_{ij} \cdot w(e^\dagger)$.

In particular $m_{ji} = m_{ij}^{-1}$, and we can set $m_{ii} = 1$.

We can see G as an undirected graph if we forget the weights, due to the existence of the bijections $E(u,v) \overset{(\cdot)^\dagger}{\to} E(v,u)$. However, the presence of weights changes the definition of the random walk on G, and in particular the stationary distribution will be different. We now want to compute this distribution.

PROPOSITION 4.2. *Let $G = (V, E)$ be a linear imbalance graph with partition $V = A_1 \sqcup \cdots \sqcup A_n$. Assume all vertices of each given class A_i have the same degree d_i, i.e., $\deg(u) = d_i$ for all $u \in A_i$.*

Suppose there exists a non-zero solution $(\alpha_1, \ldots, \alpha_n)$ to the system of equations

$$\frac{m_{ji}}{d_j}\alpha_j = \frac{1}{d_i}\alpha_i \quad \text{for every } i, j \text{ such that } E(A_i, A_j) \neq \emptyset. \tag{4.1}$$

Define the vectors $\tilde{\varphi} = (\tilde{\varphi}(u))_{u\in V}$ by $\tilde{\varphi}(u) = \alpha_i$ if $u \in A_i$, and $\varphi = \tilde{\varphi}/\|\tilde{\varphi}\|_1$.

The vector φ is a stationary distribution for the random walk on G. Moreover, the random walk on G is a reversible Markov chain.

PROOF. We need to check that

$$\tilde{\varphi}(u) = \sum_{v\in N(u),\ e\in E(u,v)} \frac{w(e^\dagger)}{\deg v}\tilde{\varphi}(v).$$

Say $u \in A_i$, and label its neighbors $v_1, \ldots, v_{t_u}$ (inside the classes $A_{j_1}, \ldots, A_{j_{t_u}}$). Then the previous equation becomes

$$\tilde{\varphi}(u) = \sum_{v\in N(u),\ e\in E(u,v)} \frac{w(e^\dagger)}{\deg v}\tilde{\varphi}(v) = \sum_{k=1}^{t_u} \frac{m_{j_k i} w_{uv_k}}{d_{j_k}}\tilde{\varphi}(v_k).$$

Substituting the values of $\tilde{\varphi}(u)$ and $\tilde{\varphi}(v_k)$, we get the equation

$$\alpha_i = \sum_{k=1}^{t_u} \frac{m_{j_k i} w_{uv_k}}{d_{j_k}}\alpha_{j_k}.$$

Using Equations (4.1), we get

$$\alpha_i = \sum_{k=1}^{t_u} \frac{w_{uv_k}}{d_i}\alpha_i = \left(\sum_{k=1}^{t_u} w_{uv_k}\right)\frac{1}{d_i}\alpha_i,$$

which is trivially true.

We say a Markov chain is reversible if, for all states u, v, we have

$$\varphi(u)P(u,v) = \varphi(v)P(v,u)$$

where $P(u,v)$ is the probability of walking from u to v. In our case, this equation becomes

$$\alpha_i \frac{w_{uv}}{d_i} = \alpha_j \frac{w_{vu}}{d_j}$$

whenever $u \in A_i, v \in A_j$, which is always satisfied (after dividing both sides by w_{vu}). This proves the reversibility of the chain. □

Proposition 4.2 imposes a total of $\binom{n}{2}$ equations, which may or may not yield a solution. However, we can reduce the number of necessary equations if the graph is connected and has composable linear imbalance.

DEFINITION 4.3. Let G and A_i be as above. Construct an undirected graph $\mathcal{G} = (\mathcal{V}, \mathcal{E})$ with vertices $\mathcal{V} = \{a_1, \ldots, a_n\}$ and with edges $\mathcal{E} = \{\{a_i, a_j\} \mid E(A_i, A_j) \neq \emptyset\}$. We say G has **composable linear imbalance**[3] if for any two neighboring vertices a_i, a_j and for any path in $\mathcal{G}$ (with distinct edges and vertices) $a_i = a_{i_0} \to a_{i_1} \to \cdots \to a_{i_k} = a_j$ from a_i to a_j we have

$$m_{ji} = m_{ji_{k-1}} m_{i_{k-1}i_{k-2}} \cdots m_{i_1 i}.$$

Every undirected graph has composable linear imbalance by defining any partition on its set of vertices. Or, alternatively, a linear imbalance graph is undirected if and only if $m_{ij} = 1$ for all i, j.

LEMMA 4.4. *Let $G = (V, E)$ be a connected graph satisfying the same conditions as in Proposition 4.2. If G has composable linear imbalance, then the set of equations*

$$\frac{m_{ji}}{d_j}\alpha_j = \frac{1}{d_i}\alpha_i \tag{4.2}$$

can be reduced to a set of $n-1$ equations, where n is the number of classes in the vertex partition of G.

PROOF. Recall $V = A_1 \sqcup \cdots \sqcup A_n$, and let $\mathcal{G}$ be the graph associated to this partition. Let $\mathcal{T}$ be any spanning tree of $\mathcal{G}$.

Consider the system of $n-1$ equations $\frac{m_{ji}}{d_j}\alpha_j = \frac{1}{d_i}\alpha_i$ whenever $\{a_i, a_j\}$ is an edge in $\mathcal{T}$. We claim this system is equivalent to the full system. Indeed, for any two vertices $a_i, a_j \in \mathcal{T}$ such that $E(A_i, A_j) \neq \emptyset$, let

$$a_i = a_{i_0} \to a_{i_1} \to \cdots \to a_{i_k} = a_j$$

be a path in $\mathcal{T}$ from a_i to a_j. Using the newly defined system, we get the equation

$$\frac{1}{d_i}\alpha_i = \frac{m_{ji_{k-1}} m_{i_{k-1}i_{k-2}} \cdots m_{i_1 i}}{d_j}\alpha_j,$$

[3]This is also known in the Markov chain literature as the Kolmogorov criterion, and it characterises chain reversibility. We use this term as it provides more meaning to our setting.

which by composability gives us the desired equation $\frac{1}{d_i}\alpha_i = \frac{m_{ji}}{d_j}\alpha_j$. □

EXAMPLE 4.5.

(1) This result can be illustrated by computing the stationary distribution for the random walk over $\Gamma_1^{SS}(\ell;p)$ with $p \equiv 11 \pmod{12}$ (the other possibilities for p are special cases of this). We partition the set of vertices V into three sets, $A_0 = V \setminus \{\mathcal{E}_0, \mathcal{E}_{1728}\}$, $A_1 = \{\mathcal{E}_0\}$, and $A_2 = \{\mathcal{E}_{1728}\}$. This partition gives the graph composable linear imbalance, with $m_{01} = 3$, $m_{02} = 2$, and $m_{12} = 2/3$. The graph $\mathcal{G}$ is a triangle,[4] which imposes three linear equations in three variables, but we get a spanning tree $\mathcal{T}$ by removing any edge. For instance, we get the equations
$$\frac{1}{\ell+1}\alpha_0 = \frac{3}{\ell+1}\alpha_1 \quad \text{and} \quad \frac{1}{\ell+1}\alpha_0 = \frac{2}{\ell+1}\alpha_2$$
which are satisfied by $(\alpha_0, \alpha_1, \alpha_2) = (1, 1/3, 1/2)$.

(2) The same procedure can be applied to the graph $\Gamma_2^{SS}(2;11)$ displayed in Figure 1. We have a disjoint partition in five one-vertex sets, and the multipliers m_{ij} between them are given by ratios of sizes of automorphism groups. By the same procedure as above, the stationary distribution is given by the vector
$$(\alpha_{\mathcal{A}_1}, \alpha_{\mathcal{A}_2}, \alpha_{\mathcal{E}_{1728}^2}, \alpha_{\mathcal{E}_0^2}, \alpha_{\Pi}) = \frac{144}{121} \cdot \left(\frac{1}{12}, \frac{1}{6}, \frac{1}{16}, \frac{1}{36}, \frac{1}{12}\right).$$

COROLLARY 4.6. *Let $G = (V, E)$ be a connected linear imbalance graph with a vertex partition $V = A_1 \sqcup \cdots \sqcup A_n$. Suppose that for each $1 \le i \le n$ there exists a positive real number g_i such that for all i, j, $m_{ij} = \frac{g_i}{g_j}$. Then G has composable linear imbalance, and it has stationary distribution $\varphi = \tilde{\varphi}/\|\tilde{\varphi}\|_1$, where*
$$\tilde{\varphi}(u) = \frac{d_i}{g_i} = \frac{\deg(u)}{g_i} \quad \textit{whenever} \quad u \in A_i.$$

PROOF. The fact that G has composable linear imbalance is trivial from the equalities $m_{ij} = \frac{g_i}{g_j}$. From Lemma 4.4, the equations $\frac{m_{ji}}{d_j}\alpha_j = \frac{1}{d_i}\alpha_i$ are satisfied for all i, j with $E(A_i, A_j) \neq \emptyset$. But these equations correspond to $\frac{g_j}{d_j}\alpha_j = \frac{g_i}{d_i}\alpha_i$ which are trivially satisfied by setting $\alpha_i = d_i/g_i$. □

We discuss now the mixing rate of a graph G satisfying the hypotheses of the last result. Let M_G be the random walk matrix. We define an inner product on $\mathbb{R}^{|V(G)|}$, denoted by $\langle\cdot,\cdot\rangle_\varphi$, by
$$\langle f, g\rangle_\varphi = \sum_{u \in V(G)} f(u)g(u)\varphi(u).$$

LEMMA 4.7 ([**28**, Lemma 12.2]). *The reversible property of the random walk on G implies:*

(1) *The inner product space $(\mathbb{R}^{|V(G)|}, \langle\cdot,\cdot\rangle_\varphi)$ has an orthonormal basis $\{f_j : 1 \le j \le |V(G)|\}$ of real-valued left eigenvectors of M_G, corresponding to real eigenvalues $\{\lambda_j : 1 \le j \le |V(G)|\}$.*

[4] It is actually a tree in many cases, but the computation is the same.

(2) *Given a random walk* $u = u_0 \to \cdots \to u_n \to \cdots$, *for all* $v \in V(G)$ *we have*

$$\frac{\Pr\left[u_n = v\right]}{\varphi(v)} = 1 + \sum_{j=2}^{|V(G)|} f_j(u) f_j(v) \lambda_j^n. \tag{4.3}$$

In particular, if the graph G is connected and aperiodic, then we know

$$1 = \lambda_1 > \lambda_2 \geq \cdots \geq \lambda_{|V(G)|} > -1.$$

Letting $\lambda_\star(G) = \max\{|\lambda| \mid \lambda \text{ is an eigenvalue of } M_G, \lambda \neq 1\}$, we have the following result bounding the mixing rate of the random walk.

PROPOSITION 4.8. *Consider a random walk* $u = u_0 \to \cdots \to u_n \to \cdots$, *and let* $v \in V(G)$ *be any vertex. If* $u \in A_i$ *and* $v \in A_j$, *we have*

$$\left|\Pr\left[u_n = v\right] - \varphi(v)\right| \leq \lambda_\star(G)^n \sqrt{\frac{\deg(v)}{\deg(u)} \frac{g_i}{g_j}}.$$

PROOF. We adapt the proof of [**28**, Theorem 12.4]. Using Eq. (4.3) and the Cauchy–Schwarz inequality we get

$$\begin{aligned} \left| \frac{\Pr\left[u_n = v\right]}{\varphi(v)} - 1 \right| &\leq \sum_{j=2}^{|V(G)|} |f_j(u) f_j(v)| \lambda_\star(G)^n \\ &\leq \lambda_\star(G)^n \left(\sum_{j=2}^{|V(G)|} f_j^2(u) \sum_{j=2}^{|V(G)|} f_j^2(v) \right)^{1/2}. \end{aligned}$$

Let δ_w be the function

$$\delta_w(u) = \begin{cases} 1 & \text{if } w = u, \\ 0 & \text{if } w \neq u. \end{cases}$$

This function can be written in the following way, using the orthonormal basis of functions $\{f_j\}_{j=1}^{|V(G)|}$:

$$\delta_w = \sum_{j=1}^{|V(G)|} \langle \delta_w, f_j \rangle_\varphi f_j = \sum_{j=1}^{|V(G)|} f_j(w) \varphi(w) f_j.$$

From this we obtain

$$\begin{aligned} \varphi(w) = \langle \delta_w, \delta_w \rangle_\varphi &= \left\langle \sum_{j=1}^{|V(G)|} f_j(w) \varphi(w) f_j, \sum_{j=1}^{|V(G)|} f_j(w) \varphi(w) f_j \right\rangle_\varphi \\ &= \varphi(w)^2 \sum_{j=1}^{|V(G)|} f_j^2(w), \end{aligned}$$

which implies $\sum_{j=2}^{|V(G)|} f_j^2(w) < \varphi(w)^{-1}$. Combining this with the first stated inequality we get

$$\left|\Pr\left[u_n = v\right] - \varphi(v)\right| \leq \lambda_\star(G)^n \sqrt{\frac{\varphi(v)}{\varphi(u)}};$$

the result follows on substituting the values of φ obtained in Corollary 4.6. □

Proposition 4.8 is the analog of classical results on random walk mixing in undirected graphs: [**30**, Theorem 5.1] for the general case, [**21**, Theorem 3.3] for regular graphs, and [**31**] and [**18**, Theorem 1] for supersingular isogeny graphs.

4.2. Isogeny graphs as linear imbalance graphs. Our results so far allow us to give the stationary distribution and convergence rate for superspecial isogeny graphs. But we can state a much more general result, and apply the same theory to interesting isogeny subgraphs.

THEOREM 4.9. *Let G be a finite, connected and aperiodic subgraph of $\Gamma_g(\ell;p)$, such that for each edge $[\phi]$ in G, its dual edge $[\phi^\dagger]$ is also in G.*

(1) *The stationary distribution of the random walk in G is given by $\varphi_G = \tilde{\varphi}_G/\|\tilde{\varphi}_G\|_1$,*
$$\tilde{\varphi}_G(\mathcal{A}) = \frac{\deg(\mathcal{A})}{\#RA(\mathcal{A})},$$
where $\deg(\mathcal{A})$ *denotes the number of isogenies in G with domain $\mathcal{A}$.*

(2) *The mixing rate is $\lambda_\star(G)$. More precisely, if $\mathcal{A}_0 \to \cdots \to \mathcal{A}_n \to \cdots$ is a random walk, and $\mathcal{A}$ is any vertex of G, then the convergence to the stationary distribution is given by*

$$|\Pr[\mathcal{A}_n \cong \mathcal{A}] - \varphi_G(\mathcal{A})| \leq \lambda_\star(G)^n \sqrt{\frac{\deg \mathcal{A}}{\deg \mathcal{A}_0}\frac{\#\mathrm{RA}(\mathcal{A}_0)}{\#\mathrm{RA}(\mathcal{A})}}. \tag{4.4}$$

PROOF. For Part (1): Lemma 3.2 tells us that G has linear imbalance, by partitioning its set of vertices according to the reduced automorphism group of each variety. Indeed, for any two neighbouring PPAVs $\mathcal{A}$ and $\mathcal{A}'$ in $\Gamma_g(\ell;p)$, we have
$$\frac{w_{\mathcal{A},\mathcal{A}'}}{w_{\mathcal{A}',\mathcal{A}}} = \frac{\#\mathrm{RA}(\mathcal{A})}{\#\mathrm{RA}(\mathcal{A}')}.$$
We can refine this partition further so that all nodes in a single class have the same degree. This way, all hypotheses of Proposition 4.2 and Corollary 4.6 are satisfied, yielding the stated distribution. Part (2) then follows from Proposition 4.8. □

Theorem 4.9 is true for all superspecial isogeny graphs $\Gamma_g^{SS}(\ell;p)$, as they are connected and non-bipartite [**25**, Corollary 18] and hence aperiodic. In fact, we can always produce a loop if g is even: if $\phi\colon \mathcal{E} \to \mathcal{E}'$ is an elliptic ℓ-isogeny, then the product $(\ell,\ldots,\ell)$-isogeny

$$(\mathcal{E}\times\mathcal{E}')^{g/2} \xrightarrow{\phi\times\phi^\dagger\times\cdots\times\phi\times\phi^\dagger} (\mathcal{E}\times\mathcal{E}')^{g/2} \tag{4.5}$$

is a loop in $\Gamma_g^{SS}(\ell;p)$. If g is odd, we let $\psi_1\colon \mathcal{E}\to\mathcal{E}$, $\psi_2\colon\mathcal{E}\to\mathcal{E}$ be two elliptic curve isogenies of respective degrees ℓ^e and ℓ^f with e and f coprime (this exists, since $\Gamma_1^{SS}(\ell;p)$ is non-bipartite [**25**, Corollary 18] and so aperiodic). Then, by constructing the previous isogeny $\phi\times\phi^\dagger\times\cdots\times\phi\times\phi^\dagger$ in genus $g-1$, we get two isogenies

$$(\phi\times\phi^\dagger)^e\times\cdots\times(\phi\times\phi^\dagger)^e\times\psi_1,$$
$$(\phi\times\phi^\dagger)^f\times\cdots\times(\phi\times\phi^\dagger)^f\times\psi_2,$$

where exponentiation means composition ($\phi\times\phi^\dagger$ is an endomorphism of $\mathcal{E}\times\mathcal{E}'$), representing two cycles of coprime lengths e and f in $\Gamma_g^{SS}(\ell;p)$.

4.3. Bounds on eigenvalues. If we fix g and ℓ, and we have a constant $\lambda = \lambda(g, \ell) < 1$ such that $\lambda_\star(\Gamma_g^{SS}(\ell; p)) \leq \lambda$ for all p, then we get a family of graphs with good expansion properties.[5] Combining this with Equation (4.4), we conclude that the diameter of each graph is $O(\log p)$, a property that also holds for regular expander graphs.

Given a d-regular undirected graph G with $\lambda_\star(G)$ as second largest eigenvalue (in absolute value), we have $d \cdot \lambda_\star(G) \geq 2\sqrt{d-1} - o_n(1)$. Here $o_n(1)$ is a quantity that tends to zero for fixed d when the number of vertices n goes to infinity. If $d \cdot \lambda_\star(G) \leq 2\sqrt{d-1}$, then G is said to be **Ramanujan** [**21**]. Ramanujan graphs have optimal expansion properties.

Isogeny graphs of supersingular elliptic curves are Ramanujan [**35**], and it was hoped that this property would extend to the more general graphs $\Gamma_g^{SS}(\ell; p)$ [**13**, Hypothesis 1]. We have shown $\Gamma_g^{SS}(\ell; p)$ does not fit into the definition of an *expander graph* for $g \geq 2$, due to the presence of non-trivial reduced automorphism groups. However, we may still ask for bounds on $\lambda_\star(\Gamma_g^{SS}(\ell; p))$, as a Ramanujan property of sorts. Now, letting $N_g(\ell)$ be the out-degree of the vertices in $\Gamma_g^{SS}(\ell; p)$, we ask a question: for which g, ℓ and p, if any, does the bound

$$N_g(\ell) \cdot \lambda_\star(\Gamma_g^{SS}(\ell; p)) \leq 2\sqrt{N_g(\ell) - 1}$$

hold?

Jordan and Zaytman [**25**] have given a first counterexample: $\Gamma_2^{SS}(2; 11)$ is not Ramanujan, as the second largest eigenvalue of the adjacency matrix is $7 + \sqrt{3}$, which is larger than $2\sqrt{N_2(2) - 1} = 2\sqrt{15 - 1}$.

We have gathered evidence that the same behaviour also occurs for (at least) all graphs $\Gamma_2^{SS}(2; p)$ for primes $11 \leq p \leq 601$. For all these primes, the superspecial Richelot isogeny graph fails to be Ramanujan, and in fact most values of $\lambda_\star$ (except for a few small primes) are very close to $11.5/15$. Giving a theoretical reason for this behaviour is left as future work.

The eigenvalues and diameters of each graph can be found in Appendix A. In Section 7 we prove that both the subgraph of Jacobians and the subgraph of elliptic products satisfy the hypotheses to have convergence to a stationary distribution, and so our data also includes their eigenvalues and diameters.

We now refine the previously stated conjectures on superspecial graphs.

CONJECTURE 4.10. *For all g and ℓ, there exists a fixed $\lambda = \lambda(g, \ell) < 1$ such that*

$$\lambda_\star(\Gamma_g^{SS}(\ell; p)) \leq \lambda \quad \textit{for every prime } p \geq 5.$$

In the case $g = 2$ and $\ell = 2$, we conjecture that

$$\frac{11}{15} \leq \lambda_\star(\Gamma_2^{SS}(2; p)) \leq \frac{12}{15} \quad \textit{for every prime } p \geq 41.$$

5. The Richelot isogeny graph

From now on, we focus on the case $g = 2$ and $\ell = 2$. Richelot [**36, 37**] gave the first explicit construction for $(2, 2)$-isogenies, so the $(2, 2)$-isogeny graph of principally polarized abelian surfaces (PPASes) is called the *Richelot isogeny graph.*

[5] Note that they should not be called *expander* graphs: this term is reserved for regular undirected graphs.

Let $\mathcal{A}_0$ be a PPAS with full rational 2-torsion. There are 15 rational Lagrangian subgroups $K_1, \ldots, K_{15}$ of $\mathcal{A}_0[2]$, and each is the kernel of a rational $(2,2)$-isogeny

$$\phi_i : \mathcal{A}_0 \to \mathcal{A}_i := \mathcal{A}_0/K_i.$$

This means that every vertex in the $(2,2)$-isogeny graph has out-degree 15. In general, none of the isogenies or codomains are isomorphic. The algorithmic construction of the isogenies and codomains depends fundamentally on whether $\mathcal{A}_0$ is a Jacobian or an elliptic product. We recall the Jacobian case in §B.1, and the elliptic product case in §B.2.

Before going further, we recall the explicit classification of (reduced) automorphism groups of PPASes. In contrast with elliptic curves, where (up to isomorphism) only two curves have nontrivial reduced automorphism group, with PPASes we see much richer structures involving many more vertices in $\Gamma_2(2;p)$.

5.1. Jacobians of genus-2 curves. Bolza [**3**] has shown that there are seven possible reduced automorphism groups for Jacobian surfaces (provided $p > 5$). Figure 2 gives Bolza's taxonomy, defining names ("types") for each of the reduced automorphism groups.

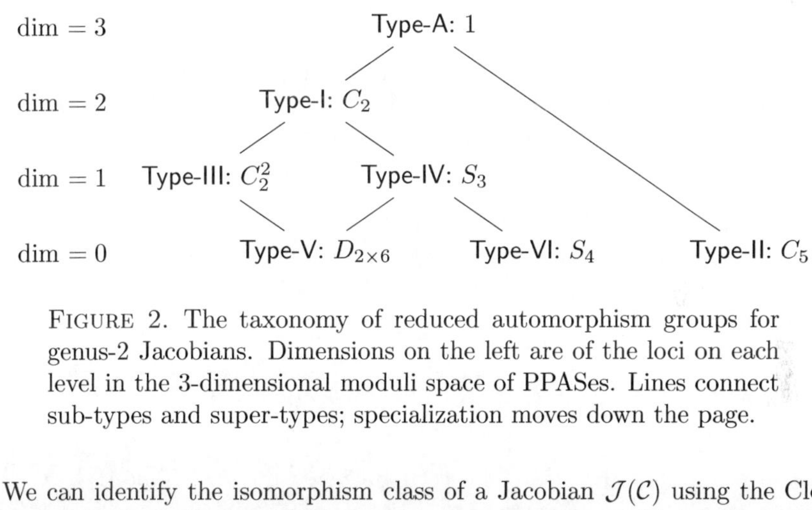

FIGURE 2. The taxonomy of reduced automorphism groups for genus-2 Jacobians. Dimensions on the left are of the loci on each level in the 3-dimensional moduli space of PPASes. Lines connect sub-types and super-types; specialization moves down the page.

We can identify the isomorphism class of a Jacobian $\mathcal{J}(\mathcal{C})$ using the Clebsch invariants A, B, C, D of $\mathcal{C}$, which are homogeneous polynomials of degree 2, 4, 6, and 10 in the coefficients of the sextic defining $\mathcal{C}$. Detailed formulæ appear in §B.3.

5.2. Products of elliptic curves. Elliptic products always have nontrivial reduced automorphism groups, because $\mathrm{RA}(\mathcal{E} \times \mathcal{E}')$ always contains the involution

$$\sigma := [1]_{\mathcal{E}} \times [-1]_{\mathcal{E}'}.$$

Note that σ fixes every Lagrangian subgroup of $(\mathcal{E} \times \mathcal{E}')[2]$ (though this is not true for $(\mathcal{E} \times \mathcal{E}')[\ell]$ if $\ell > 2$), so σ always has an impact on the Richelot isogeny graph.

Proposition 5.1 shows that there are seven possible reduced automorphism groups for elliptic product surfaces (provided $p > 3$), and Figure 3 gives a taxonomy of reduced automorphism groups analogous to that of Figure 2. We identify the isomorphism class of an elliptic product $\mathcal{E} \times \mathcal{E}'$ using the j-invariants $j(\mathcal{E})$ and $j(\mathcal{E}')$ (an unordered pair when $\mathcal{E} \not\cong \mathcal{E}'$, and a single j-invariant when $\mathcal{E} \cong \mathcal{E}'$).

PROPOSITION 5.1. *If $\mathcal{A}$ is an elliptic product surface, then (provided $p > 3$) there are seven possibilities for the isomorphism type of* $\mathrm{RA}(\mathcal{A})$.

(1) *If $\mathcal{A} \cong \mathcal{E} \times \mathcal{E}'$ for some $\mathcal{E} \not\cong \mathcal{E}'$, then one of the following holds:*
- Type-Π: $\{j(\mathcal{E}), j(\mathcal{E}')\} \cap \{0, 1728\} = \emptyset$, *and* $\mathrm{RA}(\mathcal{A}) \cong C_2$.
- Type-Π_0: $j(\mathcal{E}) = 0$ *or* $j(\mathcal{E}') = 0$, *and* $\mathrm{RA}(\mathcal{A}) \cong C_6$.
- Type-Π_{12^3}: $j(\mathcal{E}) = 1728$ *or* $j(\mathcal{E}') = 1728$, *and* $\mathrm{RA}(\mathcal{A}) \cong C_4$.
- Type-$\Pi_{0,12^3}$: $\{j(\mathcal{E}), j(\mathcal{E}')\} = \{0, 1728\}$, *and* $\mathrm{RA}(\mathcal{A}) \cong C_{12}$.

(2) *If $\mathcal{A} \cong \mathcal{E}^2$ for some $\mathcal{E}$, then one of the following holds:*
- Type-Σ: $j(\mathcal{E}) \notin \{0, 1728\}$, *and* $\mathrm{RA}(\mathcal{A}) \cong C_2^2$.
- Type-Σ_0: $j(\mathcal{E}) = 0$, *and* $\mathrm{RA}(\mathcal{A}) \cong C_6 \times S_3$.
- Type-Σ_{12^3}: $j(\mathcal{E}) = 1728$, *and* $\mathrm{RA}(\mathcal{A}) \cong C_2^2 \rtimes C_4$.

PROOF. Recall that if $\mathcal{E}$ is an elliptic curve, then: if $j(\mathcal{E}) = 0$ then $\mathrm{Aut}(\mathcal{E}) = \langle \rho \rangle \cong C_6$; if $j(\mathcal{E}) = 1728$ then $\mathrm{Aut}(\mathcal{E}) = \langle \iota \rangle \cong C_4$; and otherwise $\mathrm{Aut}(\mathcal{E}) = \langle [-1] \rangle \cong C_2$.

For Part (1): if $\mathcal{E} \not\cong \mathcal{E}'$, then $\mathrm{Aut}(\mathcal{E} \times \mathcal{E}') \cong \mathrm{Aut}(\mathcal{E}) \times \mathrm{Aut}(\mathcal{E}')$. If $\mathrm{Aut}(\mathcal{E}) = \langle \alpha \rangle$ and $\mathrm{Aut}(\mathcal{E}') = \langle \beta \rangle$, then $\mathrm{Aut}(\mathcal{E} \times \mathcal{E}') = \langle \alpha \times [1], [1] \times \beta \rangle$. Notice that $\beta^d = [-1]$ for $d = 1, 2$ or 3, so if $j(\mathcal{E}) \notin \{0, 1728\}$, then $\mathrm{RA}(\mathcal{E} \times \mathcal{E}') \cong \mathrm{Aut}(\mathcal{E}')$, which proves the first three cases. For the remaining Type-$\Pi_{0,12^3}$ case, the automorphism $[\rho] \times [\iota]$ has exact order 12, proving $\mathrm{RA}(\mathcal{E} \times \mathcal{E}') \cong C_{12}$.

For Part (2): in this case $\mathrm{Aut}(\mathcal{E}^2)$ certainly contains $\mathrm{Aut}(\mathcal{E})^2$ as a subgroup, but we also have the involution $\tau\colon (P, Q) \mapsto (Q, P)$. The existence of τ makes $\mathrm{Aut}(\mathcal{E}^2)$ non-abelian, because $(\beta \times \gamma) \circ \tau = \tau \circ (\gamma \times \beta)$ for any $\beta, \gamma \in \mathrm{Aut}(\mathcal{E})$. If $\mathrm{Aut}(\mathcal{E}) = \langle \alpha \rangle$, then $\mathrm{Aut}(\mathcal{E}^2) = \langle \alpha \times [1], [1] \times \alpha, \tau \rangle$ is the wreath product $\mathrm{Aut}(\mathcal{E}) \wr \langle \tau \rangle$, i.e., the semidirect product $(\mathrm{Aut}(\mathcal{E}) \times \mathrm{Aut}(\mathcal{E})) \rtimes \langle \tau \rangle$. More explicitly: if $\mathrm{Aut}(\mathcal{E}) = \langle \alpha \rangle$, then

$$\mathrm{Aut}(\mathcal{E}^2) \cong \langle a, b, \tau \mid a^d = b^d = \tau^2 = 1,\ ab = ba,\ a\tau = \tau b \rangle,$$

where $a = \alpha \times [1]$, $b = [1] \times \alpha$, and $d \in \{2, 4, 6\}$ is the order of α. Taking the quotient by $[-1]_{\mathcal{E}^2}$, we identify the reduced automorphism groups using GAP's `IdGroup` [19]. □

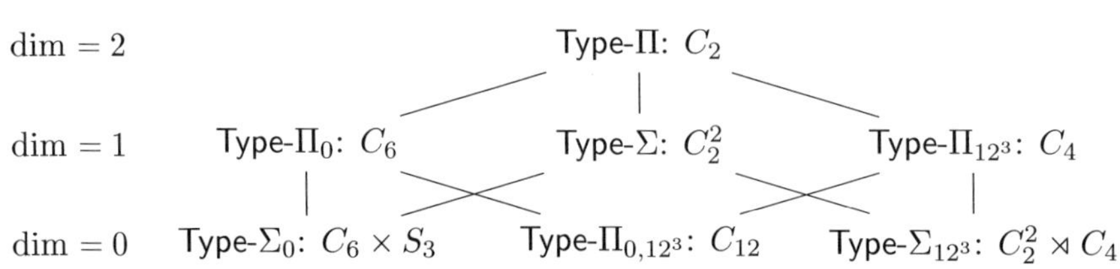

FIGURE 3. The taxonomy of reduced automorphism groups of elliptic products. Dimensions on the left are of the loci on each level in the 3-dimensional moduli space of PPASes. Lines connect sub-types and super-types; specialization moves down the page.

5.3. Implications for isogeny graphs. The vertices in $\Gamma_g(\ell; p)$ corresponding to PPAVs with nontrivial reduced automorphism groups form interesting and inter-related structures. We highlight a few of these facts for $g = 2$ and $\ell = 2$.

Katsura and Takashima observe that if we take a Jacobian vertex $[\mathcal{J}(\mathcal{C})]$ in $\Gamma_2(2; p)$, then the number of elliptic-product neighbours of $[\mathcal{J}(\mathcal{C})]$ is equal to the number of involutions α in $\mathrm{RA}(\mathcal{J}(\mathcal{C}))$ induced by involutions in $\mathrm{Aut}(\mathcal{J}(\mathcal{C}))$ (see [**26**, Proposition 6.1]). In particular: general Type-A vertices and the unique Type-II vertex have *no* elliptic product neighbours; Type-I and Type-IV vertices, and the unique Type-VI vertex, have *one* elliptic product neighbour; and the Type-III vertices and the unique Type-V vertex have *two* elliptic-square neighbours. By explicit computation of Richelot isogenies we can (slightly) extend Katsura and Takashima's results to give the complete description of weighted edges with codomain types for each of the vertex types in Table 1. The inter-relation of reduced automorphism groups and neighbourhoods of vertices and edges in the Richelot isogeny graph is further investigated (and illustrated) in [**15**].

TABLE 1. Number of edges, weights, and types of neighbours for vertices in $\Gamma_2(2;p)$ by reduced automorphism type. Observe that the edge numbers multiplied by their weights always sum to 15. Neighbour types may change under specialization (or for particular values of p), acquiring reduced automorphisms. See [**15**] for details.

Vertex	#Edges	w	Neighbour
Type-A	15	1	Type-A
Type-I	1	1	Type-Π
	6	1	Type-I
	4	2	Type-A
Type-II	3	5	Type-A
Type-III	1	1	(loop)
	2	1	Type-Σ
	4	2	Type-I
	1	4	Type-A
Type-IV	1	3	Type-Π
	3	3	Type-I
	3	1	Type-IV
Type-V	1	3	(loop)
	1	1	Type-Σ_0
	1	3	Type-Σ
	1	6	Type-I
	1	2	Type-IV
Type-VI	1	1	(loop)
	1	6	Type-Σ
	2	4	Type-IV

Vertex	#Edges	w	Neighbour
Type-Π	9	1	Type-Π
	6	1	Type-I
Type-Π_0	3	3	Type-Π
	2	3	Type-I
Type-Π_{12^3}	3	1	Type-Π_{12^3}
	3	2	Type-Π
	3	2	Type-I
Type-$\Pi_{0,12^3}$	1	3	Type-Π_{12^3}
	1	6	Type-Π
	1	6	Type-I
Type-Σ	1	1	(loop)
	3	2	Type-Π
	3	1	Type-Σ
	1	2	Type-I
	3	1	Type-III
Type-Σ_0	1	3	(loop)
	1	9	Type-Σ
	1	3	Type-V
Type-Σ_{12^3}	1	3	(loop)
	1	4	Type-Σ
	1	4	Type-Π_{12^3}
	1	4	Type-III

REMARK 5.2. Each Type-IV vertex has a triple edge to an elliptic-product neighbour. In fact, the factors of the product are always 3-isogenous (cf. [**20**, §3]). The unique Type-VI vertex is a specialization of Type-IV, and in this case the Type-Π neighbour specializes to the square of an elliptic curve with j-invariant 8000 (which has an endomorphism of degree 3). The unique Type-V vertex is also a specialization of Type-IV, and in this case the Type-Π neighbour specializes to the square of an elliptic curve of j-invariant 54000 (which as an endomorphism of degree

3); one of the Type-IV neighbours degenerates to the square of an elliptic-curve with j-invariant 0, while the other two merge, yielding a weight-2 edge; and one of the Type-I neighbours specializes to the Type-V vertex, yielding a loop, while the other two merge, yielding a weight-6 edge.

REMARK 5.3. Every Type-III vertex (and the unique Type-V vertex) has *two* elliptic-square neighbours: these are the squares of a pair of 2-isogenous elliptic curves [**20**, §4]. In this way, Type-III vertices in $\Gamma_2(2;p)$ correspond to undirected edges (i.e., edges modulo dualization of isogenies) in $\Gamma_1(2;p)$.

Ibukiyama, Katsura, and Oort have computed the precise number of superspecial genus-2 Jacobians (up to isomorphism) of each reduced automorphism type [**23**, Theorem 3.3]. We reproduce their results for $p > 5$ in Table 2, completing them with the number of superspecial elliptic products of each automorphism type (which can be easily derived from the well-known formula for the number of supersingular elliptic curves over $\mathbb{F}_{p^2}$).

TABLE 2. The number of vertices in $\Gamma_2^{SS}(\ell;p)$ of each reduced automorphism type. Here $\epsilon_{1,p} = 1$ if $p \equiv 3 \pmod 4$, 0 otherwise; $\epsilon_{2,p} = 1$ if $p \equiv 5, 7 \pmod 8$, 0 otherwise; $\epsilon_{3,p} = 1$ if $p \equiv 2 \pmod 3$, 0 otherwise; $\epsilon_{5,p} = 1$ if $p \equiv 4 \pmod 5$, 0 otherwise; and $N_p = (p-1)/12 - \epsilon_{1,p}/2 - \epsilon_{3,p}/3$ is the number of supersingular elliptic curves over $\mathbb{F}_{p^2}$ with reduced automorphism group C_2.

Type	Vertices in $\Gamma_2^{SS}(2;p)$	Type	Vertices in $\Gamma_2^{SS}(2;p)$
Type-I	$\frac{1}{48}(p-1)(p-17)$ $+\frac{1}{4}\epsilon_{1,p} + \epsilon_{2,p} + \epsilon_{3,p}$	Type-Π	$\frac{1}{2}N_p(N_p-1)$
		Type-Π_0	$\epsilon_{3,p}N_p$
Type-II	$\epsilon_{5,p}$	Type-Π_{12^3}	$\epsilon_{1,p}N_p$
Type-III	$\frac{3}{2}N_p + \frac{1}{2}\epsilon_{1,p} - \frac{1}{2}\epsilon_{2,p} - \frac{1}{2}\epsilon_{3,p}$	Type-$\Pi_{0,12^3}$	$\epsilon_{1,p} \cdot \epsilon_{3,p}$
Type-IV	$2N_p + \epsilon_{1,p} - \epsilon_{2,p}$	Type-Σ	N_p
Type-V	$\epsilon_{3,p}$	Type-Σ_0	$\epsilon_{3,p}$
Type-VI	$\epsilon_{2,p}$	Type-Σ_{12^3}	$\epsilon_{1,p}$
Type-A	$\frac{1}{2880}(p-1)(p^2-35p+346) - \frac{1}{16}\epsilon_{1,p} - \frac{1}{4}\epsilon_{2,p} - \frac{2}{9}\epsilon_{3,p} - \frac{1}{5}\epsilon_{5,p}$		

6. Random walks in the superspecial Richelot isogeny graph

We now specialize the results of §4 to the case $g = 2$, $\ell = 2$, and consider some cryptographic applications.

6.1. Random walks. Given an isogeny graph G satisfying the hypotheses of Theorem 4.9, we let

$$K_G = \max_{\mathcal{A},\mathcal{A}_0} \sqrt{\frac{\deg_G \mathcal{A}}{\deg_G \mathcal{A}_0}\,\frac{\#\mathrm{RA}(\mathcal{A}_0)}{\#\mathrm{RA}(\mathcal{A})}}.$$

If we put $G = \Gamma_2^{SS}(2;p)$ and consider the reduced automorphism groups in Proposition 5.1, then $K_G = 6$. Together with Conjecture 4.10, this gives us precise constants for the convergence of the random walk distribution on the Richelot isogeny graph. We will say that a vector $\psi \in \mathbb{R}^{|V(G)|}$ approximates the stationary distribution φ of the graph G with an error of $\varepsilon > 0$ if for each vertex $u \in V(G)$,

$|\psi(u) - \varphi(u)| \leq \varepsilon$. A random walk of length n approximates the stationary distribution with error ε if the distribution given by the walk at step n does so.

THEOREM 6.1. *Assume Conjecture* 4.10 *for* $g = 2$ *and* $\ell = 2$: *that is, assume that* $\lambda_\star(\Gamma_2^{SS}(2;p)) \leq \frac{12}{15}$ *for all* $p \geq 41$. *A random walk of length* $n \geq 4.5m \log p + 9$ *approximates the stationary distribution on* $\Gamma_2^{SS}(2;p)$ *with an error of* $\frac{1}{p^m}$. *In particular, a random walk of length*

$$n \geq 18 \log p + 9$$

approximates the stationary distribution with an error of $\frac{1}{p^4}$.

PROOF. Set $G = \Gamma_2^{SS}(2;p)$. Given a random walk $\mathcal{A}_0 \to \cdots \to \mathcal{A}_n \to \cdots$ and a vertex $\mathcal{A}$, then for all n we have

$$|\Pr[\mathcal{A}_n \cong \mathcal{A}] - \varphi_G(\mathcal{A})| \leq \lambda_\star(G)^n \sqrt{\frac{\deg_G \mathcal{A}}{\deg_G \mathcal{A}_0} \frac{\#\mathrm{RA}(\mathcal{A}_0)}{\#\mathrm{RA}(\mathcal{A})}} \leq 6\lambda_\star(G)^n.$$

The inequality $6\lambda_\star(G)^n \leq \frac{1}{p^m}$ is satisfied as long as

$$n \geq \frac{m \log p + \log 6}{\log(\lambda_\star(G)^{-1})}.$$

Since $\log 6/\log(15/12) \leq 9$ and $1/\log(15/12) \leq 4.5$, if $n \geq 4.5m \log p + 9$ then the above inequalities are satisfied. The particular case of $m = 4$ follows. □

6.2. Distributions of subgraphs. If we perform a random walk on $\Gamma_2^{SS}(2;p)$, we will encounter a certain number of products of elliptic curves along the way. We can try to predict the ratio of elliptic products to visited nodes: a first guess could be that this ratio matches the proportion of such nodes in the entire graph, which is asymptotic to $\frac{10}{p}$ (see [**9**, Proposition 2]).

However, this is not the empirical proportion that we observe in our experiment, which consists in performing 10,000 random walk steps in $\Gamma_2^{SS}(2;p)$ and counting the number N of elliptic products encountered in our path. The ratio $N/10,000$ of elliptic products to visited nodes is closer to $\frac{5}{p}$, as seen in Table 3.

TABLE 3. Number of elliptic products encountered in a 10,000-step random walk for several primes. The third row shows the proportion scaled relative to each prime.

p	101	307	503	701	907	1103
N	415	201	130	64	50	44
Ratio	$4.1915/p$	$6.1707/p$	$6.539/p$	$4.4864/p$	$4.535/p$	$4.8532/p$

Theorem 4.9, in combination with the classification of reduced automorphism groups in Proposition 5.1, gives us the true proportion of elliptic product nodes in random walks. We have $\frac{p^3}{2880} + O(p^2)$ Jacobians with trivial reduced automorphism group (this is the picture for "almost all" nodes in the graph: only $O(p^2)$ have nontrivial reduced automorphisms), and there are $\frac{p^2}{288} + O(p)$ elliptic products. However, all but $O(p)$ of those products have a reduced automorphism group of order 2, confirming that the (asymptotic) expected proportion of elliptic products in a random walk is equal to $\frac{1}{2} \times \frac{10}{p} = \frac{5}{p}$. Similarly, we could compute proportions for each abelian surface type given in Section 5.

If we combine this with the conjectured upper bound for $\lambda_\star(\Gamma_2^{SS}(2;p))$, then we can give the interpretation that elliptic products are evenly distributed in the graph, in the sense that any node is within very few steps of an elliptic product (much less than diametral distance).

6.3. The superspecial isogeny problem in genus 2 and beyond. The general problem of constructing an isogeny between two superspecial g-dimensional PPAVs $\mathcal{A}_g$ and $\mathcal{A}'_g$ over $\mathbb{F}_{p^2}$ was studied in [**13**]. The algorithm proceeds by computing isogenies $\phi : \mathcal{A}_g \to \mathcal{A}_{g-1} \times \mathcal{E}$ and $\phi' : \mathcal{A}'_g \to \mathcal{A}'_{g-1} \times \mathcal{E}'$ where $\mathcal{A}_{g-1}$ and $\mathcal{A}'_{g-1}$ have dimension $g-1$ and $\mathcal{E}$ and $\mathcal{E}'$ are elliptic curves, before computing an elliptic isogeny $\mathcal{E} \to \mathcal{E}'$ and (recursively) computing an isogeny $\mathcal{A}_{g-1} \to \mathcal{A}_{g-1}$, then combining the results to produce an isogeny $\mathcal{A}_g \to \mathcal{A}'_g$. The key step is computing the isogenies ϕ and ϕ' to product PPAVs. The expected complexity of this step is heuristic, and assumes that the isogeny graph of superspecial PPAVs has good expansion properties to ensure that $O(p)$ isogeny walks of length $O(\log p)$ will result in a walk to a product variety with probability $O(1)$. Of course, in practice one cannot simply take walks of length $O(\log p)$: we need a proper bound on the length of these walks (essentially, we need the constant hidden by the big O).

Our results show if we admit Conjecture 4.10, then the expected complexity of the algorithm in [**13**] is rigorous for $g = 2$, and we can bound the required walk lengths using the claimed eigenvalue bounds as in Theorem 6.1. In particular, for $g = 2$ and $\ell = 2$, it suffices to use walks of length $26 \log_2(p) + 8$.

6.4. Richelot isogeny hash functions. Recall the Richelot-isogeny hash function of [**9**], which is based on walks in $\Gamma_2^{SS}(2;p)$. A binary representation of the data to be hashed is broken into a series of three-bit chunks; each of the eight possible three-bit values corresponds to the choice of a step in $\Gamma_2^{SS}(2;p)$ such that the composition of the prior step with the current step is a $(4,4)$-isogeny. The hash value is (derived from) the invariants of the final vertex in the walk.

Our results show that finding an input m driving a walk into the induced subgraph $\Gamma_2^{SS}(2;p)^E$ on the elliptic product vertices would immediately yield collisions in the hash function. Indeed, looking at Table 1, we see that every vertex in $\Gamma_2^{SS}(2;p)^E$ has either outgoing edges with multiplicity greater than 1, or a Type-I neighbour with outgoing edges with multiplicity greater than 1. This means that there are multiple kernels, and thus multiple 3-bit input chunks, that produce steps to the same neighbour; in this way, given a walk to $\Gamma_2^{SS}(2;p)^E$, with at most two further steps we can construct explicit hash collisions.

Since the forward steps in these walks are restricted to a subset of eight of the fourteen possible onward edges at each vertex, the results in §4.3 do not apply directly here. Still, they give us reason to hope that these restricted random walks will approximate the uniform distribution on $\Gamma_2^{SS}(2;p)$ very quickly. If adversaries can compute walks into $\Gamma_2^{SS}(2;p)^E$ after an expected $O(p)$ steps, as they can with unrestricted walks, then they can use walks into $\Gamma_2^{SS}(2;p)^E$ to construct hash collisions in an expected $\widetilde{O}(p)$ operations, which is exponentially fewer than the $O(p^{3/2})$ required by generic attacks.

6.5. Genus 2 SIDH analogues. Our results also have constructive cryptographic applications. For example, consider the genus-2 SIDH analogue proposed by Flynn and Ti [**16**], a postquantum key exchange algorithm based on commuting random walks in $\Gamma_2^{SS}(2;p)$ and $\Gamma_2^{SS}(3;p)$. The walks involved are very short—on the order of $\frac{1}{2}\log_2 p$ steps each—and much shorter than the bound of Theorem 6.1. Our results therefore imply that this genus-2 SIDH analogue is overwhelmingly unlikely to encounter $\Gamma_2^{SS}(\ell;p)^E$, provided the base vertex is chosen sensibly.

7. Connectivity and diameters

We mentioned in §4 that Theorem 4.9 can be applied to study distributions in interesting isogeny subgraphs of the superspecial isogeny graph. Let us then distinguish three subgraphs of $\Gamma_g^{SS}(\ell;p)$, each taken to be the induced subgraph defined by its set of vertices:

- $\Gamma_g^{SS}(\ell;p)^J$, the subgraph of Jacobians;
- $\Gamma_g^{SS}(\ell;p)^P$, the subgraph of reducible PPAVs (product varieties); and
- $\Gamma_g^{SS}(\ell;p)^E$, the subgraph of products of elliptic curves.

(Observe that $\Gamma_2^{SS}(\ell;p)^P = \Gamma_2^{SS}(\ell;p)^E$). Understanding the connectivity of such subgraphs can be useful both when analysing the algorithms that work with them, and when studying the distribution of vertices in the full supersingular graph.

PROPOSITION 7.1. *The graphs $\Gamma_g^{SS}(\ell;p)^P$ and $\Gamma_g^{SS}(\ell;p)^E$ are connected and aperiodic for all g, ℓ, and p. In particular, both graphs satisfy the hypotheses of Theorem* 4.9.

PROOF. It is enough to see that $\Gamma_g^{SS}(\ell;p)^E$ is connected and aperiodic, since it is a subgraph of $\Gamma_g^{SS}(\ell;p)^P$ and given a product variety we can find a product isogeny to an elliptic product by the connectivity of $\Gamma_g^{SS}(\ell;p)$. We obtain connectivity from the fact that $\Gamma_g^{SS}(\ell;p)^E$ has a spanning subgraph which is a quotient of the tensor product of g copies of the supersingular isogeny graph $\Gamma_1^{SS}(\ell;p)$. Since $\Gamma_1^{SS}(\ell;p)$ is aperiodic, it contains an odd cycle and so $(\Gamma_1^{SS}(\ell;p))^{\otimes g}$ is connected [**45**]. We have already proved aperiodicity, since in §4.2 we constructed loops and paths of coprime lengths in $\Gamma_g^{SS}(\ell;p)^E$. □

Proposition 7.1 generalizes immediately to any connected component of the general graph $\Gamma_g(\ell;p)$ that contains elliptic products.

Conjecture 2 of [**9**] proposes that the subgraph of the superspecial Richelot isogeny graph supported on the Jacobians is connected; Theorem 7.2 confirms and proves this conjecture. (We should be able to give a similar statement for the Jacobian subgraph even without the superspecial condition, but the technique that we use only allows us to prove it for the case $g = 2$, $\ell = 2$.)

THEOREM 7.2. *The graph of Jacobians $\Gamma_2^{SS}(2;p)^J$ is connected and aperiodic. In particular, it satisfies the hypotheses of Theorem* 4.9.

PROOF. To see $\Gamma_2^{SS}(2;p)^J$ is connected, it is enough to check that the subgraph containing all Type-I Jacobians is connected. Indeed, any two Jacobians J_1 and J_2 are connected by a path in $\Gamma_2^{SS}(2;p)$, and we only need to ensure that subpaths between Type-I Jacobians can be modified to avoid elliptic products. This is always possible by Lemma 7.3 below.

The aperiodicity for primes $p \geq 13$ comes from the fact that there are always Type-III Jacobians, which always have a $(2,2)$-endomorphism. One checks easily that $\Gamma_2^{SS}(2;p)^J$ has at least one loop when p is 7 or 11. Indeed, for $p = 7$ the unique Type-VI vertex has a $(2,2)$-endomorphism ϕ with weight $w([\phi]) = 9$, while for $p = 7$ the unique Type-V vertex has a $(2,2)$-endomorphism ψ with $w([\psi]) = 3$. □

LEMMA 7.3. *Given a path* $[J_0] \to [\mathcal{E} \times \mathcal{E}'] \to [\mathcal{A}]$ *in* $\Gamma_2(2;p)$, *where* J_0 *is a Jacobian,* $\mathcal{E} \times \mathcal{E}'$ *is an elliptic product, and* $\mathcal{A}$ *is any PPAS, there exists either:*

(1) *A length-2 path*
$$[J_0] \to [J_1] \to [\mathcal{A}],$$
where J_1 *is a Jacobian, if the original path represents a* $(4,2,2)$*-isogeny, or*

(2) *A length-4 path*
$$[J_0] \to [J_1] \to [J_2] \to [J_3] \to [\mathcal{A}],$$
where each J_i *is a Jacobian, if the original walk represents a* $(4,4)$*-isogeny.*

PROOF.

CASE 1. The original path represents a $(4,2,2)$-isogeny, ϕ. Up to isomorphism, ϕ factors into a composition of two $(2,2)$-isogenies in 3 ways:

- $\phi : J_0 \to \mathcal{E} \times \mathcal{E}' \to \mathcal{A}$,
- $\phi_1 : J_0 \to \mathcal{A}_1 \to \mathcal{A}$, and
- $\phi_2 : J_0 \to \mathcal{A}_2 \to \mathcal{A}$.

The isogenies $J_0 \to \mathcal{A}_i$ each have one nontrivial kernel point in common with $J_0 \to \mathcal{E} \times \mathcal{E}'$. We know that $[J_0]$ has at most two elliptic-product neighbours (see Table 1). Recall the language of *quadratic splittings* detailed in Appendix B.1: the Lagrangian subgroups of $J_0[2]$ correspond to factorizations of $f(x)$ into three coprime quadratics, where $C_0 : y^2 = f(x)$ is a sextic model for the genus-2 curve generating J_0, and the codomain of the corresponding $(2,2)$-isogeny is an elliptic product precisely when the three quadratics are linearly dependent. After a coordinate transformation, we can suppose that $J_0 \to \mathcal{E} \times \mathcal{E}'$ is a Richelot isogeny with $\ker(J_0 \to \mathcal{E} \times \mathcal{E}') = \{x^2 - a^2, x^2 - b^2, x^2 - c^2\}$. Relabelling (a,b,c) if necessary, we can assume the point common to $\ker(J_0 \to \mathcal{E} \times \mathcal{E}')$, $\ker(J_0 \to \mathcal{A}_1)$, and $\ker(J_0 \to \mathcal{A}_2)$ corresponds to $x^2 - a^2$, and thus
$$\ker(J_0 \to \mathcal{A}_1) = \{x^2 - a^2, x^2 - (b+c)x + bc, x^2 + (b+c)x + bc\}$$
and
$$\ker(J_0 \to \mathcal{A}_2) = \{x^2 - a^2, x^2 - (b-c)x - bc, x^2 + (b-c)x - bc\}.$$
It is easy to check that the determinants of these two triples cannot both vanish unless the original curve is singular.

CASE 2. The original walk represents a $(4,4)$-isogeny, ϕ. We can always choose a neighbour $[J_2] \neq [J_0]$ of $[\mathcal{E} \times \mathcal{E}']$ such that $J_0 \to \mathcal{E} \times \mathcal{E}' \to J_2$ and $J_2 \to \mathcal{E} \times \mathcal{E}' \to \mathcal{A}$ both represent $(4,2,2)$-isogenies. Now apply Case 1 to each of these, eliminating $\mathcal{E} \times \mathcal{E}'$ from the middle of each length-2 path, and compose the results. □

REMARK 7.4. When J_0 is Type-III or Type-V in the $(4,2,2)$-isogeny case, it is possible that we obtain $[J_0] = [J_1]$, so we actually simplify to a length-1 path $[J_0] \to [\mathcal{A}]$. Further, in the $(4,4)$-isogeny case, we can even have $[J_0] = [J_2]$, and then we can simplify the original length-2 path (and the modified length-4 one) to the length-1 path $[J_0] \to [\mathcal{A}]$.

COROLLARY 7.5. *The diameters of* $\Gamma_2^{SS}(2;p)$ *and* $\Gamma_2^{SS}(2;p)^J$ *satisfy*
$$\operatorname{diam}(\Gamma_2^{SS}(2;p)) - 2 \leq \operatorname{diam}(\Gamma_2^{SS}(2;p)^J) \leq 2\operatorname{diam}(\Gamma_2^{SS}(2;p)).$$

PROOF. The first inequality comes from the fact that every elliptic product has a Richelot isogeny to a Jacobian. For the second one, apply Lemma 7.3 repeatedly to bound the distance between any two nodes in $\Gamma_2^{SS}(2;p)^J$. □

The lower bound of Corollary 7.5 is tight, as seen for $\Gamma_2^{SS}(2;521)$. Our experimental results suggest that the upper bound has some room for improvement.

8. An example: the superspecial Richelot graph for $p = 47$

We now exemplify our results on the Richelot isogeny graph for $p = 47$. The graph $\Gamma_2^{SS}(2;47)$ has an appropriate size to observe interesting behaviour. In particular, since $p \equiv 11 \bmod 12$ and $p \equiv 2 \bmod 5$, all of the vertex types described in Section 5 except Type-II appear. Table 4 lists the exact counts for each vertex type.

TABLE 4. Vertex counts for each type in the graph $\Gamma_2^{SS}(2;47)$. Here A_T denotes the subset of vertices of type T, while g_T is the corresponding value of g_i in Corollary 4.6.

Type T	A	I	II	III	IV	V	VI	Σ	Π	Π_{12^3}	Π_0	Σ_{12^3}	$\Pi_{0,12^3}$	Σ_0
$\#A_T$	14	31	0	4	6	1	1	3	3	3	3	1	1	1
g_T	1	2	–	4	6	12	23	4	2	4	6	16	12	36

Let us compute the stationary distribution for the full graph $\Gamma_2^{SS}(2;47)$. First, we partition the vertex set according to each type: $A_{\mathsf{Type-A}}$ contains the 14 Type-A vertices, $A_{\mathsf{Type-I}}$ the 31 Type-I vertices, and so on. In the notation of Corollary 4.6, if $A_i = A_T$ for a type T, then the values of g_i are the g_T in Table 4. (In general, we would also have $g_{II} = 1/5$.) Since all vertices have 15 Lagrangian subgroups in their two-torsion, Corollary 4.6 says that (after normalization) the stationary distribution is given by

$$\tilde{\varphi}(\mathcal{A}) = \frac{1}{g_T} \qquad \text{whenever } \mathcal{A} \text{ is of type } T.$$

We can observe this partially in Figure 4. The picture lacks the edge weights, which we have omitted for the sake of clarity. Nevertheless, we see clearly that vertices with larger reduced automorphism groups are more isolated, because lots of isogenies are identified through automorphisms. This makes these vertices harder to reach in a random walk, so they have a smaller value in the stationary distribution.

We may also compute the stationary distributions of $\Gamma_2^{SS}(2;47)^J$ and $\Gamma_2^{SS}(2;47)^E$. Recall from Table 1 that the degrees in these graphs are no longer regular: for example, a Type-A varieties have 15 isogenies to other Jacobians, while Type-I varieties have 14 isogenies to other Jacobians and a single isogeny to a product of elliptic curves. The stationary probability for a vertex $\mathcal{A}$ of type T is

$$\tilde{\varphi}(\mathcal{A}) = \frac{\deg \mathcal{A}}{g_T} \qquad \text{whenever } \mathcal{A} \text{ is of type } T,$$

where $\deg \mathcal{A}$ is now the number of isogenies from $\mathcal{A}$ to vertices *in the same graph*, and g_T is defined as above.

In this setting, the vertices which are not of Type-A in $\Gamma_2^{SS}(2;47)^J$ get more isolated, because they all have out-degree less than 15. On the other hand, the stationary distribution is uniformized slightly in $\Gamma_2^{SS}(47;p)^E$, because the vertices with

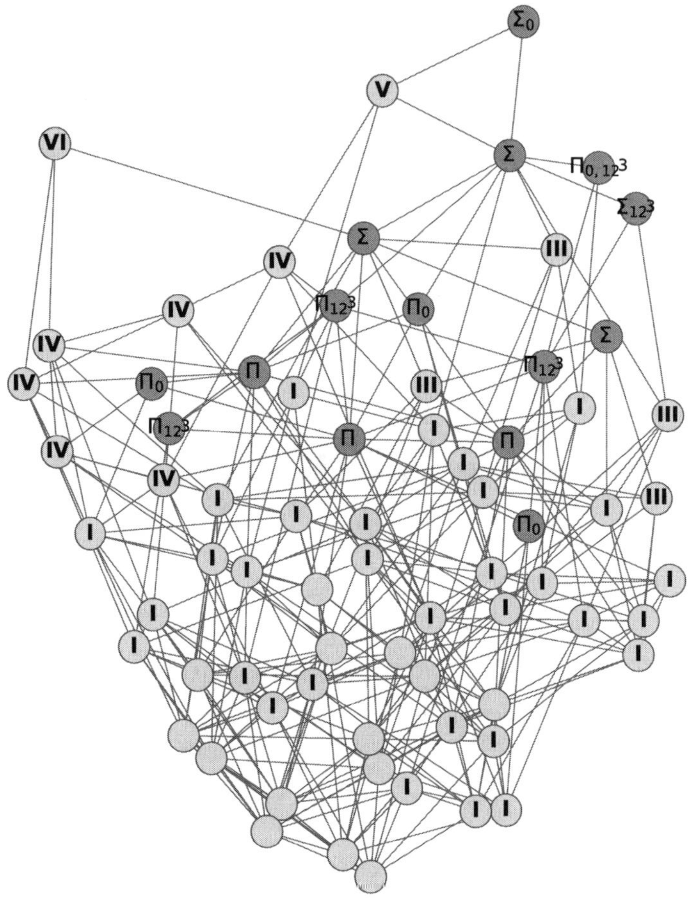

FIGURE 4. The superspecial Richelot isogeny graph for $p = 47$. Vertices are labeled with their types; unlabeled vertices are Type-A, with trivial reduced automorphism group. Loops are omitted.

larger automorphism groups have one, two or three fewer isogenies to Jacobians. This can be seen in Figure 5.

These phenomena generalize immediately to $\Gamma_2^{SS}(\ell; p)$ for all primes $\ell \neq p$, due to the generality achieved in Theorem 4.9.

Appendix A. Experimental diameters and $\lambda_\star$ for $\Gamma_2^{SS}(2; p)$

The following table consists of experimental data computed for the graphs $G = \Gamma_2^{SS}(2; p)$, $J = \Gamma_2^{SS}(2; p)^J$ and $E = \Gamma_2^{SS}(2; p)^E$. The computed values are the diameters $d(G)$, $d(J)$ and $d(E)$, and the (scaled) second-largest eigenvalues of each graph. In particular, the second eigenvalues of $\Gamma_2^{SS}(2; p)$ support Conjecture 4.10. We use the notation $\tilde{\lambda}_\star = 15\lambda_\star$.

p	$d(G)$	$d(J)$	$d(E)$	$\tilde{\lambda}_\star(G)$	$\tilde{\lambda}_\star(J)$	$\tilde{\lambda}_\star(E)$
17	3	3	2	10.671	9.203	3.000

19	3	3	2	11.072	10.016	1.833
23	3	4	2	10.241	8.993	4.102
29	4	4	4	10.472	9.522	6.460
31	3	4	2	11.183	10.516	5.748
37	4	4	2	10.797	10.025	5.372
41	5	5	6	11.436	10.098	7.837
43	4	4	2	11.153	10.650	5.495
47	4	5	4	11.131	10.526	7.580
53	5	5	4	11.060	10.769	6.145
59	5	5	5	11.475	10.447	7.927
61	5	6	3	11.451	11.037	6.978
67	5	4	4	11.563	11.210	7.537
71	5	5	4	11.341	10.885	7.183
73	5	5	4	11.577	11.129	7.575
79	5	5	3	11.216	10.774	6.576
83	6	6	5	11.262	11.023	8.241
89	6	6	6	11.307	10.681	8.418
97	5	5	6	11.494	11.089	7.973
101	6	6	7	11.192	10.817	8.474
103	6	6	5	11.217	10.980	8.644
107	6	6	6	11.379	11.203	7.344
109	6	6	4	11.168	10.985	6.549
113	6	6	6	11.386	11.156	7.593
127	6	6	4	11.612	11.383	7.522
131	7	6	8	11.525	11.373	8.179
137	6	6	6	11.648	11.440	7.193
139	6	6	5	11.528	11.424	7.682
149	7	7	8	11.534	11.407	8.131
151	6	6	4	11.387	11.285	7.338
157	6	7	6	11.508	11.291	8.489
163	6	6	6	11.638	11.376	8.012
167	7	7	6	11.494	11.359	8.116
173	7	7	7	11.631	11.408	8.077
179	7	7	8	11.586	11.459	8.075
181	6	7	6	11.347	11.267	8.270
191	7	7	7	11.461	11.348	8.307
193	6	6	5	11.537	11.431	7.754
197	7	7	8	11.295	11.207	7.789
199	7	7	6	11.361	11.261	8.041
211	7	7	6	11.610	11.522	7.933
223	7	7	7	11.484	11.339	8.334
227	7	7	7	11.480	11.397	8.110
229	7	7	6	11.605	11.486	8.076
233	7	7	6	11.523	11.420	7.672
239	8	7	8	11.581	11.431	8.246
241	7	7	6	11.507	11.342	8.233
251	8	7	8	11.568	11.371	8.585
257	8	7	8	11.636	11.462	8.315

263	7	7	7	11.539	11.433	7.640
269	8	7	8	11.448	11.337	8.405
271	7	7	6	11.537	11.482	8.037
277	7	8	6	11.530	11.396	7.935
281	7	7	8	11.479	11.366	8.297
283	7	7	7	11.582	11.504	8.272
293	8	8	8	11.582	11.430	8.390
307	7	7	7	11.614	11.535	8.244
311	8	8	7	11.507	11.383	8.411
313	8	7	7	11.645	11.480	8.439
317	8	8	7	11.543	11.495	7.922
331	7	7	7	11.505	11.450	8.018
337	7	7	7	11.613	11.542	8.005
347	8	8	8	11.520	11.457	8.185
349	8	8	8	11.465	11.407	8.485
353	8	8	8	11.561	11.490	8.143
359	8	8	8	11.556	11.500	8.311
367	8	8	7	11.553	11.463	8.352
373	8	8	7	11.475	11.411	8.259
379	8	7	7	11.474	11.408	8.202
383	8	8	7	11.548	11.492	8.351
389	8	8	9	11.582	11.544	8.280
397	8	8	7	11.593	11.523	8.368
401	8	8	8	11.558	11.492	8.315
409	8	8	8	11.626	11.575	8.354
419	9	8	10	11.555	11.472	8.552
421	8	8	6	11.614	11.569	8.015
431	8	8	8	11.585	11.512	8.276
433	8	8	9	11.615	11.532	8.516
439	8	8	8	11.509	11.459	8.389
443	8	8	9	11.501	11.458	8.287
449	8	8	8	11.546	11.499	8.178
457	8	8	8	11.539	11.460	8.429
461	9	8	9	11.588	11.513	8.452
463	8	8	9	11.514	11.458	8.394
467	8	8	8	11.608	11.561	8.332
479	8	8	9	11.579	11.524	8.202
487	8	8	8	11.546	11.512	8.320
491	8	8	8	11.606	11.529	8.217
499	8	8	8	11.492	11.457	8.168
503	9	8	8	11.606	11.529	8.209
509	9	9	9	11.607	11.542	8.431
521	10	8	10	11.618	11.566	8.295
523	8	8	8	11.596	11.545	8.338
541	8	8	8	11.518	11.469	8.255
547	8	8	8	11.591	11.555	8.282
557	9	8	10	11.528	11.490	8.277
563	9	9	8	11.542	11.486	8.360

569	9	8	10	11.573	11.525	8.366
571	8	8	8	11.605	11.560	8.262
577	8	8	8	11.612	11.490	8.438
587	9	9	9	11.628	11.565	8.362
593	9	8	10	11.642	11.565	8.446
599	9	9	9	11.535	11.481	8.449
601	8	8	8	11.553	11.518	8.219

Appendix B. Explicit formulæ for genus-2 computations

This appendix collects useful formulæ for computing explicit Richelot isogenies, and identifying the reduced automorphism groups of abelian surfaces.

B.1. Richelot isogenies. Let $\mathcal{C} : y^2 = F(x)$ be a genus-2 curve, with F squarefree of degree 5 or 6. The Lagrangian subgroups of $\mathcal{J}(\mathcal{C})[2]$ correspond to factorizations of F into quadratics (of which one may be linear, if $\deg(F) = 5$):

$$\mathcal{C} : y^2 = F(x) = F_1(x)F_2(x)F_3(x),$$

up to permutation of the F_i and constant multiples. We call such factorizations *quadratic splittings*.

Fix one such quadratic splitting $\{F_1, F_2, F_3\}$; then the corresponding subgroup $K \subset \mathcal{J}(\mathcal{C})[2]$ is the kernel of a $(2,2)$-isogeny $\phi : \mathcal{J}(\mathcal{C}) \to \mathcal{J}(\mathcal{C})/K$. For each $1 \leq i \leq 3$, we write $F_i(x) = F_{i,2}x^2 + F_{i,1}x + F_{i,0}$. Now let

$$\delta = \delta(F_1, F_2, F_3) := \begin{vmatrix} F_{1,0} & F_{1,1} & F_{1,2} \\ F_{2,0} & F_{2,1} & F_{2,2} \\ F_{3,0} & F_{3,1} & F_{3,2} \end{vmatrix}.$$

If $\delta(F_1, F_2, F_3) \neq 0$, then $\mathcal{J}(\mathcal{C})/K$ is isomorphic to a Jacobian $\mathcal{J}(\mathcal{C}')$, which we can compute using Richelot's algorithm (see [**5**] and [**39**, §8]). First, let

$$G_1(x) := \delta^{-1} \cdot (F_2'(x)F_3(x) - F_3'(x)F_2(x)),$$
$$G_2(x) := \delta^{-1} \cdot (F_3'(x)F_1(x) - F_1'(x)F_3(x)),$$
$$G_3(x) := \delta^{-1} \cdot (F_1'(x)F_2(x) - F_2'(x)F_1(x)).$$

Now the isogenous Jacobian is $\mathcal{J}(\mathcal{C}')$, where $\mathcal{C}'$ is the curve

$$\mathcal{C}' : y^2 = G(x) = G_1(x)G_2(x)G_3(x)$$

and the quadratic splitting $\{G_1, G_2, G_3\}$ corresponds to the kernel of the dual isogeny $\phi^\dagger : \mathcal{J}(\mathcal{C}') \to \mathcal{J}(\mathcal{C})$. The F_i and G_i are related by the identity

$$F_1(x_1)G_1(x_2) + F_2(x_1)G_2(x_2) + F_3(x_1)G_3(x_2) + (x_1 - x_2)^2 = 0.$$

Bruin and Doerksen present a convenient form for a divisorial correspondence $\mathcal{R} \subset \mathcal{C} \times \mathcal{C}'$ inducing the isogeny ϕ (see [**7**, §4]):

$$\text{(B.1)} \qquad \mathcal{R} : \begin{cases} F_1(x_1)G_1(x_2) + F_2(x_1)G_2(x_2) = 0, \\ F_1(x_1)G_1(x_2)(x_1 - x_2) = y_1 y_2, \\ F_2(x_1)G_2(x_2)(x_1 - x_2) = -y_1 y_2. \end{cases}$$

If $\delta(F_1, F_2, F_3) = 0$, then $\mathcal{J}(\mathcal{C})/K$ is isomorphic to an elliptic product $\mathcal{E} \times \mathcal{E}'$. Let $D(\lambda)$ be the discriminant of the quadratic polynomial $F_1 + \lambda F_2$, and let λ_1 and λ_2 be the roots of $D(\lambda)$; then $F_1 + \lambda_1 F_2 = U^2$ and $F_1 + \lambda_2 F_2 = V^2$ for some linear polynomials U and V. Now $F_1 = \alpha_1 U^2 + \beta_1 V^2$ and $F_2 = \alpha_2 U^2 + \beta_2 V^2$ for some

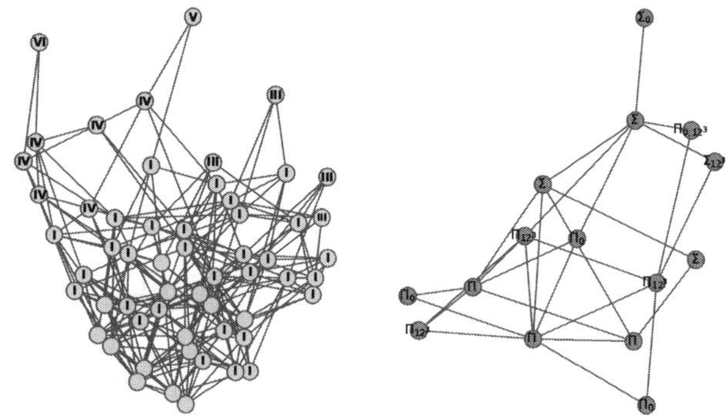

(A) Jacobian subgraph (B) Elliptic product subgraph

FIGURE 5. The subgraphs $\Gamma_2^{SS}(2;47)$ supported on Jacobians (left) and elliptic products (right). Vertex positions are mantained with respect to Figure 4.

α_1, β_1, α_2, and β_2, and since in this case F_3 is a linear combination of F_1 and F_2, we must have $F_3 = \alpha_3 U^2 + \beta_3 V^2$ for some α_3 and β_3. Now, rewriting the defining equation of $\mathcal{C}$ as

$$\mathcal{C} : Y^2 = \prod_{i=1}^{3} (\alpha_i U^2 + \beta_i V^2),$$

it is clear that the elliptic curves

$$\mathcal{E} : Y^2 = \prod_{i=1}^{3} (\alpha_i X + \beta_i Z) \quad \text{and} \quad \mathcal{E}' : Y^2 = \prod_{i=1}^{3} (\beta_i X + \alpha_i Z)$$

are the images of double covers $\pi : \mathcal{C} \to \mathcal{E}$ and $\pi' : \mathcal{C} \to \mathcal{E}'$ defined by $\pi((X : Y : Z)) = (U : Y : V)$ and $\pi'((X : Y : Z)) = (V : Y : U)$, respectively. The product of these covers induces the isogeny $\phi : \mathcal{J}(\mathcal{C}) \to \mathcal{E} \times \mathcal{E}'$.

B.2. Isogenies from elliptic products. Consider a generic pair of elliptic curves over $\Bbbk$, defined by

$$\mathcal{E} : y^2 = (x - s_1)(x - s_2)(x - s_3)$$

and

$$\mathcal{E}' : y^2 = (x - s_1')(x - s_2')(x - s_3').$$

We have $\mathcal{E}[2] = \{0_\mathcal{E}, P_1, P_2, P_3\}$ and $\mathcal{E}'[2] = \{0_{\mathcal{E}'}, P_1', P_2', P_3'\}$ where $P_i := (s_i, 0)$ and $P_i' := (s_i', 0)$. For each $1 \leq i \leq 3$, we let

$$\psi_i : \mathcal{E} \longrightarrow \mathcal{E}_i := \mathcal{E}/\langle P_i \rangle \quad \text{and} \quad \psi_i' : \mathcal{E}' \to \mathcal{E}_i' := \mathcal{E}'/\langle P_i' \rangle$$

be the quotient 2-isogenies. These can be computed using Vélu's formulæ [**44**].

The fifteen Lagrangian subgroups of $(\mathcal{E} \times \mathcal{E}')[2]$ fall naturally into two kinds. Nine of the kernels correspond to products of 2-isogeny kernels in $\mathcal{E}[2]$. Namely, for each $1 \leq i, j \leq 3$ we have a subgroup

$$K_{i,j} := \langle (P_i, 0_{\mathcal{E}'}), (0_\mathcal{E}, P_i') \rangle \subset (\mathcal{E} \times \mathcal{E}')[2],$$

and a quotient isogeny

$$\phi_{i,j} : \mathcal{E} \times \mathcal{E}' \to (\mathcal{E} \times \mathcal{E}')/K_{i,j} \cong \mathcal{E}_i \times \mathcal{E}'_j.$$

Of course, $\phi_{i,j} = \psi_i \times \psi_j$; we can thus compute $\phi_{i,j}$, and the codomains $\mathcal{E}_i \times \mathcal{E}'_j$, using Vélu's formulæ as above.

The other six kernels correspond to 2-Weil anti-isometries $\mathcal{E}[2] \cong \mathcal{E}'[2]$: they are

$$K_\pi := \{(0_{\mathcal{E}}, 0_{\mathcal{E}'}), (P_1, P'_{\pi(1)}), (P_2, P'_{\pi(2)}), (P_3, P'_{\pi(3)})\} \quad \text{for } \pi \in \operatorname{Sym}(\{1,2,3\}),$$

with quotient isogenies

$$\phi_\pi : \mathcal{E} \times \mathcal{E}' \to \mathcal{A}_\pi := (\mathcal{E} \times \mathcal{E}')/K_\pi.$$

If the anti-isometry $P_i \mapsto P'_{\pi(i)}$ is induced by an isomorphism $\mathcal{E} \to \mathcal{E}'$, then $\mathcal{A}_\pi$ is isomorphic to $\mathcal{E} \times \mathcal{E}'$; otherwise, it is the Jacobian of a genus-2 curve $\mathcal{C}_\pi$, which we can compute using the formulæ below (taken from [**22**, Proposition 4]).

Writing $\alpha_i := x(P_i)$ and $\beta_i := x(P'_{\pi(i)})$ for $1 \le i \le 3$, let

$$\begin{aligned}
a_1 &:= \frac{(\alpha_3 - \alpha_2)^2}{\beta_3 - \beta_2} + \frac{(\alpha_2 - \alpha_1)^2}{\beta_2 - \beta_1} + \frac{(\alpha_1 - \alpha_3)^2}{\beta_1 - \beta_3}, \\
b_1 &:= \frac{(\beta_3 - \beta_2)^2}{\alpha_3 - \alpha_2} + \frac{(\beta_2 - \beta_1)^2}{\alpha_2 - \alpha_1} + \frac{(\beta_1 - \beta_3)^2}{\alpha_1 - \alpha_3}, \\
a_2 &:= \alpha_1(\beta_3 - \beta_2) + \alpha_2(\beta_1 - \beta_3) + \alpha_3(\beta_2 - \beta_1), \\
b_2 &:= \beta_1(\alpha_3 - \alpha_2) + \beta_2(\alpha_1 - \alpha_3) + \beta_3(\alpha_2 - \alpha_1), \\
A &:= \Delta' \cdot a_1/a_2 \text{ where } \Delta' := (\beta_2 - \beta_3)^2(\beta_1 - \beta_3)^2(\beta_1 - \beta_2)^2, \\
B &:= \Delta \cdot b_1/b_2 \text{ where } \Delta := (\alpha_2 - \alpha_3)^2(\alpha_1 - \alpha_3)^2(\alpha_1 - \alpha_2)^2,
\end{aligned}$$

and finally

$$\begin{aligned}
F_1 &:= A(\alpha_2 - \alpha_1)(\alpha_1 - \alpha_3)X^2 + B(\beta_2 - \beta_1)(\beta_1 - \beta_3)Z^2, \\
F_2 &:= A(\alpha_3 - \alpha_2)(\alpha_2 - \alpha_1)X^2 + B(\beta_3 - \beta_2)(\beta_2 - \beta_1)Z^2, \\
F_3 &:= A(\alpha_1 - \alpha_3)(\alpha_3 - \alpha_2)X^2 + B(\beta_1 - \beta_3)(\beta_3 - \beta_2)Z^2.
\end{aligned}$$

Now the curve $\mathcal{C}_\pi$ may be defined by

$$\mathcal{C}_\pi : Y^2 = -F_1(X,Z)F_2(X,Z)F_3(X,Z).$$

The dual isogeny $\phi_\pi^\dagger : \mathcal{J}(\mathcal{C}_\pi) \to \mathcal{E} \times \mathcal{E}'$ corresponds to the quadratic splitting $\{F_1, F_2, F_3\}$.

B.3. Identifying reduced automorphism types of Jacobians. We can identify the isomorphism class of a Jacobian $\mathcal{J}(\mathcal{C})$ using the Clebsch invariants A, B, C, D of $\mathcal{C}$, which are homogeneous polynomials of degree 2, 4, 6, and 10 in the coefficients of the sextic defining $\mathcal{C}$. These invariants should be seen as coordinates on the weighted projective space $\mathbb{P}(2,4,6,10)$: that is,

$$(A : B : C : D) = (\lambda^2 A : \lambda^4 B : \lambda^6 C : \lambda^{10} D)$$

for all nonzero λ in $\Bbbk$. The Clebsch invariants can be computed using a series of transvectants involving the sextic (see [**33**, §1]), but it is more convenient to use (for example) `ClebschInvariants` in Magma [**4**] or `clebsch_invariants` from the `sage.schemes.hyperelliptic_curves.invariants` library of Sage [**43**]. If $\mathcal{C}/\overline{\mathbb{F}}_p$ is superspecial, then $(A : B : C : D)$ are in $\mathbb{F}_{p^2}$.

To determine $\mathrm{RA}(\mathcal{J}(\mathcal{C}))$ for a given genus-2 $\mathcal{C}$, we use necessary and sufficient conditions on the Clebsch invariants derived by Bolza [**3**, §11], given here in Table 6. These criteria involve some derived invariants: following Mestre's notation [**33**], let

$$A_{11} = 2C + \frac{1}{3}AB, \qquad A_{12} = \frac{2}{3}(B^2 + AC), \qquad A_{23} = \frac{1}{2}B \cdot A_{12} + \frac{1}{3}C \cdot A_{11},$$
$$A_{22} = D, \qquad A_{31} = D, \qquad A_{33} = \frac{1}{2}B \cdot A_{22} + \frac{1}{3}C \cdot A_{12}$$

(recall again that $\operatorname{char} \mathbb{k}$ is not 2 or 3). Finally, the R-invariant is defined by

$$R^2 = \frac{1}{2}\begin{vmatrix} A_{11} & A_{12} & A_{31} \\ A_{12} & A_{22} & A_{23} \\ A_{31} & A_{23} & A_{33} \end{vmatrix}.$$

TABLE 6. The classification of reduced automorphism groups of Jacobian surfaces, with necessary and sufficient conditions on the Clebsch invariants for each type.

Type	$\mathrm{RA}(\mathcal{J}(\mathcal{C}))$	Conditions on Clebsch invariants
Type-A	1	$R \neq 0$, $(A:B:C:D) \neq (0:0:0:1)$
Type-I	C_2	$R = 0$ and $A_{11}A_{22} \neq A_{12}$
Type-II	C_5	$(A:B:C:D) = (0:0:0:1)$
Type-III	C_2^2	$BA_{11} - 2AA_{12} = -6D$, $D \neq 0$, $CA_{11} + 2BA_{12} = AD$, $6C^2 \neq B^3$
Type-IV	S_3	$6C^2 = B^3$, $3D = 2BA_{11}$, $2AB \neq 15C$, $D \neq 0$
Type-V	$D_{2\times 6}$	$6B = A^2$, $D = 0$, $A_{11} = 0$, $A \neq 0$
Type-VI	S_4	$(A:B:C:D) = (1:0:0:0)$

References

[1] Sinan Aksoy, Fan Chung, and Xing Peng, *Extreme values of the stationary distribution of random walks on directed graphs*, Adv. in Appl. Math. **81** (2016), 128–155, DOI 10.1016/j.aam.2016.06.012. MR3551666

[2] Reza Azarderakhsh, Brian Koziel, Matt Campagna, Brian LaMacchia, Craig Costello, Patrick Longa, Luca De Feo, Michael Naehrig, Basil Hess, Joost Renes, Amir Jalali, Vladimir Soukharev, David Jao, and David Urbanik, *Supersingular Isogeny Key Encapsulation*, `http://sike.org`, 2017.

[3] Oskar Bolza, *On binary sextics with linear transformations into themselves*, Amer. J. Math. **10** (1887), no. 1, 47–70, DOI 10.2307/2369402. MR1505464

[4] Wieb Bosma, John J. Cannon, Claus Fieker, and Allan Steel, *Handboook of Magma functions*, 2.25 ed., January 2020.

[5] Jean-Benoît Bost and Jean-François Mestre, *Moyenne arithmético-géométrique et périodes des courbes de genre* 1 *et* 2 (French), Gaz. Math. **38** (1988), 36–64. MR970659

[6] Bradley Wayne Brock, *Superspecial curves of genera two and three*, ProQuest LLC, Ann Arbor, MI, 1993. Thesis (Ph.D.)–Princeton University. MR2689446

[7] Nils Bruin and Kevin Doerksen, *The arithmetic of genus two curves with* (4, 4)*-split Jacobians*, Canad. J. Math. **63** (2011), no. 5, 992–1024, DOI 10.4153/CJM-2011-039-3. MR2866068

[8] Reinier Bröker, Kristin Lauter, and Andrew V. Sutherland, *Modular polynomials via isogeny volcanoes*, Math. Comp. **81** (2012), no. 278, 1201–1231, DOI 10.1090/S0025-5718-2011-02508-1. MR2869057

[9] Wouter Castryck, Thomas Decru, and Benjamin Smith, *Hash functions from superspecial genus-2 curves using Richelot isogenies*, J. Math. Cryptol. **14** (2020), no. 1, 268–292, DOI 10.1515/jmc-2019-0021. MR4134760

[10] Wouter Castryck, Tanja Lange, Chloe Martindale, Lorenz Panny, and Joost Renes, *CSIDH: an efficient post-quantum commutative group action*, Advances in cryptology—ASIACRYPT 2018. Part III, Lecture Notes in Comput. Sci., vol. 11274, Springer, Cham, 2018, pp. 395–427, DOI 10.1007/978-3-030-03332-3_15. MR3897883

[11] Denis X. Charles, Eyal Z. Goren, and Kristin E. Lauter, *Families of Ramanujan graphs and quaternion algebras*, Groups and symmetries, CRM Proc. Lecture Notes, vol. 47, Amer. Math. Soc., Providence, RI, 2009, pp. 53–80, DOI 10.1090/crmp/047/05. MR2500554

[12] Denis X. Charles, Kristin E. Lauter, and Eyal Z. Goren, *Cryptographic hash functions from expander graphs*, J. Cryptology **22** (2009), no. 1, 93–113, DOI 10.1007/s00145-007-9002-x. MR2496385

[13] Craig Costello and Benjamin Smith, *The supersingular isogeny problem in genus 2 and beyond*, Post-quantum cryptography, Lecture Notes in Comput. Sci., vol. 12100, Springer, Cham, [2020] ©2020, pp. 151–168, DOI 10.1007/978-3-030-44223-1_9. MR4139650

[14] Luca De Feo, David Jao, and Jérôme Plût, *Towards quantum-resistant cryptosystems from supersingular elliptic curve isogenies*, J. Math. Cryptol. **8** (2014), no. 3, 209–247, DOI 10.1515/jmc-2012-0015. MR3259113

[15] Enric Florit and Benjamin Smith, *An atlas of the superspecial richelot isogeny graph*, Preprint: `https://hal.inria.fr/hal-03094296`, 2020.

[16] E. V. Flynn and Yan Bo Ti, *Genus two isogeny cryptography*, Post-quantum cryptography, Lecture Notes in Comput. Sci., vol. 11505, Springer, Cham, 2019, pp. 286–306, DOI 10.1007/978-3-030-25510-7_16. MR3989010

[17] Mireille Fouquet and François Morain, *Isogeny volcanoes and the SEA algorithm*, Algorithmic number theory (Sydney, 2002), Lecture Notes in Comput. Sci., vol. 2369, Springer, Berlin, 2002, pp. 276–291, DOI 10.1007/3-540-45455-1_23. MR2041091

[18] Steven D. Galbraith, Christophe Petit, and Javier Silva, *Identification protocols and signature schemes based on supersingular isogeny problems*, Advances in cryptology—ASIACRYPT 2017. Part I, Lecture Notes in Comput. Sci., vol. 10624, Springer, Cham, 2017, pp. 3–33, DOI 10.1007/978-3-319-70694-8_1. MR3747691

[19] The GAP Group, *GAP – Groups, Algorithms, and Programming, Version 4.11.0*, 2020.

[20] P. Gaudry and É. Schost, *On the invariants of the quotients of the Jacobian of a curve of genus 2*, Applied algebra, algebraic algorithms and error-correcting codes (Melbourne, 2001), Lecture Notes in Comput. Sci., vol. 2227, Springer, Berlin, 2001, pp. 373–386, DOI 10.1007/3-540-45624-4_39. MR1913484

[21] Shlomo Hoory, Nathan Linial, and Avi Wigderson, *Expander graphs and their applications*, Bull. Amer. Math. Soc. (N.S.) **43** (2006), no. 4, 439–561, DOI 10.1090/S0273-0979-06-01126-8. MR2247919

[22] Everett W. Howe, Franck Leprévost, and Bjorn Poonen, *Large torsion subgroups of split Jacobians of curves of genus two or three*, Forum Math. **12** (2000), no. 3, 315–364, DOI 10.1515/form.2000.008. MR1748483

[23] Tomoyoshi Ibukiyama, Toshiyuki Katsura, and Frans Oort, *Supersingular curves of genus two and class numbers*, Compositio Math. **57** (1986), no. 2, 127–152. MR827350

[24] David Jao and Luca De Feo, *Towards quantum-resistant cryptosystems from supersingular elliptic curve isogenies*, Post-quantum cryptography, Lecture Notes in Comput. Sci., vol. 7071, Springer, Heidelberg, 2011, pp. 19–34, DOI 10.1007/978-3-642-25405-5_2. MR2931459

[25] Bruce W. Jordan and Yevgeny Zaytman, *Isogeny graphs of superspecial abelian varieties and generalized Brandt matrices*, preprint, `arXiv:2005.09031` [math.NT], 2020.

[26] Toshiyuki Katsura and Katsuyuki Takashima, *Counting Richelot isogenies between superspecial abelian surfaces*, ANTS XIV—Proceedings of the Fourteenth Algorithmic Number Theory Symposium, Open Book Ser., vol. 4, Math. Sci. Publ., Berkeley, CA, 2020, pp. 283–300, DOI 10.2140/obs.2020.4.283. MR4235119

[27] David Russell Kohel, *Endomorphism rings of elliptic curves over finite fields*, ProQuest LLC, Ann Arbor, MI, 1996. Thesis (Ph.D.)–University of California, Berkeley. MR2695524

[28] David A. Levin and Yuval Peres, *Markov chains and mixing times*, American Mathematical Society, Providence, RI, 2017. Second edition of [MR2466937]; With contributions by Elizabeth L. Wilmer; With a chapter on "Coupling from the past" by James G. Propp and David B. Wilson, DOI 10.1090/mbk/107. MR3726904

[29] Ke-Zheng Li and Frans Oort, *Moduli of supersingular abelian varieties*, Lecture Notes in Mathematics, vol. 1680, Springer-Verlag, Berlin, 1998, DOI 10.1007/BFb0095931. MR1611305

[30] L. Lovász, *Random walks on graphs: a survey*, Combinatorics, Paul Erdős is eighty, Vol. 2 (Keszthely, 1993), Bolyai Soc. Math. Stud., vol. 2, János Bolyai Math. Soc., Budapest, 1996, pp. 353–397. MR1395866

[31] Ricardo Menares, *Equidistribution of Hecke points on the supersingular module*, Proc. Amer. Math. Soc. **140** (2012), no. 8, 2687–2691, DOI 10.1090/S0002-9939-2011-11148-1. MR2910756

[32] J.-F. Mestre, *La méthode des graphes. Exemples et applications* (French), Proceedings of the international conference on class numbers and fundamental units of algebraic number fields (Katata, 1986), Nagoya Univ., Nagoya, 1986, pp. 217–242. MR891898

[33] Jean-François Mestre, *Construction de courbes de genre* 2 *à partir de leurs modules* (French), Effective methods in algebraic geometry (Castiglioncello, 1990), Progr. Math., vol. 94, Birkhäuser Boston, Boston, MA, 1991, pp. 313–334. MR1106431

[34] Frans Oort, *A stratification of a moduli space of abelian varieties*, Moduli of abelian varieties (Texel Island, 1999), Progr. Math., vol. 195, Birkhäuser, Basel, 2001, pp. 345–416, DOI 10.1007/978-3-0348-8303-0_13. MR1827027

[35] Arnold K. Pizer, *Ramanujan graphs and Hecke operators*, Bull. Amer. Math. Soc. (N.S.) **23** (1990), no. 1, 127–137, DOI 10.1090/S0273-0979-1990-15918-X. MR1027904

[36] Friedrich Julius Richelot, *Essai sur une méthode générale pour déterminer les valeurs des intégrales ultra-elliptiques, fondée sur des transformations remarquables de ces trnscendates*, Comptes Rendus Mathématique. Académie des Sciences. Paris **2** (1836), 622–627.

[37] Fried. Jul. Richelot, *De transformatione integralium Abelianorum primi ordinis commentatio* (Latin), J. Reine Angew. Math. **16** (1837), 221–284, DOI 10.1515/crll.1837.16.221. MR1578134

[38] Alexander Rostovtsev and Anton Stolbunov, *Public-key cryptosystem based on isogenies*, Cryptology ePrint Archive, Report 2006/145, April 2006.

[39] Benjamin Smith, *Explicit endomorphisms and correspondences*, Ph.D. thesis, University of Sydney, 2005.

[40] Anton Stolbunov, *Constructing public-key cryptographic schemes based on class group action on a set of isogenous elliptic curves*, Adv. Math. Commun. **4** (2010), no. 2, 215–235, DOI 10.3934/amc.2010.4.215. MR2654134

[41] Andrew V. Sutherland, *Identifying supersingular elliptic curves*, LMS J. Comput. Math. **15** (2012), 317–325, DOI 10.1112/S1461157012001106. MR2988819

[42] Katsuyuki Takashima, *Efficient algorithms for isogeny sequences and their cryptographic applications*, Mathematical modelling for next-generation cryptography, Math. Ind. (Tokyo), vol. 29, Springer, Singapore, 2018, pp. 97–114. MR3586863

[43] The Sage Developers, *Sagemath, the Sage Mathematics Software System (Version 9.1)*, 2020, `https://www.sagemath.org`.

[44] Jacques Vélu, *Isogénies entre courbes elliptiques* (French), C. R. Acad. Sci. Paris Sér. A-B **273** (1971), A238–A241. MR294345

[45] Paul M. Weichsel, *The Kronecker product of graphs*, Proc. Amer. Math. Soc. **13** (1962), 47–52, DOI 10.2307/2033769. MR133816

Departament de Matemàtiques i Infomàtica, Universitat de Barcelona(UB), Gran Via de les Corts Catalanes 585, 08007 Barcelona, Spain
Email address: `enricflorit@ub.edu`

Inria and Laboratoire d'Informatique (LIX), CNRS, École polytechnique, Institut Polytechnique de Paris, 91120 Palaiseau, France
Email address: `smith@lix.polytechnique.fr`

Contemporary Mathematics
Volume **779**, 2022
https://doi.org/10.1090/conm/779/15673

Frobenius structures on hypergeometric equations

Kiran S. Kedlaya

ABSTRACT. We give an exposition of Dwork's construction of Frobenius structures associated to generalized hypergeometric equations via the interpretation of the latter due to Gelfand–Kapranov–Zelevinsky in the language of A-hypergeometric systems. As a consequence, we extract some explicit formulas for the degeneration at 0 in terms of the Morita p-adic gamma function.

1. Introduction

Hypergeometric differential equations, of arbitrary order, provide some key examples of Picard–Fuchs equations and of rigid local systems. As such, they admit p-adic analytic *Frobenius structures* which interpolate the zeta functions associated to certain motives over finite fields.

The purpose of this note is to extract from Dwork's book [**16**] an explicit construction of Frobenius structures on hypergeometric equations (see Theorem 4.1.2), and in particular a formula for the residue at 0 (see Corollary 4.3.3), using A-hypergeometric systems in the sense of Gelfand–Kapranov–Zelevinsky [**22**] (which we introduce in very little detail in §3). We also give a brief indication of how this knowledge can be used as the basis for an efficient algorithm to compute the action of Frobenius on the (rational) crystalline realizations of hypergeometric motives, in the style of Lauder's *deformation method* [**36**]. We have implemented this method in SAGEMATH [**33**] and gotten good results in practice; however, some further analysis is needed on the tradeoff between rigor and efficiency caused by the choice of working precision for certain power series and p-adic coefficients (see Remark 5.3.1).

2. Generalities

We first recall some general facts and definitions concerning ordinary differential equations, including the definition of a Frobenius structure.

2020 *Mathematics Subject Classification.* Primary 33C20, 12H25.

The author was supported by NSF (grants DMS-1501214, DMS-1802161, DMS-2053473), UCSD (Warschawski Professorship), and the IAS School of Mathematics (Visiting Professorship 2018–2019). Additional funding/hospitality was provided by ICTP (September 2017), HRIM–Bonn (March 2018), KIAS–Seoul (August 2019), AIM (August 2019), EPSRC (grant EP/K034383/1), and the Simons Collaboration on Arithmetic Geometry, Number Theory, and Computation.

2.1. Ordinary differential equations. We first recall some standard concepts in order to set notation for them.

DEFINITION 2.1.1. Let D be a differential operator acting on a field F of characteristic zero. By a *D-differential equation*, we will always mean a homogeneous linear differential equation in the variable y of the form

$$D^n(y) + a_{n-1}D^{n-1}(y) + \cdots + a_0 y = 0 \tag{2.1.1.1}$$

with $a_0, \dots, a_{n-1} \in F$. For uniformity of notation, we set $a_n = 1$.

By a *D-differential system* of rank n, we will mean an equation in the variable $\mathbf{v}$ (a column vector of length n) of the form

$$N\mathbf{v} + D(\mathbf{v}) = 0, \tag{2.1.1.2}$$

where N is an $n \times n$ matrix over F. This is the same structure as a *connection* over F whose underlying module is equipped with a distinguished basis.

REMARK 2.1.2. Given the equation (2.1.1.1), let N be the companion matrix

$$N = \begin{pmatrix} 0 & -1 & \cdots & 0 & 0 \\ 0 & 0 & & 0 & 0 \\ \vdots & & \ddots & & \vdots \\ 0 & 0 & & 0 & -1 \\ a_0 & a_1 & \cdots & a_{n-2} & a_{n-1} \end{pmatrix};$$

then the solutions of (2.1.1.2) are precisely the vectors of the form

$$\mathbf{v} = \begin{pmatrix} y \\ D(y) \\ \vdots \\ D^{n-1}(y) \end{pmatrix}$$

where y is a solution of (2.1.1.1).

Conversely, given the equation (2.1.1.2), note that for U an invertible $n \times n$ matrix over F, the equation

$$N_U\mathbf{w} + D(\mathbf{w}) = 0, \qquad N_U := U^{-1}NU + U^{-1}D(U)$$

is equivalent to the original equation via the substitutions

$$\mathbf{v} \mapsto U\mathbf{w}, \qquad \mathbf{w} \mapsto U^{-1}\mathbf{v}.$$

The *cyclic vector theorem* (see for example [**32**, Theorem 5.4.2]) then implies that for any choice of N, there exists some U for which N_U is a companion matrix. However, there is typically no natural choice of U.

DEFINITION 2.1.3. Let X be a locally ringed space over $\operatorname{Spec}\mathbb{Q}$. Let Ω be a coherent sheaf on X equipped with a derivation $d : \mathcal{O}_X \to \Omega$. A *connection* on X (with respect to d) consists of a pair $(\mathcal{E}, \nabla)$ in which $\mathcal{E}$ is a vector bundle (locally free coherent sheaf) $\mathcal{E}$ on X and $\nabla : \mathcal{E} \to \mathcal{E} \otimes_{\mathcal{O}_X} \Omega$ is an additive morphism satisfying the Leibniz rule with respect to d: for $U \subseteq X$ open, $f \in \Gamma(U, \mathcal{O})$, $\mathbf{v} \in \Gamma(U, \mathcal{E})$, we have

$$d(f\mathbf{v}) = f\nabla(\mathbf{v}) + \mathbf{v} \otimes d(f).$$

We also refer to such a pair as being a connection on $\mathcal{E}$. The elements of the kernel of ∇ on $\mathcal{E}(U)$ are called the *horizontal sections* of $\mathcal{E}$, or more precisely of $(\mathcal{E}, \nabla)$, over U.

Given two connections $(\mathcal{E}_1, \nabla_1)$, $(\mathcal{E}_2, \nabla_2)$, the *tensor product* is the connection $(\mathcal{E}_1 \otimes_{\mathcal{O}_X} \mathcal{E}_2, \nabla)$ given by

$$\nabla(f\mathbf{v} \otimes \mathbf{w}) = f\nabla_1(\mathbf{v}) \otimes \mathbf{w} + f\mathbf{v} \otimes \nabla_2(\mathbf{w}) + d(f) \otimes \mathbf{v} \otimes \mathbf{w}.$$

Given a connection $(\mathcal{E}, \nabla)$, the *dual* is the unique connection whose underlying bundle is the modulo-theoretic dual $\mathcal{E}^\vee$ for which the canonical pairing $\mathcal{E} \otimes \mathcal{E}^\vee \to \mathcal{O}_X$ is a morphism of connections.

REMARK 2.1.4. In the case where $X = \operatorname{Spec} F$, $\Omega = \mathcal{O}_X$, $d = D$, and $\mathcal{E} = \mathcal{O}_X^{\oplus n}$, any connection on $\mathcal{E}$ has the form $\mathbf{v} \mapsto N\mathbf{v} + D(\mathbf{v})$ for some $n \times n$ matrix N over F (and conversely any such matrix defines a connection). The solutions of the equation (2.1.1.1) then correspond to the horizontal sections of $\mathcal{E}$ over X. The dual connection (with the dual basis) corresponds to the matrix $-N^T$.

DEFINITION 2.1.5. Let $F\{D\}$ denote the *Ore polynomial ring* in D; it is a noncommutative F-algebra whose underlying set coincides with that of $F[D]$, but whose multiplication is characterized by the identity

$$Dx - xD = D(x) \qquad (x \in F).$$

Then a connection on $\operatorname{Spec} F$ is the same as a left $F\{D\}$-module whose underlying F-vector space is identified with the set of length-n column vectors over F, with the action of D given by $\mathbf{v} \mapsto N\mathbf{v} + D(\mathbf{v})$; passing from N to N_U amounts to changing basis on this vector space via the matrix U.

Given a D-differential system defined by a D-differential equation (2.1.1.1), the dual of the corresponding connection is the left $F\{D\}$-module $F\{D\}/F\{D\}(D^n + a_{n-1}D^{n-1} + \cdots + a_0)$.

2.2. Regular singularities. Throughout §2.2, let K be a field of characteristic 0.

DEFINITION 2.2.1. In the notation of §2.1, take $F = K(z)$ to be equipped with the derivation $D = z\frac{d}{dz}$. We then say that the equation (2.1.1.1) is *regular* at 0 if $\operatorname{ord}_0(a_i) \geq 0$ for $i = 0, \ldots, n-1$. We say that (2.1.1.2) is *regular* at 0 if $\operatorname{ord}_0(N_{ij}) \geq 0$ for $i, j = 1, \ldots, n$.

DEFINITION 2.2.2. With notation as in Definition 2.2.1, fix an algebraic closure of K. Define the *local exponents* at 0 of the equation (2.1.1.2) to be the negations of the roots of the characteristic polynomial of $N|_{z=0}$. By the classical theory of regular (Fuchsian) singularities, the images of the local exponents under $\exp(2\pi i \bullet)$ compute the eigenvalues of local monodromy around $z = 0$. Note that this only uses the values of the exponents modulo $\mathbb{Z}$; in fact it is only these residues that are intrinsic under meromorphic changes of coordinates, as one can make integral shifts using *shearing transformations* [**32**, Proposition 7.3.10].

DEFINITION 2.2.3. Now in the notation of §2.1, take $F = K(z)$ to be equipped with the derivation $D = \frac{d}{dz}$. For $z_0 \in \mathbf{P}^1_K$, the equation (2.1.1.2) is *regular* at z_0 if the entries of N have at worst simple poles at $z = z_0$; for $z = 0$, this is consistent with Definition 2.2.1. The equation (2.1.1.1) is *regular* at z_0 if the corresponding matrix equation is; for $z_0 \in \mathbf{A}^1_K$, this translates into the condition

$$\operatorname{ord}_{z_0}(a_i) \geq i - n \qquad (i = 0, \ldots, n-1).$$

2.3. Frobenius structures on differential equations.

HYPOTHESIS 2.3.1. *Throughout §2.3, fix a prime p. Let X be an open subspace of* $\mathbf{P}^1_{\mathbb{Q}_p}$. *Let Z be the complement of X in* $\mathbf{P}^1_{\mathbb{Q}_p}$; *to simplify notation, we assume that* $\{0, \infty\} \subseteq Z$.

DEFINITION 2.3.2. By a *Frobenius lift*, we will mean a $\mathbb{Q}_p$-linear map $\sigma : \mathcal{O}(X) \to \mathcal{O}(X)$ such that $\sigma(z) - z^p \in p\mathbb{Z}_p[z]_{(p)}$. For instance, we may take $\sigma(z) = z^p$; we call this the *standard Frobenius lift* (with respect to the coordinate z).

DEFINITION 2.3.3. Let $\mathbf{P}^{1,\mathrm{an}}_{\mathbb{Q}_p}$ be the analytification of $\mathbf{P}^1_{\mathbb{Q}_p}$ in the sense of rigid analytic geometry. (For the purposes of this discussion, we use Tate's model of p-adic analytic geometry; however any of the equivalent models of p-adic analytic geometry may be used instead, such as Berkovich spaces or Huber adic spaces.)

Let $(\mathcal{E}, \nabla)$ be a connection on X. We define a *Frobenius structure* on $(\mathcal{E}, \nabla)$ with respect to the Frobenius lift σ as an isomorphism $\sigma^*\mathcal{E} \cong \mathcal{E}$ of vector bundles with connection on some subspace V of $\mathbf{P}^{1,\mathrm{an}}_{\mathbb{Q}_p}$ whose complement consists of a union of closed discs, each contained in the open unit disc around some point of Z.

More generally, for $(\mathcal{E}', \nabla')$ another connection on X, we define a *Frobenius intertwiner* from $(\mathcal{E}, \nabla)$ to $(\mathcal{E}', \nabla')$ with respect to the Frobenius lift σ to be an isomorphism $\sigma^*\mathcal{E} \cong \mathcal{E}'$ of vector bundles with connection on some subspace V as above.

REMARK 2.3.4. In the context of Remark 2.1.4, a Frobenius intertwiner corresponds to an invertible $n \times n$ matrix Φ with entries in the ring $\mathcal{O}(V)$ satisfying

$$(2.3.4.1) \qquad N'\Phi - c_\sigma \sigma(N) + D(\Phi) = 0, \qquad c_\sigma = \frac{\sigma(dz/z)}{dz/z} = \frac{D(\sigma(z))}{\sigma(z)}.$$

The effect of changing basis by two invertible matrices U, U' is to replace Φ with

$$\Phi_{U,U'} := U^{-1}\Phi\sigma(U'),$$

which defines a Frobenius intertwiner from N_U to $N'_{U'}$.

REMARK 2.3.5. When a Frobenius intertwiner exists, one can always rescale it by an invertible elements of $\mathbb{Q}_p$. In many cases, one can show that there can be at most one Frobenius structure up to rescaling (see Lemma 2.3.6 below); however, we will need some extra information in order to normalize for this scalar ambiguity.

LEMMA 2.3.6. *Let* $(\mathcal{E}, \nabla)$ *and* $(\mathcal{E}', \nabla')$ *be two connections on X satisfying the following conditions.*

(a) *The restriction of* $(\mathcal{E}, \nabla)$ *to some open unit disc is trivial.*
(b) *The points of Z are pairwise noncongruent modulo p.*
(c) *At each* $z \in Z$, $(\mathcal{E}', \nabla')$ *is regular with exponents in* $\mathbb{Z}_{(p)}$.
(d) *The connection* $(\mathcal{E}', \nabla')$ *is irreducible over* $\mathbb{Q}_p(z)$.

Then up to $\mathbb{Q}_p^\times$*-scalar multiplication, there exists at most one Frobenius interwiner from* $(\mathcal{E}, \nabla)$ *to* $(\mathcal{E}', \nabla')$.

PROOF. By Baldassari's theorem on continuity of the radius of convergence of p-adic differential equations [**3**], condition (a) implies triviality of $(\mathcal{E}, \nabla)$ also on the restriction to a generic open unit disc. With this, we may apply [**14**] to conclude. □

REMARK 2.3.7. While the definition of a Frobenius intertwiner was made in terms of the chosen Frobenius lift σ, there is a certain independence from this choice: for any other Frobenius lift $\tilde{\sigma}$, there is a functorial way to transform Frobenius intertwiners with respect to σ into Frobenius intertwiners with respect to $\tilde{\sigma}$ using the Taylor isomorphism. As we will mostly be concerned with Frobenius defined with respect to a fixed Frobenius lift $z \mapsto z^p$, we will not develop this point here; see for example [**32**, §17.3].

LEMMA 2.3.8. *Let D_0 denote the open unit disc around 0, and suppose that $Z \cap D_0 = \{0\}$. Let $(\mathcal{E}, \nabla), (\mathcal{E}', \nabla')$ be connections on X which are regular at 0 with exponents in $\mathbb{Q} \cap \mathbb{Z}_{(p)}$. Suppose that there exists a Frobenius intertwiner Φ from $(\mathcal{E}, \nabla)$ to $(\mathcal{E}', \nabla')$ with respect to the standard Frobenius lift σ.*

(a) *As multisets of $\mathbb{Q}/\mathbb{Z}$, the local exponents of $(\mathcal{E}', \nabla')$ at 0 correspond to p times the local exponents of $(\mathcal{E}, \nabla)$.*

(b) *On D_0, we have decompositions*

$$\mathcal{E} \cong \bigoplus_{\lambda \in \mathbb{Z}_{(p)} \cap [0,1)} \mathcal{E}_\lambda, \qquad \mathcal{E}' \cong \bigoplus_{\mu \in \mathbb{Z}_{(p)} \cap [0,1)} \mathcal{E}'_\mu$$

of connections such that $\mathcal{E}_\lambda$ (resp. $\mathcal{E}'_\mu$) admits a basis on which $D = z\frac{d}{dz}$ acts by multiplication by λ (resp. μ) plus a nilpotent scalar matrix.

(c) *Any Frobenius structure Φ on $(\mathcal{E}, \nabla)$ extends holomorphically to the punctured open unit disc around 0 and meromorphically across 0. More precisely, with bases as in (b), for $\lambda, \mu \in \mathbb{Z}_{(p)} \cap [0,1)$ with $p\lambda \equiv \mu \pmod{\mathbb{Z}}$, Φ carries $\sigma^*\mathcal{E}_\lambda$ into $\mathcal{E}'_\mu$ and $t^{p\mu-\lambda}\Phi$ acts holomorphically on the chosen bases.*

PROOF. Suppose first that the exponents at 0 are all in $\mathbb{Z}$. In this case, (a) is trivial, (b) follows from [**32**, Proposition 17.5.1], and (c) follows from (b) by logic as in Remark 2.3.10 below.

To treat the general case, let m be the least common denominator of the exponents; then pulling back along $z \mapsto z^m$ gives another pair of connections admitting a Frobenius intertwiner, to which we may apply the previous argument to deduce the claim. Compare the proof of [**34**, Lemma 2.3]. □

REMARK 2.3.9. By making the substitution $z \mapsto z^{-1}$, we may immediately infer that Lemma 2.3.8 holds with the point 0 replaced by ∞. The same does not apply directly to other points of $\mathbf{P}^1_{\mathbb{Q}_p}$ because the relevant substitutions change the Frobenius lift; however, by Remark 2.3.7 we may still infer that Lemma 2.3.8(a) holds at any point of $\mathbf{P}^1_{\mathbb{Q}_p}$.

We next introduce the idea that one can compute a Frobenius structure by solving a differential equation and imposing an initial condition.

REMARK 2.3.10. Assuming that a given pair of connections given by matrices N, N' admits a Frobenius intertwiner Φ for the standard Frobenius lift σ, one can attempt to compute it by first finding formal solution matrices U, U' of N, N' at 0, i.e., finding invertible matrices U, U' over $\mathbb{Q}_p[\![z]\!]$ for which N_U and $N'_{U'}$ are scalar matrices. In the context of hypergeometric equations, we will even have explicit formulas for U, U' in terms of hypergeometric series and their derivatives.

We may further ensure that $N_U, N'_{U'}$ are block diagonal matrices with blocks indexed by $\lambda \in \mathbb{Z}_{(p)} \cap [0,1)$, in which each of the blocks N_λ, N'_λ equals λ plus a nilpotent matrix. In this case, Φ_U is itself a block permutation matrix with nonzero (λ, μ)-block whenever $p\lambda \equiv \mu \pmod{\mathbb{Z}}$. If we call this block Φ_λ, as per (2.3.4.1) we have

$$N_\mu \Phi_\lambda + D(\Phi_\lambda) = p\Phi_\lambda N_\lambda.$$

(Here we have replaced $\sigma(N_\lambda)$ with N_λ because N_λ has entries in $\mathbb{Q}_p$, which are fixed by σ.) Since N_λ and N_μ are scalar matrices, we may write $\Phi_\lambda = \sum_{n=-\infty}^{\infty} \Phi_n z^n$ and see that

$$N_\mu \Phi_n + n\Phi_n = p\Phi_n N_\lambda;$$

since $N_\lambda - \lambda$ and $N_\mu - \mu$ are nilpotent, this implies that $\Phi_n = 0$ unless $\mu + n = p\lambda$; that is, Φ_λ equals $t^{p\lambda-\mu}$ times an invertible matrix over $\mathbb{Q}_p$.

REMARK 2.3.11. Keeping notation as in Remark 2.3.10, by writing Φ in the form $U\Phi_U\sigma(U')^{-1}$, we can express the entries of Φ as elements of $\mathbb{Q}_p[\![z]\!]$. In order to be a Frobenius structure, these series have to also represent entries of $\mathcal{O}(V)$ for some V; this in particular implies that the series in $\mathbb{Q}_p[\![z]\!]$ we are considering have bounded coefficients, that is, they belong to the subring $\mathbb{Z}_p[\![z]\!][p^{-1}]$ of $\mathbb{Q}_p[\![z]\!]$.

This containment generally does not hold "by accident." For a typical differential equation, there is no choice of the scalar matrices $\Phi_{\lambda,0} := t^{\mu-p\lambda}\Phi_\lambda$ for which this last containment holds; in this case, no Frobenius structure can exist. When a Frobenius structure does exist, typically the values of $\Phi_{\lambda,0}$ are uniquely determined, up to a joint scalar multiplication, by the fact that they give rise to entries of F having bounded coefficients. This can be used as a mechanism for discovering the entries of $\Phi_{\lambda,0}$ empirically without any prior knowledge; see [**43**] for some examples of this and [**8**] for a more comprehensive treatment.

By contrast, in the case of hypergeometric equations, we will give a computable formula for the matrices $\Phi_{\lambda,0}$. (Since the entries are elements of $\mathbb{Q}_p$ which are in general transcendental over $\mathbb{Q}$, this means that for any fixed integer N, we can compute rational numbers which differ from the entries of $\Phi_{\lambda,0}$ by values in $p^N\mathbb{Z}_p$.)

REMARK 2.3.12. Keeping notation as in Remark 2.3.11, suppose that there exists a Frobenius structure Φ for which we have a computable formula for matrices $\Phi_{\lambda,0}$. The entries of Φ are elements of $\mathcal{O}(V)$; this ring is a certain completion of $\mathcal{O}(X)$ contained in the p-adic completion. We may thus represent the entries of Φ as sums of the form

$$P(z) + \sum_{i=1}^{\infty} \frac{c_i}{Q(z)^i} \qquad (P(z) \in \mathbb{Q}_p[z^\pm], c_i \in \mathbb{Q}_p, \lim_{i\to\infty} c_i = 0)$$

where $Q(z)$ is the monic polynomial with simple zeroes at $Z \setminus \{0, \infty\}$. (In the case of hypergeometric equations, we will have $Q(z) = z - 1$.)

In order to obtain a representation of Φ which is accurate to some prescribed p-adic accuracy, we need an effective bound on the decay rate of the c_i; this amounts to identifying a choice of the subspace V and a bound on Φ over V. In the case where the points of Z have pairwise distinct images under specialization, this can be done by studying the effect of changing the Frobenius lift (Remark 2.3.7).

3. Hypergeometric equations and the GKZ construction

We now describe the generalized hypergeometric equation that we consider, the Gelfand–Kapranov–Zelevinsky construction of A-hypergeometric systems, and how the two are related.

3.1. Hypergeometric differential equations.

DEFINITION 3.1.1. Define the differential operator $D := z\frac{d}{dz}$ on the field $K(z)$ as in Definition 2.2.1. The *generalized hypergeometric equation* with parameters in K given by

$$\underline{\alpha}; \underline{\beta} = \alpha_1, \dots, \alpha_m; \beta_1, \dots, \beta_n$$

is the linear differential equation of the form

$$P(\underline{\alpha};\underline{\beta})(y) = 0, \quad P(\underline{\alpha};\underline{\beta}) := z\prod_{i=1}^{m}(D+\alpha_i) - \prod_{j=1}^{n}(D+\beta_j-1). \tag{3.1.1.1}$$

(We conflate this equation with the equivalent equation in terms of the operator $\frac{d}{dz}$, which is somewhat less compact to express.) The case $m = n = 2$ recovers the classical (Gaussian) hypergeometric equation. We will primarily be interested in the case $K = \mathbb{Q}$, but in this section we treat the case $K = \mathbb{C}$ following Beukers–Heckman [**5**].

REMARK 3.1.2. Under the substitution $z \mapsto (-1)^{m-n}z^{-1}$, solutions of (3.1.1.1) correspond to solutions of $P(\underline{\alpha}';\underline{\beta}')(y) = 0$ for

$$\underline{\alpha}'; \underline{\beta}' := 1-\beta_1, \dots, 1-\beta_n; 1-\alpha_1, \dots, 1-\alpha_m.$$

REMARK 3.1.3. As in [**5**, Proposition 2.3], one has

$$(D+\delta-1)P(\underline{\alpha};\underline{\beta}) = P(\underline{\alpha},\delta;\underline{\beta},\delta)$$
$$P(\underline{\alpha};\underline{\beta})(D+\delta) = P(\underline{\alpha},\delta;\underline{\beta},\delta+1).$$

As per [**5**, Corollary 2.4], it follows that for $i = 1, \dots, m$ and $j = 1, \dots, n$,

$$P(\underline{\alpha};\underline{\beta})(D+\alpha_i-1) = (D+\alpha_i-1)P(\alpha_1,\dots,\alpha_i-1,\dots,\alpha_m;\beta_1,\dots,\beta_n)$$
$$(D+\beta_j-1)P(\underline{\alpha};\underline{\beta}) = P(\alpha_1,\dots,\alpha_m;\beta_1,\dots,\beta_j-1,\dots,\beta_n)(D+\beta_j).$$

This has the consequence that for all practical purposes, the analysis of the hypergeometric equation is insensitive to integer shifts in the parameters. In particular, there is no real loss of generality in normalizing the parameters so that

$$0 \le \mathrm{Re}(\alpha_1) \le \dots \le \mathrm{Re}(\alpha_m) < 1, \qquad 0 \le \mathrm{Re}(\beta_1) \le \dots \le \mathrm{Re}(\beta_n) < 1;$$

this will become convenient when we start manipulating series solutions of (3.1.1.1).

REMARK 3.1.4. For $n = 1$, (3.1.1.1) becomes

$$(z-1)D + (z-1)(1-\beta_1) + z(\alpha_1-\beta_1+1) = 0$$

with formal solutions

$$y = cz^{1-\beta_1}(z-1)^{\alpha_1-\beta_1+1}.$$

We next recall the explicit description of formal solutions of (3.1.1.1) at $z = 0$. The formal solutions at $z = \infty$ may be described similarly by interchanging the roles of the α and the β. The formal solutions at $z = 1$ behave somewhat differently; see [**5**, Proposition 2.8].

DEFINITION 3.1.5. For n a nonnegative integer, define the rising Pochhammer symbol

$$(\alpha)_n := \alpha(\alpha+1)\cdots(\alpha+n-1).$$

Define the *Clausen–Thomae hypergeometric series*

$$ {}_mF_{n-1}\left(\begin{matrix}\alpha_1,\dots,\alpha_m\\ \beta_1,\dots,\beta_{n-1}\end{matrix}\middle| z\right) := \sum_{k=0}^{\infty} \frac{(\alpha_1)_k\cdots(\alpha_m)_k}{(\beta_1)_k\cdots(\beta_{n-1})_k}\frac{z^k}{k!}. $$

The case $m = n = 3$ was first considered by Clausen [**12**]; the general case was first considered by Thomae [**42**].

PROPOSITION 3.1.6. *In* (3.1.1.1), *suppose that* $\beta_n = 1$ *(so that* $(\beta_n)_k = k!$*) and that no* β_i *is a nonpositive integer (which is to say that* $(\beta_i)_k \neq 0$ *for all* $k \geq 0$*). Then*

$$ {}_mF_{n-1}\left(\begin{matrix}\alpha_1,\dots,\alpha_m\\ \beta_1,\dots,\beta_{n-1}\end{matrix}\middle| z\right) $$

is a solution of (3.1.1.1) *in* $\mathbb{C}[\![z]\!]$.

PROOF. This may be seen by a direct calculation: applying the operator $z(D+\alpha_1)\cdots(D+\alpha_m)$ to the given series yields

$$ \sum_{k=0}^{\infty} \frac{(\alpha_1)_{k+1}\cdots(\alpha_m)_{k+1}}{(\beta_1)_k\cdots(\beta_{n-1})_k}\frac{z^{k+1}}{k!} $$

while applying $(D+\beta_1-1)\cdots(D+\beta_n-1) = (D+\beta_1-1)\cdots(D+\beta_{n-1}-1)D$ yields the equivalent expression

$$ \sum_{k=0}^{\infty} \frac{(\alpha_1)_k\cdots(\alpha_m)_k}{(\beta_1)_{k-1}\cdots(\beta_{n-1})_{k-1}}\frac{kz^k}{k!}. \qquad \square $$

COROLLARY 3.1.7. *In* (3.1.1.1), *suppose that* $m \leq n$ *and that* $\beta_1, \dots, \beta_n \in \mathbb{Q}$ *are pairwise distinct modulo* $\mathbb{Z}$. *Then the sums*

(3.1.7.1)

$$ z^{1-\beta_i}\,{}_mF_{n-1}\left(\begin{matrix}\alpha_1-\beta_i+1,\dots,\alpha_m-\beta_i+1\\ \beta_1-\beta_i+1,\dots,\widehat{\beta_i-\beta_i+1},\dots,\beta_n-\beta_i+1\end{matrix}\middle| z\right) \qquad (i = 1,\dots,n) $$

form a $\mathbb{C}$*-basis of the solutions of* (3.1.1.1) *in the Puiseux field* $\bigcup_{l=1}^{\infty}\mathbb{C}((z^{1/l}))$.

By formally differentiating with respect to parameters, we see what happens when some of the β's come together modulo $\mathbb{Z}$.

COROLLARY 3.1.8. *In* (3.1.1.1), *suppose that no two of* $\beta_1, \dots, \beta_n \in \mathbb{Q}$ *differ by* a nonzero *integer (e.g., because they all belong to* $[0,1)$*). For each* $\beta \in \{\beta_1,\dots,\beta_n\}$ *occurring with multiplicity* μ, *for* $= 1, \dots, \mu-1$, *consider the sums*

(3.1.8.1)

$$ z^{1-\beta}\sum_{i=0}^{j}\frac{j!(\log z)^{j-i}}{(j-i)!}\sum_{k=0}^{\infty}[\epsilon^i]\left(\frac{(\alpha_1-\beta+1+\epsilon)_k\cdots(\alpha_m-\beta+1+\epsilon)_k}{(\beta_1-\beta+1+\epsilon)_k\cdots(\beta_n-\beta+1+\epsilon)_k}\right)z^k $$

where $[\epsilon^i](*)$ *means the coefficient of* ϵ^i *of the expansion of* $*$ *as a formal power series in* ϵ. *These then form a* $\mathbb{C}$*-basis of the solutions of* (3.1.1.1) *in the ring* $\bigcup_{m=1}^{\infty}\mathbb{C}((z^{1/m}))[\log z]$.

PROOF. For $i = 0$, Proposition 3.1.6 implies that (3.1.8.1) is a solution for $\epsilon = 0$. We obtain $\mu - 1$ additional linearly independent solutions by formally differentiating with respect to $-\beta$; noting that the derivative of $z^{1-\beta}$ with respect to $-\beta$ is $(\log z)z^{1-\beta}$, we obtain the claimed formula. $\square$

COROLLARY 3.1.9. *In (3.1.1.1), suppose that* $\beta_1, \ldots, \beta_n \in \mathbb{Q}$ *and* $0 \leq \beta_1 \leq \cdots \leq \beta_n < 1$. *Let* $i_1 < \cdots < i_l$ *be the sequence of indices* $i \in \{1, \ldots, n\}$ *for which either* $i = 1$, *or* $i > 1$ *and* $\beta_{i-1} < \beta_i$. *For* $h = 1, \ldots, l$, *let* μ_h *denote the multiplicity of* β_{i_h} *(so that* $\mu_j = i_{h+1} - i_h$ *if* $h < l$ *and* $n+1-i_h$ *otherwise). Define the series* $f_1, \ldots, f_n \in \mathbb{C}[\![z]\!]$ *by the following formula: for* $h = 1, \ldots, l$ *and* $j = 0, \ldots, \mu_h - 1$,

$$f_{i_h+j} := \frac{1}{j!}\sum_{k=0}^{\infty}[\epsilon^j]\left(\frac{(\alpha_1-\beta+1+\epsilon)_k \cdots (\alpha_m-\beta+1+\epsilon)_k}{(\beta_1-\beta+1+\epsilon)_k \cdots (\beta_n-\beta+1+\epsilon)_k}\right)z^k.$$

Let U *be the matrix over* $\mathbb{C}[\![z]\!]$ *given by the following formula: for* $h = 1, \ldots, l$; $i = 1, \ldots, n$; $j = 0, \ldots, \mu_h - 1$,

$$U_{i(i_h+j)} = \sum_{k=\max\{0,j-i+1\}}^{j} \frac{j!(i-1)!}{k!(j-k)!(i-1-j+k)!}(D+1-\beta_{i_h})^{i-1-j+k}(f_{i_h+k}).$$

Then U *is invertible and* N_U *is a block matrix with block lengths* $\mu_1, \ldots, \mu_m$ *in which*

$$(N_U)_{(i_h+i)(i_h+j)} = \begin{cases} \beta_{i_h} - 1 & i = j \\ -j & j = i+1 \\ 0 & \text{otherwise} \end{cases} \qquad (h = 1, \ldots, m; 0 \leq i, j \leq \mu_h - 1).$$

PROOF. In the ring $\mathbb{C}((z))[\log z]$, we may define the elements $g_1, \ldots, g_n$ so that for $h = 1, \ldots, m$, $j = 0, \ldots, \mu_h - 1$, the series g_{i_h+j} is given by (3.1.8.1) for $\beta = \beta_{i_h}$, omitting the factor of $z^{1-\beta}$. Define the invertible $n \times n$ matrix V over $\mathbb{C}((z))[\log z]$ by setting $V_{ij} = (D+1-\beta_j)^{i-1}(g_j)$; then N_V is the diagonal matrix with entries $\beta_1 - 1, \ldots, \beta_n - 1$.

By construction, we have

$$g_{i_h+j} = \sum_{k=0}^{j}\binom{j}{k}(\log z)^k f_{i_h+j-k} \qquad (j = 0, \ldots, \mu_h - 1);$$

consequently, for $i = 1, \ldots, n$ we have

$$\begin{aligned}(D+1-\beta_{i_h})^{i-1}(g_{i_h+j}) &= \sum_{l=0}^{j}\binom{j}{l}(D+1-\beta_{i_h})^{i-1}((\log z)^l f_{i_h+j-l}) \\ &= \sum_{k=0}^{j}\binom{j}{k}(\log z)^{j-k}\sum_{l=j-k}^{j} *(D+1-\beta_{i_h})^{i-1-l+j-k}(f_{i_h+j-l}), \\ * &= \frac{k!(i-1)!}{(j-l)!(l-j+k)!(i-1-l+j-k)!}.\end{aligned}$$

That is, we have $V = UW$ where W is the block matrix with block lengths $\mu_1, \ldots, \mu_m$ in which

$$W_{(i_h+i)(i_h+j)} = \binom{j}{i}(\log z)^{j-i} \qquad (0 \leq i, j \leq \mu_h - 1);$$

it follows that $N_U = WN_VW^{-1} + WD(W^{-1})$. Since each block of N_V is a scalar matrix, we have $WN_VW^{-1} = N_V$; meanwhile, an elementary computation shows that the h-th block of $WD(W^{-1})$ is nilpotent with superdiagonal entries $-1, -2, \ldots, -\mu_h + 1$. □

We recall the local structure of the singularities of (3.1.1.1) in the case $m = n$.

PROPOSITION 3.1.10. *For $m = n$, the equation (3.1.1.1) is regular with singularities at $0, 1, \infty$ having local exponents as follows:*

$$\begin{aligned} z = 0: &\quad 1-\beta_1, \dots, 1-\beta_n \\ z = \infty: &\quad \alpha_1, \dots, \alpha_n \\ z = 1: &\quad 0, \dots, n-2, \gamma, \qquad \gamma := \sum_{i=1}^n \beta_i - \sum_{i=1}^n \alpha_i. \end{aligned}$$

PROOF. See [**5**, §2]. □

Although we will not use this overtly, for context we recall the explicit description of the monodromy representation of (3.1.1.1).

PROPOSITION 3.1.11. *Suppose that $m = n$ and that $\alpha_i - \beta_j \notin \mathbb{Z}$ for $i, j = 1, \dots, n$.*

(a) *Put $a_i := \exp(2\pi i \alpha_i), b_i := \exp(2\pi i \beta_i)$ and define the polynomials*

$$\prod_{i=1}^n (T - a_i) = T^n + A_1 T^{n-1} + \cdots + A_n, \quad \prod_{i=1}^n (T - b_i) = T^n + B_1 T^{n-1} + \cdots + B_n.$$

Then in a suitable basis (see Remark 3.1.12), the local monodromy operators (3.1.1.1) may taken to be

$$h_0 := B^{-1}, \qquad h_1 := A^{-1}B, \qquad h_\infty := A \in \mathrm{GL}_n(\mathbb{C}),$$

$$A := \begin{pmatrix} 0 & 0 & \cdots & 0 & -A_n \\ 1 & 0 & \cdots & 0 & -A_{n-1} \\ \vdots & & \ddots & & \vdots \\ 0 & 0 & \cdots & 1 & -A_1 \end{pmatrix}, \qquad B := \begin{pmatrix} 0 & 0 & \cdots & 0 & -B_n \\ 1 & 0 & \cdots & 0 & -B_{n-1} \\ \vdots & & \ddots & & \vdots \\ 0 & 0 & \cdots & 1 & -B_1 \end{pmatrix}.$$

(b) *The representation described in (a) is irreducible.*

(c) *The matrix h_1 is a complex reflection with special eigenvalue $c := \exp(2\pi i \gamma)$, meaning that $h_1 - 1$ has rank 1.*

PROOF. Part (a) is a theorem of Levelt [**5**, Theorem 3.5]. Parts (b) and (c) are immediate corollaries; see [**5**, Proposition 3.3] for (b) and [**5**, Proposition 2.10] for (c). □

REMARK 3.1.12. In Proposition 3.1.11, if one further assumes that the α_i and β_j are all distinct mod $\mathbb{Z}$, one can make the choice of a "suitable basis" quite explicit in terms of the local solutions given by Corollary 3.1.7. This was originally shown by Golyshev–Mellit [**23**].

REMARK 3.1.13. In case $m \neq n$, the local structure of the singularities of (3.1.1.1) is rather different; to simplify notation, we assume that $m < n$. In this case, (3.1.1.1) is of order n and its local monodromy at 0 is as described above; however, we no longer have a singularity at $z = 1$, and the singularity at $z = \infty$ is now irregular. This can be understood in terms of *confluence*, where the regular singularities at 1 and ∞ have coalesced into an irregular singularity upon degeneration of one of the parameters. To make this more explicit, consider the one-parameter family of hypergeometric equations

$$P(\alpha_1, \dots, \alpha_m, 1/t, \dots, 1/t; \beta_1, \dots, \beta_n)$$

indexed by a parameter t. This is equivalent via the substitution $z \mapsto t^{n-m}z$ to the equation

$$z\prod_{i=1}^{m}(D+\alpha_i)\prod_{i=m+1}^{n}(tD+1)-\prod_{j=1}^{n}(D+\beta_j-1)$$

with a regular singularity at $z=t^{m-n}$. Taking the limit as $t\to 0$ yields the operator $P(\alpha_1,\dots,\alpha_m;\beta_1,\dots,\beta_n)$.

3.2. The GKZ interpretation. In preparation for adopting the point of view of Dwork [**16**], we recall the description of the hypergeometric equation (3.1.1.1) in terms of a GKZ (Gelfand–Kapranov–Zelevinsky) A-hypergeometric system, following [**18**] (see also [**1**], [**10**, §1.4], and [**9**, §2]).

We begin by rewriting the hypergeometric equation to simplify the dependence on the parameters $\underline{\alpha},\underline{\beta}$ at the expense of replacing the original series with a function of multiple variables. (Warning: the use of the letter Φ here has nothing to do with the Frobenius intertwiners discussed in §2.3.)

LEMMA 3.2.1. *Consider a function $\Phi(\underline{x},\underline{y})$ of indeterminates $\underline{x}=x_1,\dots,x_m$ and $\underline{y}=y_1,\dots,y_n$. (For the moment, we leave it unspecified what sort of function we have in mind.)*

(a) *The function Φ is annihilated by the operators*

$$x_j\frac{\partial}{\partial x_j}+y_k\frac{\partial}{\partial y_k}+\alpha_j-\beta_k+1 \qquad (j=1,\dots,m;k=1,\dots,n) \tag{3.2.1.1}$$

if and only if there exists a univariate function $f(z)$ such that

$$\Phi(\underline{x},\underline{y})=x_1^{-\alpha_1}\cdots x_m^{-\alpha_m}y_1^{\beta_1-1}\cdots y_n^{\beta_n-1}f((-1)^m x_1^{-1}\cdots x_m^{-1}y_1\cdots y_n). \tag{3.2.1.2}$$

(b) *For Φ, f satisfying (3.2.1.2), Φ is annihilated by the operator*

$$\prod_{j=1}^{m}\frac{\partial}{\partial x_j}-\prod_{j=1}^{n}\frac{\partial}{\partial y_j}. \tag{3.2.1.3}$$

if and only if f is a solution of the hypergeometric equation (3.1.1.1).

PROOF. For Φ as in (3.2.1.2) and $z=(-1)^m x_1^{-1}\cdots x_m^{-1}y_1\cdots y_n$, we have

$$x_j\frac{\partial}{\partial x_j}(\Phi)(\underline{x},\underline{y})=((-D-\alpha_j)(f))(z) \tag{3.2.1.4}$$

$$y_j\frac{\partial}{\partial y_j}(\Phi)(\underline{x},\underline{y})=((D+\beta_j-1)(f))(z). \tag{3.2.1.5}$$

In particular, any such Φ satisfies (3.2.1.1). Conversely, to check that any Φ satisfying (3.2.1.1) satisfies (3.2.1.2) for some f, we may formally reduce to the case where $\alpha_i=0,\beta_i=1$ for all i. In this case, (3.2.1.1) implies that Φ remains constant under any substitution of the form

$$x_j\mapsto cx_j, \qquad y_k\mapsto cy_k$$

(for some j,k, with the other variables left unchanged); consequently, (3.2.1.2) holds for

$$f(z):=\Phi(1,\dots,1,(-1)^m z).$$

This proves (a).

For z as above, the operator (3.2.1.3) may be rewritten as

$$y_1^{-1}\cdots y_n^{-1}\left(z\prod_{j=1}^m\left(-x_j\frac{\partial}{\partial x_j}\right)-\prod_{j=1}^n\left(y_j\frac{\partial}{\partial y_j}\right)\right).$$

This makes it clear that from (3.2.1.4), (3.2.1.5), we immediately deduce (b). □

COROLLARY 3.2.2. *Suppose that $\beta_1, \dots, \beta_n$ are pairwise distinct modulo $\mathbb{Z}$. In terms of the indeterminates $\underline{x}$, $\underline{y} = x_1, \dots, x_m, y_1, \dots, y_n$, for $i = 1, \dots, n$ define (formally)*

$$f_i(z) := z^{1-\beta_i}{}_mF_{n-1}\left(\begin{matrix}\alpha_1-\beta_i+1,\dots,\alpha_m-\beta_i+1\\ \beta_1-\beta_i+1,\dots,\widehat{\beta_i-\beta_i+1},\dots,\beta_n-\beta_i+1\end{matrix}\middle| z\right)$$
$$\Phi_i(\underline{x},\underline{y}) := x_1^{-\alpha_1}\cdots x_m^{-\alpha_m}y_1^{\beta_1-1}\cdots y_n^{\beta_n-1}f_i((-1)^m x_1^{-1}\cdots x_m^{-1}y_1\cdots y_n).$$

Then $\Phi_1, \dots, \Phi_n$ are all annihilated by the operators (3.2.1.1) *and* (3.2.1.3).

PROOF. Combine Lemma 3.2.1 with Corollary 3.1.7. □

DEFINITION 3.2.3. For m a positive integer, let $W_m := \mathbb{C}\langle x_1, \dots, x_m, \partial_1, \dots, \partial_m\rangle$ denote the *Weyl algebra*, i.e., the quotient of the noncommutative polynomial algebra in $x_1, \dots, x_m, \partial_1, \dots, \partial_m$ by the two-sided ideal generated by

$$x_ix_j - x_jx_i,\ \partial_i\partial_j - \partial_j\partial_i,\ \partial_ix_i - x_i\partial_i - 1 \qquad (i, j = 1, \dots, m).$$

We write θ_i as shorthand for $x_i\partial_i$.

For d a nonnegative integer, let A be a $d \times m$ matrix over $\mathbb{Z}$. (In the notation of [**1**, §2], our d is n therein, our m is N therein, and the columns of A correspond to the lattice points therein.) The *toric ideal* associated to A is the ideal

$$I_A = \{\partial^u - \partial^v : u, v \in \mathbb{Z}^m_{\geq 0}, Au = Av\} \subseteq \mathbb{C}[\partial_1, \dots, \partial_m].$$

For $\delta \in \mathbb{C}^d$ a column vector, for $i = 1, \dots, d$ we may define an *Euler operator*

$$A_{i1}\theta_1 + \cdots + A_{im}\theta_m - \delta_i \in W_m.$$

The *GKZ ideal* (or *hypergeometric ideal*) defined by A and δ is the left ideal $J_{A,\delta}$ of W_m generated by I_A and the Euler operators.

EXAMPLE 3.2.4. Define the $(m+n-1) \times (m+n)$ matrix A over $\mathbb{Z}$ by the block expression

$$A = \begin{pmatrix} I_m & 0 & 1 \\ 0 & -I_{n-1} & 1 \end{pmatrix};$$

the toric ideal is generated by $\partial_1 \cdots \partial_m - \partial_{m+1} \cdots \partial_{m+n}$. Let $\delta \in \mathbb{C}^{m+n}$ be the column vector

$$(\alpha_1 - \beta_n + 1, \dots, \alpha_m - \beta_n + 1, \beta_1 - \beta_n, \dots, \beta_{n-1} - \beta_n);$$

the Euler operators then have the form

$$\theta_j + \theta_{m+n} + \alpha_j - \beta_n + 1 \quad (j = 1, \dots, m)$$
$$-\theta_{m+j} + \theta_{m+n} + \beta_j - \beta_n \quad (j = 1, \dots, n-1).$$

By Lemma 3.2.1, the formula

(3.2.4.1)
$$\Phi(x_1, \dots, x_{m+n}) = x_1^{-\alpha_1} \cdots x_m^{-\alpha_m} x_{m+1}^{\beta_1-1} \cdots x_{m+n}^{\beta_n-1} f((-1)^m x_1^{-1} \cdots x_m^{-1} x_{m+1} \cdots x_{m+n})$$

defines a bijection between the functions $f(z)$ satisfying (3.1.1.1) and the functions $\Phi(x_1, \dots, x_{m+n})$ annihilated by $J_{A,\delta}$.

It will be useful to also have a symmetric variant of Example 3.2.4.

EXAMPLE 3.2.5. Define an $mn \times (m+n)$ matrix A over $\mathbb{Z}$, using the index set $\{1, \dots, m\} \times \{1, \dots, n\}$ in place of $\{1, \dots, mn\}$, by

$$A_{(i_1,i_2)j} = \begin{cases} 1 & j \in \{i_1, m+i_2\} \\ 0 & \text{otherwise.} \end{cases}$$

and a column vector $\delta \in \mathbb{C}^{mn}$ by

$$\delta_{(i_1,i_2)} = \alpha_{i_1} - \beta_{i_2} + 1.$$

The Euler operators then have the form

$$\theta_{i_1} + \theta_{i_2} + \alpha_{i_1} - \beta_{i_2} + 1 \quad (i_1 = 1, \dots, m; i_2 = 1, \dots, n).$$

This GKZ system is isomorphic to the previous one, in a sense to be made explicit in §3.4.

REMARK 3.2.6. Let d', m' be two more positive integers, let A be a $d' \times m'$ matrix over $\mathbb{Z}$, and let $\delta' \in \mathbb{C}^{d'}$. We then have a canonical isomorphism of $\mathbb{C}$-vector spaces

$$W_m/J_{A,\delta} \otimes_{\mathbb{C}} W_{m'}/J_{A',\delta'} \cong W_{m+m'}/J_{A\oplus A',\delta\oplus\delta'}$$

which promotes to an isomorphism of left $W_{m+m'}$-modules if we identify the variables of $W_{m'}$ with the variables $x_{m+1}, \dots, x_{m+m'}, \partial_{m+1}, \dots, \partial_{m+m'}$ of $W_{m+m'}$.

REMARK 3.2.7. In Example 3.2.4, if we drop the last column, the toric ideal becomes the zero ideal. In this case, the functions annihilated by $J_{A,\delta}$ are just the constant multiples of $x_1^{-\alpha_1+\beta_n-1} \cdots x_m^{-\alpha_m+\beta_n-1} x_{m+1}^{\beta_1-\beta_n} \cdots x_{m+n-1}^{\beta_{n-1}-\beta_n}$; this can be viewed as an instance of the product construction described in Remark 3.2.6.

REMARK 3.2.8. A comment related to Remark 3.2.7 is that the definition of a GKZ system in Example 3.2.4 is insensitive to an overall translation

$$\alpha_i \mapsto \alpha_i + c, \qquad \beta_i \mapsto \beta_i + c;$$

the value of c only appears in the comparison with the hypergeometric equation in (3.2.4.1) (and specifically in the exponents of the leading powers).

3.3. Dwork's exponential module. Returning to the general GKZ setup, we now introduce Dwork's construction of the exponential module (compare [**17**, §4]).

DEFINITION 3.3.1. Retain notation as in Definition 3.2.3. Let R_A be the $\mathbb{C}$-subalgebra of $\mathbb{C}[X_1^{\pm}, \dots, X_d^{\pm}]$ generated by the monomials $X^{(j)} := X_1^{A_{1j}} \cdots X_d^{A_{dj}}$ for $j = 1, \dots, m$. Define also $R_A[x] := R_A[x_1, \dots, x_m]$. Define the element

$$g_A := \lambda \sum_{j=1}^{m} x_j X^{(j)} \in R_A[x].$$

(In the original construction one takes $\lambda = 1$; since we can absorb λ by rescaling x_j there is no extra generality in varying λ, but this will be convenient for the construction of Frobenius structures.)

There are obvious "natural" actions of the derivations

$$\partial_1, \dots, \partial_m, \qquad \Theta_1, \dots, \Theta_d := X_1 \frac{\partial}{\partial X_1}, \dots, X_d \frac{\partial}{\partial X_d}$$

on $R_A[x]$ (but not $\frac{\partial}{\partial X_i}$ in general). Define the twisted operators

$$\partial_{A,j} := \partial_j + \partial_j(g_A) = \partial_j + \lambda x_j X^{(j)}$$

$$D_{A,\delta,i} := \Theta_i + \Theta_i(g_A) + \delta_i = X_i \frac{\partial}{\partial X_i} + \delta_i + \lambda \sum_{j=1}^{m} A_{ij} x_j X^{(j)}.$$

We give $R_A[x]$ the structure of a left W_m-module by specifying that ∂_j acts via $\partial_{A,j}$.

REMARK 3.3.2. In the setting of Example 3.2.4, we have

$$\begin{aligned} g_A &= \lambda(x_1 X_1 + \cdots + x_m X_m + x_{m+1} X_{m+1}^{-1} + \cdots \\ &\qquad + x_{m+n-1} X_{m+n-1}^{-1} + x_{m+n} X_1 \cdots X_{m+n-1}) \\ D_{A,\delta,i} &= X_i \frac{\partial}{\partial X_i} + \lambda x_i X_i + \lambda x_{m+n} X_1 \cdots X_{m+n-1} + \alpha_i - \beta_n + 1 \\ D_{A,\delta,i} &= X_i \frac{\partial}{\partial X_i} - \lambda x_i X_i^{-1} + \lambda x_{m+n} X_1 \cdots X_{m+n-1} + \beta_{i-m} - \beta_n. \end{aligned}$$

where the second and third equations are for $i = 1, \dots, m$ and $i = m+1, \dots, m+n-1$ respectively.

LEMMA 3.3.3. *The formula*

$$x_1 \mapsto x_1, \dots, x_m \mapsto x_m, \partial_1 \mapsto X^{(1)}, \dots, \partial_m \mapsto X^{(m)}$$

defines a surjective homomorphism $\phi : W_m \to R_A[x]$ of left W_m-modules (for the exotic module structure on $R_A[x]$ from Definition 3.3.1) which induces the following isomorphisms of left W_m-modules:

$$W_m / W_m I_A \cong R_A[x]$$

$$W_m / J_{A,\delta} \cong R_A[x] / \sum_{i=1}^{d} D_{A,\delta,i} R_A[x].$$

PROOF. See [**1**, Theorem 4.4]. (Compare also [**18**, Theorem 6.8] and [**16**, Corollary 11.1.3].) □

REMARK 3.3.4. Even beyond the setting of Example 3.2.4, one can give a good "toric" description of $W_m/J_{A,\delta}$. As this is not necessary for our purposes, we defer to [**1**] for details.

3.4. Morphisms of A-hypergeometric systems.

DEFINITION 3.4.1. Let A' be a $d' \times m'$ matrix over $\mathbb{Z}$ and let $\delta' \in \mathbb{C}^{d'}$ be a column vector. By a *morphism* from the GKZ hypergeometric system with parameters (A, δ) to the GKZ hypergeometric system with parameters (A', δ'), we will mean a homomorphism $\psi : R_A[x] \to R_{A'}[x]$ of $\mathbb{C}$-modules which induces a homomorphism

$$\overline{\psi} : R_A[x] / \sum_{i=1}^{d} D_{A,\delta,i} R_A[x] \to R_{A'}[x] / \sum_{i'=1}^{d'} D_{A',\delta',i'} R_{A'}[x].$$

In order to make this meaningful, we must also have some compatibility with the ∂_j; we will describe this on a case-by-case basis.

CONSTRUCTION 3.4.2. Let B be a $d' \times d$ matrix over $\mathbb{Z}$ and let B' be a $d \times d'$ matrix over $\mathbb{Z}$ satisfying

$$BA = A', \qquad B'A' = A, \qquad B\delta = \delta', \qquad B'\delta' = \delta.$$

Consider the $\mathbb{C}$-linear ring homomorphisms $\psi : R_A[x] \to R_{A'}[x]$, $\psi' : R_{A'}[x] \to R_A[x]$ of $\mathbb{C}$-modules given by

$$\psi : \quad x_j \mapsto x_j, \qquad X_i \mapsto \prod_{i'=1}^{d'} X_{i'}^{B_{i'i}},$$

$$\psi' : \quad x_j \mapsto x_j, \qquad X_{i'} \mapsto \prod_{i=1}^{d} X_i^{B'_{ii'}}.$$

These satisfy the following identities:

$$\begin{aligned} \psi' \circ \psi = \mathrm{id}_{R_A[x]}, &\qquad \psi \circ \psi' = \mathrm{id}_{R_{A'}[x]}, \\ \psi(g_A) = g_{A'}, &\qquad \psi'(g_{A'}) = A, \\ D_{A',\delta',i'} \circ \psi &= \sum_i B_{i'i} \psi \circ D_{A,\delta,i}, \\ D_{A,\delta,i} \circ \psi' &= \sum_{i'} B'_{ii'} \psi' \circ D_{A',\delta',i'}. \end{aligned}$$

Consequently, ψ and ψ' define morphisms $(A, \delta) \to (A', \delta')$, $(A', \delta') \to (A, \delta)$ which are inverses of each other and manifestly commute with $\partial_1, \ldots, \partial_m$.

EXAMPLE 3.4.3. In Example 3.2.5, we have obvious isomorphisms as in Construction 3.4.2 corresponding to the permutations of $\alpha_1, \ldots, \alpha_n$ and of $\beta_1, \ldots, \beta_n$; however, these are not automorphisms because they change δ. We may similarly construct an isomorphism effecting the interchange of parameters from Remark 3.1.2.

EXAMPLE 3.4.4. We construct an isomorphism, in the sense of Construction 3.4.2, between the minimal GKZ system corresponding to a hypergeometric equation (Example 3.2.4) and the more symmetric version (Example 3.2.5). This uses the matrices

$$B_{(i_1,i_2)i} = \begin{cases} 1 & i = i_1 \\ -1 & i = m + i_2 \\ 0 & \text{otherwise.} \end{cases} \qquad B'_{i(i_1,i_2)} = \begin{cases} 1 & (i_1, i_2) = (i, n) \\ 1 & (i_1, i_2) = (i - m, n) \\ -1 & (i_1, i_2) = (i - m, i - m) \\ 0 & \text{otherwise.} \end{cases}$$

CONSTRUCTION 3.4.5. Let $T \in \mathbb{Z}^d$ be a vector in the column span of A and put $A' := A$, $\delta' := \delta - T$. Let $\psi : R_A[x] \to R_{A'}[x]$ be the map given by multiplication by $X_1^{T_1} \cdots X_d^{T_d}$; it satisfies

$$D_{A',\delta',i} \circ \psi = \psi \circ D_{A,\delta,i} \qquad (i = 1, \ldots, d)$$

and therefore defines a morphism $(A, \delta) \to (A', \delta')$ which manifestly commutes with $\partial_1, \ldots, \partial_m$.

We now consider some cases where the interaction with $\partial_1, \dots, \partial_m$ is a bit more subtle.

CONSTRUCTION 3.4.6. Let A' be the $d \times (m-1)$ matrix obtained from A by omitting the last column, and put $\delta' := \delta$. The ring homomorphism $\psi : R_A[x] \to R_{A'}[x]$ specializing x_m to 0 then satisfies

$$D_{A',\delta,i} \circ \psi = \psi \circ D_{A,\delta,i} \qquad (i = 1, \dots, d);$$

consequently, it defines a morphism $(A, \delta) \to (A', \delta')$ which commutes with the operators $\partial_1, \dots, \partial_{m-1}$. This does not extend to ∂_m because no such operator has been defined on $R_{A'}[x]$.

CONSTRUCTION 3.4.7. Put $A' = A$, $\delta' := p\delta$, and consider the morphism $\varphi : R_A[x] \to R_A[x]$ given by the substitution $x_j \mapsto x_j^p, X_i \mapsto X_i^p$. If we define

$$h := \lambda \sum_{j=1}^{m} (x_j X^{(j)} - (x_j X^{(j)})^p),$$

then

$$\begin{aligned} (D_{A,p\delta,i} - \Theta_i(h)) \circ \varphi &= p\varphi \circ D_{A,\delta,i} \qquad (i = 1, \dots, d) \\ (x_j \partial_{A,j} - x_j \partial_j(h)) \circ \varphi &= p\varphi \circ (x_j \partial_{A,j}) \qquad (j = 1, \dots, m). \end{aligned}$$

Formally, this means that $\exp(h)\varphi$ is a morphism which defines a Frobenius intertwiner (because of the factor of p in the second relation). In the p-adic context, this becomes not merely formal because of the convergence properties of the Dwork exponential series (for a suitable choice of λ).

REMARK 3.4.8. Somewhat tangentially to our current discussion, we note that one could also make the Frobenius intertwiner nonformal by working over a base ring equipped with a topology in which λ, $x_j - 1$, and $X_i - 1$ are small enough to make the series $\exp(h)$ convergent. This hints towards a potential connection with q-de Rham cohomology in the sense of Scholze [**39**] and prismatic cohomology in the sense of Bhatt–Scholze [**6**].

4. Hypergeometric Frobenius intertwiners

We now give our interpretation of Dwork's construction of Frobenius intertwiners for hypergeometric equations, based on morphisms of A-hypergeometric systems.

4.1. Existence of Frobenius intertwiners.

DEFINITION 4.1.1. Fix a choice of π in an algebraic closure of $\mathbb{Q}_p$ satisfying $\pi^{p-1} = -p$. Define the *Dwork exponential series* to be the series

$$E_\pi(t) := \sum_{j=0}^{\infty} c_j t^j = \exp(\pi(t - t^p));$$

it has radius of convergence $p^{(p-1)/p^2} > 1$ [**37**, §VII.2.4].

THEOREM 4.1.2 (Dwork). *Let $\underline{\alpha}; \underline{\beta}$ and $\underline{\alpha}', \underline{\beta}'$ be two sequences in $\mathbb{Z}_p$ such that $p\underline{\alpha}; p\underline{\beta}$ are congruent modulo $\mathbb{Z}$ to some permutations of $\underline{\alpha}', \underline{\beta}'$. Then over $\mathbb{Q}_p(\pi)$, there exists a Frobenius intertwiner between the connections corresponding to $P(\underline{\alpha}, \underline{\beta})$ and $P(\underline{\alpha}', \underline{\beta}')$.*

PROOF. We construct the desired intertwiner as follows.

- Take A, δ as in Example 3.2.5, then apply Construction 3.4.7 (taking λ there to be our chosen π) to replace $\underline{\alpha}; \underline{\beta}$ with $p\underline{\alpha}; p\underline{\beta}$.
- Use Construction 3.4.5 to replace $p\underline{\alpha}; p\underline{\beta}$ with a permutation of $\underline{\alpha}'; \underline{\beta}'$.
- Use Example 3.4.3 to undo the permutation of $\underline{\alpha}'; \underline{\beta}'$.

Note that the convergence property of the Dwork exponential is needed in the first step. □

REMARK 4.1.3. In Theorem 4.1.2, if $m = n$ and $\alpha_i - \beta_j \notin \mathbb{Z}$ for all i, j, then we may combine Lemma 2.3.6 and Proposition 3.1.11 to deduce that the Frobenius intertwiner is unique up to scalar multiplication. On the other hand, we can resolve the ambiguity completely by observing that the construction given by Theorem 4.1.2 has the following properties.

(a) In case $\underline{\alpha} = \underline{\alpha}', \underline{\beta} = \underline{\beta}'$, the Frobenius intertwiner is the identity.
(b) The construction of the Frobenius intertwiner is compatible (in a natural sense which we decline to notate) with permutations of each of $\underline{\alpha}, \underline{\beta}, \underline{\alpha}', \underline{\beta}'$.
(c) Suppose that

$$\alpha_i' = p\alpha_i + \mu_i, \qquad \beta_j' = p\alpha_j + \nu_j \qquad (\mu_i, \nu_j \in \mathbb{Z}). \tag{4.1.3.1}$$

Then the restriction of the Frobenius intertwiner to any fixed point of X varies p-adically continuously as we vary $\underline{\alpha}, \underline{\beta}$ while maintaining (4.1.3.1) and fixing μ_i, ν_j.

REMARK 4.1.4. For $n = 2$, an alternate construction of the Frobenius intertwiner has been given by Salinier [**38**] using rigidity; this has been generalized to all n by Vargas Montoya [**44**]. While this approach is technically simpler than Dwork's method, the latter is more useful for our ultimate aim of making explicit computations.

4.2. Gamma factors and the Dwork exponential series. In order to make use of Construction 3.4.7, we recall the description due to Dwork[1] [**7**], [**15**, §1] of the relationship between the Morita p-adic gamma function and Gauss sums provided by the Gross–Koblitz formula [**25**]. See Remark 5.1.2 for the geometric interpretation of this.

DEFINITION 4.2.1. Recall (or see [**37**, §VII.1.1]) that there exists a unique continuous function $\Gamma_p : \mathbb{Z}_p \to \mathbb{Z}_p^\times$ characterized by the properties

$$\Gamma_p(0) = 1 \tag{4.2.1.1}$$

$$\frac{\Gamma_p(x+1)}{\Gamma_p(x)} = \begin{cases} -x & x \notin p\mathbb{Z}_p \\ -1 & x \in \mathbb{Z}_p. \end{cases} \tag{4.2.1.2}$$

This function is the *Morita p-adic gamma function.*

DEFINITION 4.2.2. For $a, b \in \mathbb{Z}_{(p)} \setminus \mathbb{Z}$ with $pb - a = \mu \in \mathbb{Z}$, Dwork defines the symbol $\gamma_p(a, b) \in \mathbb{Q}_p(\pi)$ by the formula

$$\gamma_p(a, b) = \sum_{i \in \mathbb{Z}} c_{pi+\mu}(b)_i / (-\pi)^i.$$

[1] The attribution is predicated on the fact that [**7**] was written by Dwork under the pseudonym Maurizio Boyarsky [**31**, p. 341, first sidebar].

Equivalently, writing

$$\psi(f)(x) = \frac{1}{p} \sum_{z^p = x} f(x),$$

we have

$$\psi(x^{a-pb} E_\pi(x)) \equiv \gamma_p(a,b) \mod \left(x\frac{d}{dx} + b + \pi x \right) \mathbb{Q}_p(\pi)[\![x]\!]. \tag{4.2.2.1}$$

For fixed $\mu \in \mathbb{Z}$, using the series representation we may extend $\gamma_p(pb - \mu, b)$ to a continuous function of $b \in \mathbb{Z}_p$; note that $\gamma_p(0,0) = 1$. For $s, t \in \mathbb{Z}$, we have the functional equation [**15**, (1.7)]

$$\gamma_p(a+s, b+t) = \gamma_p(a,b)(-\pi)^{t-s} \frac{(a)_s}{(b)_t}. \tag{4.2.2.2}$$

THEOREM 4.2.3 (Dwork). *For $a, b \in \mathbb{Z}_{(p)}$ with $pb - a = \mu \in \{0, \dots, p-1\}$, we have*

$$\gamma_p(a,b) = \pi^\mu \Gamma_p(a).$$

PROOF. Using the above discussion, one checks that $\gamma(a,b)/\pi^\mu$ satisfies the defining properties (4.2.1.1), (4.2.1.2) of $\Gamma_p(a)$; this proves the claim. □

As indicated in [**7**], Theorem 4.2.3 can be viewed as an equivalent form of the Gross–Koblitz formula for Gauss sums [**25**]. In other words, we immediately compute the Frobenius intertwiners for hypergeometric equations of order 1.

COROLLARY 4.2.4. *Let $\{x\} := x - \lfloor x \rfloor$ denote the fractional part of x. In the case*

$$m = n = 1, \qquad \alpha_1, \alpha_1', \beta_1, \beta_1' \in [0,1), \qquad \alpha_1 \neq \beta_1,$$

for

$$\mu = p(\alpha_1 - \beta_1) - (\alpha_1' - \beta_1') \in \mathbb{Z},$$

the Frobenius interwiner of Theorem 4.1.2 is given by multiplication by

$$\gamma(\alpha_1' - \beta_1' + 1, \alpha_1 - \beta_1 + 1) := \pi^\mu \Gamma_p(\{\alpha_1' - \beta_1'\}) \times \left(\left(\begin{cases} \frac{1}{\alpha_1 - \beta_1} & \alpha_1 > \beta_1 \\ p & \alpha_1 < \beta_1 \end{cases} \right) \times \left(\begin{cases} \alpha_1' - \beta_1' & \alpha_1' > \beta_1' \\ -1 & \alpha_1' < \beta_1' \end{cases} \right) \right).$$

PROOF. We first make some auxiliary calculations in order to prepare for the use of Theorem 4.2.3. Note that

$$p\alpha_1 - \alpha_1', p\beta_1 - \beta_1' \in \mathbb{Z} \cap (-1, p) = \{0, \dots, p-1\}$$

and so

$$\mu = (p\alpha_1 - \alpha_1') - (p\beta_1 - \beta_1') \in \{1-p, \dots, p-1\}.$$

If $\alpha_1 > \beta_1$, then we also have $p(\alpha_1 - \beta_1) \in (0, p)$, $\alpha_1' - \beta_1' \in (-1,1)$ and so

$$\mu \in \{1-p, \dots, p-1\} \cap (-1, p+1) = \{0, \dots, p-1\}. \tag{4.2.4.1}$$

Similarly, if $\alpha_1 < \beta_1$, then $p(\alpha_1 - \beta_1) \in (-p, 0)$, $\alpha_1' - \beta_1' \in (-1,1)$ and so

$$\mu \in \{1-p, \dots, p-1\} \cap (-p-1, 1) = \{1-p, \dots, 0\}$$

and

$$p(\alpha_1 - \beta_1 + 1) - (\alpha_1' - \beta_1' + 1) = (p-1) + \mu \in \{0, \dots, p-1\}. \tag{4.2.4.2}$$

In particular, μ is either zero or has the same sign as $\alpha_1 - \beta_1$.

By (4.2.2.1) and (4.2.2.2), the Frobenius intertwiner is given by multiplication by

$$\gamma(\alpha_1' - \beta_1' + 1, \alpha_1 - \beta_1 + 1) = \gamma(\alpha_1' - \beta_1', \alpha_1 - \beta_1)\frac{\alpha_1' - \beta_1'}{\alpha_1 - \beta_1}.$$

In case $\alpha_1 < \beta_1$, we apply Theorem 4.2.3 and (4.2.4.2) to write

$$\gamma(\alpha_1' - \beta_1' + 1, \alpha_1 - \beta_1 + 1) = \pi^{(p-1)+\mu}\Gamma_p(\alpha_1' - \beta_1' + 1).$$

In case $\alpha_1 > \beta_1$, we may apply Theorem 4.2.3 and (4.2.4.1) to write

$$\gamma(\alpha_1' - \beta_1', \alpha_1 - \beta_1) = \pi^{\mu}\Gamma_p(\alpha_1' - \beta_1').$$

We can thus write the intertwiner as

$$\begin{cases} \pi^{\mu}\frac{\alpha_1'-\beta_1'}{\alpha_1-\beta_1}\Gamma_p(\alpha_1' - \beta_1') & \alpha_1 > \beta_1 \\ -p\pi^{\mu}\Gamma_p(\alpha_1' - \beta_1' + 1) & \alpha_1 < \beta_1. \end{cases} \tag{4.2.4.3}$$

Now note that if $\alpha_1' - \beta_1'$ and $\alpha_1 - \beta_1$ are of opposite sign, we cannot have $\mu = 0$, and so we can rewrite $(\alpha_1' - \beta_1')\Gamma_p(\alpha_1' - \beta_1')$ as $-\Gamma_p(\alpha_1' - \beta_1' + 1)$ or vice versa. This yields the stated formula. □

4.3. Specialization and factorization. Using the GKZ interpretation, we may immediately extend the previous computation to arbitrary rank.

HYPOTHESIS 4.3.1. *Throughout §4.3, suppose that $m \leq n$; $\alpha_i, \beta_j \in \mathbb{Z}_{(p)} \cap [0,1)$ for i, $j = 1, \ldots, n$; and $\alpha_i \neq \beta_j$ for i, $j = 1, \ldots, n$. Define $\alpha_i' := \{p\alpha_i\}, \beta_j' := \{p\beta_j\}$.*

THEOREM 4.3.2. *Suppose that $k \in \{1, \ldots, n\}$ is such that $\beta_j \neq \beta_k$ for $j \neq k$. Then the matrix Φ_λ for $\lambda = \beta_k$ is the 1×1 scalar*

$$\prod_{i=1}^{m} \gamma(\alpha_i' - \beta_k' + 1, \alpha_i - \beta_k + 1) \prod_{j=1}^{n} \gamma(\beta_j' - \beta_k' + 1, \beta_j - \beta_k + 1)^{-1}$$

(Note that the factor $j = k$ contributes 1 to the product.)

PROOF. For ease of notation we treat only the case $k = n$. In this case, under the GKZ interpretation, we may read off Φ_λ by specializing x_{m+n} to 0 via the morphism from Construction 3.4.6. In this case, as per Remark 3.2.7 we obtain the specified factorization. □

By combining Theorem 4.3.2 with Corollary 4.2.4, we get an explicit formula for the initial condition for the Frobenius intertwiner in the case where $\beta_1, \ldots, \beta_n$ are pairwise distinct mod $\mathbb{Z}$.

COROLLARY 4.3.3. *In addition to Hypothesis 4.3.1, suppose that $\beta_1, \ldots, \beta_n$ are pairwise distinct. Consider the formal solution matrix obtained by multiplying the function (3.1.7.1) corresponding to β_k by the scalar factor*

$$\frac{\prod_{i=1}^{m}(\alpha_i - \beta_k)_+}{\prod_{j=1}^{n}(\beta_j - \beta_k)_+}, \qquad (x)_+ := \begin{cases} x & x > 0 \\ 1 & x \leq 0. \end{cases} \tag{4.3.3.1}$$

Define the zigzag function *associated to $\underline{\alpha}, \underline{\beta}$ as the function $Z : \mathbb{R} \to \mathbb{R}$ given by*

$$Z(x) = \#\{i \in \{1, \ldots, m\} : \alpha_i < x\} - \#\{j \in \{1, \ldots, n\} : \beta_i < x\}.$$

Then the sole entry of Φ_λ for $\lambda = \beta_k$ can be written as

$$(-1)^{Z(\beta'_k)} p^{Z(\beta_k)} \mu^c \frac{\prod_{i=1}^m \Gamma_p(\{\alpha'_i - \beta'_k\})}{\prod_{j=1}^n \Gamma_p(\{\beta'_j - \beta'_k\})}$$

for

$$c := \sum_{i=1}^n (p\alpha_i - \alpha'_i) - \sum_{j=1}^n (p\beta_j - \beta'_j).$$

REMARK 4.3.4. In Corollary 4.3.3, the factor μ^c does not depend on k. We may thus eliminate it at the expense that our normalization no longer matches that of Theorem 4.2.3.

REMARK 4.3.5. In applications, we will typically be interested in the case where, in addition to the conditions of Hypothesis 4.3.1, one has that $m = n$ and $\underline{\alpha}, \underline{\beta} \subset \mathbb{Z}_{(p)} \cap [0,1)$ are *Galois-stable*, meaning that any two elements of $\mathbb{Z}_{(p)} \cap [0,1)$ with the same denominator occur with the same multiplicity in $\underline{\alpha}$ and $\underline{\beta}$. These conditions ensure the existence of a family of *hypergeometric motives* with this hypergeometric equation as associated Picard–Fuchs equation.

In this situation, a further renormalization beyond that of Remark 4.3.4 is sometimes warranted in order to ensure that the Frobenius structure correctly computes the characteristic polynomials of the p-Frobenius of the associated hypergeometric motives. This is achieved by taking the entry of Φ_λ to be

$$(-1)^{Z(\beta'_k)} p^{Z(\beta_k) - \min\{Z(\beta_*)\}} \frac{\prod_{i=1}^m \Gamma_p(\{\alpha'_i - \beta'_k\})/\Gamma_p(\alpha'_i)}{\prod_{j=1}^n \Gamma_p(\{\beta'_j - \beta'_k\})/\Gamma_p(\beta'_j)}.$$

The net effect of the factors $\Gamma_p(\alpha'_i)$ and $\Gamma_p(\beta'_j)$ is limited by the identity

$$\Gamma_p(x)\Gamma_p(1-x) = (-1)^y, \qquad y \in \{1, \dots, p\}, \quad y \equiv x \pmod{p}$$

and its special case

$$\Gamma_p\left(\frac{1}{2}\right)^2 = \left(\frac{-1}{p}\right) \qquad (p \neq 2).$$

4.4. An example with repeated parameters. In lieu of extending Theorem 4.3.2 to the case where the β_j are not all distinct (which would create some notational headaches), we sketch an example originally due to Shapiro [**40, 41**].

EXAMPLE 4.4.1. Consider the case

$$m = n = 4, \qquad \underline{\alpha}; \underline{\beta} = \left(\frac{1}{5}, \frac{2}{5}, \frac{3}{5}, \frac{4}{5}\right); (1,1,1,1).$$

This example is well-known; the corresponding hypergeometric equation is a Picard–Fuchs equation for the Dwork pencil of quintic threefolds.

Assume $p \neq 2, 5$. (The restriction $p \neq 5$ is essential; the restriction $p = 2$ is probably not, but is made in [**41**].) For $\lambda = 0$, the matrix $\Phi_{\lambda,0}$ is upper-triangular with eigenvalues $1, p, p^2, p^3$. To compute the off-diagonal entries, we use p-adic interpolation: consider the statement of Corollary 4.3.3 for

$$\underline{\beta} = (1, 1+\epsilon, 1+2\epsilon, 1+3\epsilon), \qquad \epsilon := \frac{p^n}{3p^n+1}.$$

For U the formal solution matrix, the matrix Φ_U equals the diagonal matrix whose k-diagonal entry (for $k = 0, \dots, 3$) equals

$$z^{(p-1)k\epsilon}(-p)^k \frac{\Gamma_p(-1/5 - k\epsilon)\Gamma_p(-2/5 - k\epsilon)\Gamma_p(-3/5 - k\epsilon)\Gamma_p(-4/5 - k\epsilon)}{\Gamma_p(-k\epsilon)\Gamma_p((1-k)\epsilon)\Gamma_p((2-k)\epsilon)\Gamma_p((3-k)\epsilon)}.$$

Using Definition 4.2.2 and Theorem 4.2.3, one may compute coefficients of the Taylor series for Γ_p (see for example [**41**, Proposition 3.1]); we may thus rewrite Φ_0 truncated at ϵ^4, and the formal solution matrix U truncated at ϵ^4 and z^4. Taking the limit of $\Phi = U\Phi_U\sigma(U^{-1})$ as $\epsilon \to 0^+$, and using the relationship between derivatives of Γ_p and p-adic zeta values (e.g., see [**11**, Proposition 11.5.19]), one may recover Shapiro's formula

$$\Phi_\lambda = \begin{pmatrix} 1 & 0 & 0 & \frac{2^3}{5^2}(p^3-1)\zeta_p(3) \\ 0 & p & 0 & 0 \\ 0 & 0 & p^2 & 0 \\ 0 & 0 & 0 & p^3 \end{pmatrix}.$$

We leave further details to the interested reader.

5. Applications to computation of L-functions

The formula of Dwork can be used as part of an efficient algorithm for computing Euler factors of L-functions associated to hypergeometric motives. We sketch this here. (In the case $n = 2$, an alternate approach has been described by Asakura [**2**].)

5.1. Hypergeometric motives.

Definition 5.1.1. Suppose that $m = n$ and that $\underline{\alpha}, \underline{\beta} \subset \mathbb{Q}$ are both Galois-stable. Then there exists a family of motives $H(\underline{\alpha}; \underline{\beta}; t)$ over $\mathbb{Q}(t)$ which for $t \neq \{0, 1\}$ is pure of dimension n and weight

$$w = \max(Z) - \min(Z) - 1$$

where Z denotes the zigzag function defined in Corollary 4.3.3. For example, this motive can be found inside the family of varieties considered in [**4**].

If we specialize to a value of t in $\overline{\mathbb{Q}}$, then the motive $H(\underline{\alpha}; \underline{\beta}; t)$ has good reduction at all places of the number field $\mathbb{Q}(t)$ at which $\alpha_1, \dots, \alpha_n, \beta_1, \dots, \beta_n$ have nonnegative valuation and $t, t^{-1}, t-1$ have nonnegative valuation. An excluded prime is said to be *wild* if the first condition fails (note that this does not depend on t) and *tame* otherwise.

Remark 5.1.2. When the Galois-stable condition holds and the β_j are pairwise distinct, the specialization of $H(\underline{\alpha}; \underline{\beta}; t)$ at $t = 0$ is a CM motive, whose associated L-function is therefore given by certain Jacobi sums. The formula given in Corollary 4.3.3 can also be derived by applying the Gross-Koblitz formula to these Jacobi sums.

When the β_j are not pairwise distinct, the specialization of $H(\underline{\alpha}; \underline{\beta}; t)$ at $t = 0$ becomes a mixed motive, whose L-function then includes a contribution from extension classes. Again, it should be possible to make an explicit link with degenerations of Corollary 4.3.3; for example, in Example 4.4.1, the appearance of $\zeta_p(3)$ should be related via motivic considerations to a corresponding appearance of $\zeta(3)$ in mirror symmetry [**35**].

REMARK 5.1.3. We expect that there are corresponding families of motives associated to GKZ systems associated to parameters (A, δ) with $\delta \in \mathbb{Q}^d$, under a suitable analogue of the Galois-stable condition: for each prime p for which $\delta \in \mathbb{Z}_{(p)}^d$, the GKZ system with parameters $(A, p\delta)$ should be isomorphic to the original one.

REMARK 5.1.4. The families $H(\underline{\alpha}; \underline{\beta}; t)$ and $H(\underline{\beta}; \underline{\alpha}; t^{-1})$ are isomorphic. This can be used in certain cases where one wants to make an asymmetric restriction on $\underline{\alpha}$ and $\underline{\beta}$, as in our computation of Frobenius structures.

5.2. The approach via trace formulas. Before describing the approach we have in mind, we describe an alternate approach for purposes of comparison.

REMARK 5.2.1. Suppose that $t \in \mathbb{Q} \setminus \{0, 1\}$ and p is a prime at which $H(\underline{\alpha}; \underline{\beta}; t)$ has good reduction. For f a positive integer, let H_{p^f} be the trace of the f-th power of the p-Frobenius acting on $H(\underline{\alpha}; \underline{\beta}; t)$; note that this depends only on the residue of t modulo p. By combining [**4**] with the Gross-Koblitz formula, one can obtain a highly practical formula for H_{p^f}; this is a poorly documented result of Cohen–Rodriguez Villegas–Watkins, but the formula can be found in the documentation of the MAGMA package on hypergeometric motives:
`http://magma.maths.usyd.edu.au/magma/handbook/hypergeometric_motives.`
The same formula is also implemented in SAGEMATH.

5.3. The approach via Frobenius structures. To simplify this discussion, we assume that $\beta_1, \dots, \beta_n$ are pairwise distinct. Recall that via Remark 5.1.4, we can swap $\underline{\alpha}$ with $\underline{\beta}$ to achieve these conditions in some cases where it is not initially satisfied.

Let N denote the companion matrix for the differential operator $P(\underline{\alpha}, \underline{\beta})$. Let U denote the formal solution matrix obtained from the matrix U of Corollary 3.1.9 by multiplying its k-th column by the factor (4.3.3.1) for $k = 1, \dots, n$. By Theorem 4.2.3 and Corollary 4.3.3, there is a Frobenius structure on N with $\Phi = \Phi_0 \sigma(U^{-1})$, where Φ_0 is the matrix with

$$(\Phi_0)_{i,j} = (-1)^{Z(\beta_i)} p^{Z(\beta_j) - \min\{Z(\beta_*)\}} \frac{\prod_{k=1}^n \Gamma_p(\{\alpha_k - \beta_i\})/\Gamma_p(\alpha_k)}{\prod_{k=1}^n \Gamma_p(\{\beta_k - \beta_i\})/\Gamma_p(\beta_k)} t^{1-p+\lfloor p\beta_j \rfloor}$$

whenever $\beta_i \equiv p\beta_j \pmod{\mathbb{Z}}$ and $(\Phi_0)_{i,j} = 0$ otherwise. Note that this computation nominally takes place in $\mathbb{Q}_p((t))$; in order to represent the elements of Φ as rigid analytic functions, we must multiply by a suitable power of $t - 1$, then truncate modulo suitable powers of p and t. One can then specialize t to any $(p-1)$-st root of unity to obtain a matrix whose characteristic polynomial gives the Euler factor of $H(\underline{\alpha}; \underline{\beta}; p)$. (Beware that we have not yet checked that the scalar normalization is correct. One way to do this would be to use this formula to reprove the Beukers–Cohen–Mellit trace formula.)

We have an experimental SAGEMATH implementation of this algorithm, and have done numerous tests to confirm its agreement with Beukers–Cohen–Mellit (albeit without fixing the precision estimates; see below). See [**33**].

REMARK 5.3.1. In order to make the previous algorithm rigorous, one must bound the p-adic and t-adic precision requirements. The power of $t - 1$ can be estimated using the method of [**34**]. This depends on estimating the p-adic valuation of Φ_0; this appears to be controlled by the p-adic valuations of the differences $\alpha_k - \beta_i$ and $\beta_k - \beta_i$. In any case, it appears that for a fixed p-adic truncation (which suffices for the computation of Euler factors), the power of $t-1$ is bounded independently of p; this means that the t-adic truncation can be bounded by cp for some constant c independently of p.

This has the following consequences for an average polynomial time algorithm. One is trying to evaluate the entries of the matrix

$$(t-1)^e U \Phi_0 \sigma(U^{-1}),$$

modulo some fixed power of p; they look like polynomial of degree bounded by cp where c is independent of p. This means that for the purposes of evaluation $\sigma(U^{-1})$, we need only a constant number of terms of U^{-1}; these coefficients are moreover rational numbers with no dependence at all on p. We may thus frame the problem as that of computing, for various primes p, a certain $\mathbb{Q}$-linear combination of coefficients of terms of U of the form t^{ap+b} for certain fixed pairs (a,b), then reducing the result modulo a fixed power of p.

5.4. Comparison of approaches. When comparing the relative efficacy of the trace formula and Frobenius structures, it is important to separate different use cases in which the relative strengths of the approaches play different roles. In the following discussion, we mostly ignore constants and logarithmic factors.

REMARK 5.4.1. Suppose we wish to compute $H_p(\underline{\alpha};\underline{\beta};t)$ for a single choice of $\underline{\alpha},\underline{\beta},t$ and a single prime p. Both approaches have complexity linear in p; however, the trace formula carries less overhead in this context and thus is preferable in practice. Moreover, if one repeats the computation for the same p and different values of $\underline{\alpha},\underline{\beta}$, one can cache the Mahler expansion of Γ_p for additional savings.

REMARK 5.4.2. Suppose we fix $\underline{\alpha},\underline{\beta},p$, and wish to compute either $H_p(\underline{\alpha};\underline{\beta};t)$ for all values of t (or equivalently, for $t \in \{2,\dots,p-1\}$). In this case, the trace formula can be computed as a polynomial in a variable (running over $(p-1)$-st roots of unity); alternatively, the Frobenius structure can be computed and then specialized repeatedly.

REMARK 5.4.3. Suppose we wish to compute the full Euler factor of the L-function associated to $H(\underline{\alpha};\underline{\beta};t)$ at a prime p. In this case, the trace formula approach requires computing $H_{p^f}(\underline{\alpha};\underline{\beta};t)$ for f ranging from 1 to half the degree of the associated L-function; the formula is a sum over p^f-1 terms. By contrast, the Frobenius structure computation gives the entire Euler factor at once, with complexity linearly in p.

REMARK 5.4.4. Suppose we wish to compute the first X Dirichlet coefficients of the L-function associated to $H(\underline{\alpha};\underline{\beta};t)$; this is the relevant use case when making numerical computations with the L-function. Using the trace formula directly scales quadratically in X; however, it should be possible to develop an *average polynomial time* algorithm in the sense of Harvey [**26**, **27**] (see also [**28**, **29**]). A partial result has been given by Costa–Kedlaya–Roe [**13**], who compute $H_p(\underline{\alpha};\underline{\beta};t) \pmod p$ for all primes $p \le X$ with complexity linear in X. As remarked upon in [**13**], it should be possible to adapt this approach to compute $H_p(\underline{\alpha};\underline{\beta};t)$ exactly for all primes $p \le X$ with similar complexity.

It is less clear how to include higher prime powers into this approach. However, one can use Frobenius structures to circumvent this difficulty, by directly computing full Euler factors for all primes $p \le X^{1/2}$, then using these to recover $H_{p^f}(\underline{\alpha};\underline{\beta};t)$

for all prime powers $p^f \leq X$ with $f > 1$. Since this involves $O(X^{1/2})$ computations each of complexity linear in $X^{1/2}$, this does not dominate the computation of prime Dirichlet coefficients.

REMARK 5.4.5. Suppose we wish to compute the full Euler factors of the L-function associated to $H(\underline{\alpha}; \underline{\beta}; t)$ at all primes $p \leq X$; this is the relevant use case when studying statistical properties of the Euler factors (e.g., the generalized Sato-Tate conjecture). In this case, it should be possible (and relatively straightforward) to give an average polynomial time computation using Frobenius structures; however, we do not develop this point further here.

6. Towards A-hypergeometric motives

In this paper, we have used only a restricted form of the theory of A-hypergeometric systems. However, it is likely that the circle of ideas giving rise to hypergeometric motives and L-functions can be extended beyond this special case; this question was posed at the end of the introduction of [**13**]. We record here some references that point towards such an extension.

The story of hypergeometric motives begins with Greene's construction of finite hypergeometric sums [**24**]. This was generalized to A-hypergeometric systems by Gelfand–Graev [**21**].

Greene's sums were reinterpreted in terms of ℓ-adic cohomology by Katz [**30**]. This interpretation was extended to the Gelfand–Graev construction by Lei Fu [**19**]. The p-adic construction described in this paper has been generalized by Fu–Wan–Zhang [**20**]. However, we know of no analogue of the Beukers–Cohen–Mellit construction.

References

[1] Alan Adolphson, *Hypergeometric functions and rings generated by monomials*, Duke Math. J. **73** (1994), no. 2, 269–290, DOI 10.1215/S0012-7094-94-07313-4. MR1262208

[2] M. Asakura, *An algorithm of computing special values of Dwork's p-adic hypergeometric functions in polynomial time*, arXiv:1909.02700v3, 2020.

[3] Francesco Baldassarri, *Continuity of the radius of convergence of differential equations on p-adic analytic curves*, Invent. Math. **182** (2010), no. 3, 513–584, DOI 10.1007/s00222-010-0266-7. MR2737705

[4] Frits Beukers, Henri Cohen, and Anton Mellit, *Finite hypergeometric functions*, Pure Appl. Math. Q. **11** (2015), no. 4, 559–589, DOI 10.4310/PAMQ.2015.v11.n4.a2. MR3613122

[5] F. Beukers and G. Heckman, *Monodromy for the hypergeometric function* ${}_nF_{n-1}$, Invent. Math. **95** (1989), no. 2, 325–354, DOI 10.1007/BF01393900. MR974906

[6] B. Bhatt and P. Scholze, *Prisms and prismatic cohomology*, arXiv:1905.08229v3, 2021.

[7] Maurizio Boyarsky, *p-adic gamma functions and Dwork cohomology*, Trans. Amer. Math. Soc. **257** (1980), no. 2, 359–369, DOI 10.2307/1998301. MR552263

[8] P. Candelas, X. de la Ossa, and D. van Straten, *Local zeta functions from Calabi–Yau differential equations*, arXiv:2104.07816v1, 2021.

[9] Alberto Castaño Domínguez and Christian Sevenheck, *Irregular Hodge filtration of some confluent hypergeometric systems*, J. Inst. Math. Jussieu **20** (2021), no. 2, 627–668, DOI 10.1017/S1474748019000288. MR4223435

[10] E. Cattani, *Three lectures on hypergeometric functions*, https://people.math.umass.edu/~cattani/hypergeom_lectures.pdf.

[11] Henri Cohen, *Number theory. Vol. II. Analytic and modern tools*, Graduate Texts in Mathematics, vol. 240, Springer, New York, 2007. MR2312338

[12] Th. Clausen, *Ueber die Fälle, wenn die Reihe von der Form*

$$y = 1 + \frac{\alpha}{1} \cdot \frac{\beta}{\gamma} x + \frac{\alpha . \alpha + 1}{1.2} \cdot \frac{\beta . \beta + 1}{\gamma \cdot \gamma + 1} x^2 + \text{etc.}$$

ein Quadrat von der Form

$$z = 1 + \frac{\alpha'}{1} \cdot \frac{\beta'}{\gamma'} \cdot \frac{\delta'}{\epsilon'} x + \frac{\alpha'.\alpha'+1}{1.2} \cdot \frac{\beta'.\beta'+1}{\gamma' \cdot \gamma'+1} \cdot \frac{\delta'.\delta'+1}{\epsilon'.\epsilon'+1} x^2 + \text{etc.}$$

hat (German), J. Reine Angew. Math. **3** (1828), 89–91, DOI 10.1515/crll.1828.3.89. MR1577682

[13] Edgar Costa, Kiran S. Kedlaya, and David Roe, *Hypergeometric L-functions in average polynomial time*, ANTS XIV—Proceedings of the Fourteenth Algorithmic Number Theory Symposium, Open Book Ser., vol. 4, Math. Sci. Publ., Berkeley, CA, 2020, pp. 143–159, DOI 10.2140/obs.2020.4.143. MR4235111

[14] B. Dwork, *On the uniqueness of Frobenius operator on differential equations*, Algebraic number theory, Adv. Stud. Pure Math., vol. 17, Academic Press, Boston, MA, 1989, pp. 89–96, DOI 10.2969/aspm/01710089. MR1097612

[15] Bernard Dwork, *On the Boyarsky principle*, Amer. J. Math. **105** (1983), no. 1, 115–156, DOI 10.2307/2374383. MR692108

[16] Bernard Dwork, *Generalized hypergeometric functions*, Oxford Mathematical Monographs, The Clarendon Press, Oxford University Press, New York, 1990. Oxford Science Publications. MR1085482

[17] B. Dwork, *Cohomological interpretation of hypergeometric series*, Rend. Sem. Mat. Univ. Padova **90** (1993), 239–263. MR1257141

[18] B. Dwork and F. Loeser, *Hypergeometric series*, Japan. J. Math. (N.S.) **19** (1993), no. 1, 81–129, DOI 10.4099/math1924.19.81. MR1231511

[19] Lei Fu, *ℓ-adic GKZ hypergeometric sheaves and exponential sums*, Adv. Math. **298** (2016), 51–88, DOI 10.1016/j.aim.2016.04.021. MR3505737

[20] L. Fu, D. Wan, and H. Zhang, *The p-adic Gelfand-Kapranov-Zelevinsky hypergeometric complex*, `arXiv:1804.05297v1`, 2018.

[21] I. M. Gel′fand and M. I. Graev, *Hypergeometric functions over finite fields* (Russian), Dokl. Akad. Nauk **381** (2001), no. 6, 732–737. MR1892519

[22] I. M. Gelfand, M. M. Kapranov, and A. V. Zelevinsky, *Discriminants, resultants and multidimensional determinants*, Modern Birkhäuser Classics, Birkhäuser Boston, Inc., Boston, MA, 2008. Reprint of the 1994 edition. MR2394437

[23] Vasily Golyshev and Anton Mellit, *Gamma structures and Gauss's contiguity*, J. Geom. Phys. **78** (2014), 12–18, DOI 10.1016/j.geomphys.2013.12.007. MR3170307

[24] John Greene, *Hypergeometric functions over finite fields*, Trans. Amer. Math. Soc. **301** (1987), no. 1, 77–101, DOI 10.2307/2000329. MR879564

[25] Benedict H. Gross and Neal Koblitz, *Gauss sums and the p-adic Γ-function*, Ann. of Math. (2) **109** (1979), no. 3, 569–581, DOI 10.2307/1971226. MR534763

[26] David Harvey, *Counting points on hyperelliptic curves in average polynomial time*, Ann. of Math. (2) **179** (2014), no. 2, 783–803, DOI 10.4007/annals.2014.179.2.7. MR3152945

[27] David Harvey, *Computing zeta functions of arithmetic schemes*, Proc. Lond. Math. Soc. (3) **111** (2015), no. 6, 1379–1401, DOI 10.1112/plms/pdv056. MR3447797

[28] David Harvey and Andrew V. Sutherland, *Computing Hasse-Witt matrices of hyperelliptic curves in average polynomial time*, LMS J. Comput. Math. **17** (2014), no. suppl. A, 257–273, DOI 10.1112/S1461157014000187. MR3240808

[29] David Harvey and Andrew V. Sutherland, *Computing Hasse-Witt matrices of hyperelliptic curves in average polynomial time, II*, Frobenius distributions: Lang-Trotter and Sato-Tate conjectures, Contemp. Math., vol. 663, Amer. Math. Soc., Providence, RI, 2016, pp. 127–147, DOI 10.1090/conm/663/13352. MR3502941

[30] Nicholas M. Katz, *Exponential sums and differential equations*, Annals of Mathematics Studies, vol. 124, Princeton University Press, Princeton, NJ, 1990, DOI 10.1515/9781400882434. MR1081536

[31] Nicholas M. Katz and John Tate, *Bernard Dwork (1923–1998)*, Notices Amer. Math. Soc. **46** (1999), no. 3, 338–343. MR1669973

[32] Kiran S. Kedlaya, *p-adic differential equations*, Cambridge Studies in Advanced Mathematics, vol. 125, Cambridge University Press, Cambridge, 2010, DOI 10.1017/CBO9780511750922. MR2663480

[33] K.S. Kedlaya, GitHub repository, `https://github.com/kedlaya/hgm-frobstruct`.

[34] Kiran S. Kedlaya and Jan Tuitman, *Effective convergence bounds for Frobenius structures on connections*, Rend. Semin. Mat. Univ. Padova **128** (2012), 7–16 (2013), DOI 10.4171/RSMUP/128-2. MR3076829

[35] M. Kim and W. Yang, *Mirror symmetry, mixed motives, and* $\zeta(3)$, `arXiv:1710.02344v3`, 2019.

[36] Alan G. B. Lauder, *Deformation theory and the computation of zeta functions*, Proc. London Math. Soc. (3) **88** (2004), no. 3, 565–602, DOI 10.1112/S0024611503014461. MR2044050

[37] Alain M. Robert, *A course in p-adic analysis*, Graduate Texts in Mathematics, vol. 198, Springer-Verlag, New York, 2000, DOI 10.1007/978-1-4757-3254-2. MR1760253

[38] Alain Salinier, *Structure de Frobenius forte de l'équation différentielle hypergéométrique* (French, with English summary), C. R. Acad. Sci. Paris Sér. I Math. **305** (1987), no. 10, 393–396. MR916337

[39] Peter Scholze, *Canonical q-deformations in arithmetic geometry* (English, with English and French summaries), Ann. Fac. Sci. Toulouse Math. (6) **26** (2017), no. 5, 1163–1192, DOI 10.5802/afst.1563. MR3746625

[40] I. Shapiro, *Frobenius map for quintic threefolds*, Int. Math. Res. Not. IMRN **13** (2009), 2519–2545, DOI 10.1093/imrn/rnp024. MR2520788

[41] Ilya Shapiro, *Frobenius map and the p-adic gamma function*, J. Number Theory **132** (2012), no. 8, 1770–1779, DOI 10.1016/j.jnt.2012.03.005. MR2922344

[42] J. Thomae, *Ueber die höheren hypergeometrischen Reihen, insbesondere über die Reihe:* $1 + \frac{a_0 a_1 a_2}{1.b_1 b_2} x + \frac{a_0(a_0+1)a_1(a_1+1)a_2(a_2+1)}{1.2.b_1(b_1+1)b_2(b_2+1)} x^2 + \cdots\cdots\cdot$ (German), Math. Ann. **2** (1870), no. 3, 427–444, DOI 10.1007/BF01448236. MR1509670

[43] D. van Straten, *CY-operators and L-functions*, in *2017 MATRIX Annals*, Springer, 2019, 491–503.

[44] Daniel Vargas-Montoya, *Algébricité modulo p, séries hypergéométriques et structures de Frobenius fortes* (French, with English and French summaries), Bull. Soc. Math. France **149** (2021), no. 3, 439–477, DOI 10.24033/bsmf.283. MR4349570

Department of Mathematics, University of California San Diego, La Jolla, California 92093

Email address: `kedlaya@ucsd.edu`

URL: `https://kskedlaya.org`

Contemporary Mathematics
Volume **779**, 2022
https://doi.org/10.1090/conm/779/15674

The regulator dominates the rank

Fabien Pazuki

We believe that the results presented here would have been to the liking of Alexey Zykin, who is deeply missed. In loving memory of Alexey and Tanya.

ABSTRACT. After noticing that the regulator of a number field dominates the rank of its group of units, we bound from below the regulator of the Mordell-Weil group of elliptic curves over global function fields of characteristic $p \geq 5$. The lower bound is an increasing function of the rank and of the height. This partially answers Question 7.1 and Question 7.2 in Autissier et al [Int. Math. Res. Not., 7, 2021, 4976-4993].

1. Introduction

Regulators of number fields and regulators of Mordell-Weil groups of abelian varieties have attracted a lot of attention, both for their own sake, and for the role they play in the Class Number Formula and in the strong form of the Birch and Swinnerton-Dyer conjecture, respectively. When studying families of number fields or families of abelian varieties, it is sometimes necessary to estimate the size of the regulator in terms of easier invariants, like the discriminant and degree of the number fields, or like the height of the abelian varieties and the rank of their Mordell-Weil group, respectively. In this note, we propose a new lower bound on the regulator of elliptic curves defined over global function fields. This lower bound is an increasing function of the rank of the elliptic curve (when the height is big enough), which is a new phenomenon, and which mirrors a similar situation taking place between the regulator of a number field and its rank of units. We describe both results in the rest of this introduction.

1.1. Regulators and ranks of units of number fields. Let us start with the following theorem, which has been an important motivation for this work. In the sequel, if F is a number field, we denote by d its degree over $\mathbb{Q}$. Let r_1 be the number of real embeddings of F, and r_2 be the number of pairs of complex conjugate embeddings of F. The group of units of F is a $\mathbb{Z}$-module of finite rank,

2020 *Mathematics Subject Classification.* Primary 11G50, 14G40.

Key words and phrases. Heights, elliptic curves, regulators, Mordell-Weil.

We thank the Swedish Research Council under grant no. 2016-06596, as this work was finalized while the author was in residence at Institut Mittag-Leffler in Djursholm, Sweden during the fall of 2021. The author was supported by ANR-17-CE40-0012 Flair and ANR-20-CE40-0003 Jinvariant.

we denote this rank by r_F. Let R_F be the regulator of F, and let w_F be the number of roots of unity in F.

THEOREM 1.1 (Friedman, [**Fri89**, page 620, Corollary]). *Let F be a number field. Then*

$$\frac{R_F}{w_F} \geq 0.0031 \exp(0.241d + 0.497r_1). \tag{1}$$

By Dirichlet's unit theorem, we know that $r_F = r_1 + r_2 - 1$. We also know that $d = r_1 + 2r_2$. This has the following easy consequence when used in inequality (1).

COROLLARY 1.2. *Let F be a number field. Then*

$$R_F \geq 0.0062 \exp(0.241 r_F). \tag{2}$$

So the story begins with the following fact given by inequality (2): the regulator R_F of a number field F dominates the rank r_F of its group of units. This triggers questions about other contexts, for instance: to what extent would the regulator of the Mordell-Weil group of an abelian variety over a global field dominate the rank of this Mordell-Weil group?

1.2. Elliptic curves and ranks of Mordell-Weil groups. Our goal is to prove that the regulator $\mathrm{Reg}(E/K)$ of an elliptic curve E defined over a function field K of characteristic $p \geq 5$ dominates the rank of its Mordell-Weil group. In doing so we partially answer Question 7.1 and Question 7.2 of [**AHP21**] in the case where $K = \mathbb{F}_q(\mathcal{C})$ is a function field of characteristic $p \geq 5$, where $\mathcal{C}$ is a smooth projective and geometrically connected curve defined over its constant field $\mathbb{F}_q$ and of genus $g \geq 0$. Note that the rank of elliptic curves over function fields of positive characteristic is not bounded [**Ulm02, Gri20**], hence this improvement is non-trivial. Let us state the result.

THEOREM 1.3. *Let $K = \mathbb{F}_q(\mathcal{C})$ be a function field of characteristic $p \geq 5$ and genus g. Let E be an elliptic curve over K of discriminant $\Delta(E/K)$, of trace zero, and let p^s denote the inseparability degree of the j-map of E. Let r denote the rank of $E(K)$. There exists a positive real number $c_0 = c_0(q, g, p^s)$ such that*

$$\mathrm{Reg}(E/K) \geq \Big(c_0 \log 12h(E)\Big)^r, \tag{3}$$

where $h(E) = \frac{1}{12} \deg \Delta(E/K)$, and the inequality holds with the explicit value

$$c_0 = \Big(p^{2s} 12\sqrt{q}(\log q)^2(5g+9)10^{15.5+23g}\Big)^{-1}.$$

We can now deduce the following corollary, which can be seen as a refined Northcott property for the regulators of elliptic curves over function fields in characteristic $p \geq 5$.

COROLLARY 1.4. *Let $K = \mathbb{F}_q(\mathcal{C})$ be a function field of characteristic $p \geq 5$ and genus g. The set of elliptic curves of trace zero over K, with positive rank, bounded inseparability degree and bounded regulator is finite.*

REMARK 1.5. Under the ABC conjecture, the BSD conjecture, and the GRH, one obtains an inequality for elliptic curves over number fields similar to the inequality (3) using [**Mes86**]. This would lead to an improvement of Theorem 4 page 1124 of [**Paz16**], as the regulator would bound the rank from above.

The rest of the text presents a proof of Theorem 1.3 and of Corollary 1.4. After giving the prerequisites in the next section, we prove inequality (3). The proof relies on the Minkowski successive minima inequality, combined with a lower bound on the canonical height of non-torsion points on elliptic curves. This is not enough, though, we need extra input to obtain the correct dependance in the rank. We are then able to give an explicit estimate on the analytic rank in Lemma 3.1, following Brumer's work, and transfer this estimate on the algebraic rank via Tate's work. The estimate is of sufficient quality to yield the result.

2. Definitions and prerequisites

Here we gather the basic definitions –function fields, heights, regulators of elliptic curves– and the key results used later in the proof of Theorem 1.3.

2.1. Function fields. Let $K = k(\mathcal{C})$ be the function field of a smooth projective and geometrically connected curve $\mathcal{C}$ defined over its constant field k and of genus $g \geq 0$. Let M_K stand for a complete set of inequivalent valuations $v(.)$. The set M_K is in bijection with the set of closed point in $\mathcal{C}$. Given a place $v \in M_K$, the residue field k_v of K at v is a finite extension of k: the degree $n_v := [k_v : k]$ of this extension will be called the degree of v.

This gives a normalization such that for any element $x \in K$, $x \neq 0$, the following product formula holds

$$\sum_{v \in M_K} n_v v(x) = 0.$$

A divisor I on the field K is a formal sum $\sum_{v \in M_K} a_v \cdot v$ where $a_v \in \mathbb{Z}$ is zero for all but finitely many places v. We pose

$$\deg(I) = \sum_{v \in M_K} n_v a_v.$$

We define the height on K by $h(0) = 0$ and for any non-zero $x \in K$, by

$$h(x) = \sum_{v \in M_K} n_v \max\{0, -v(x)\}.$$

If we now consider E to be an elliptic curve defined over the function field K, we define the Néron-Tate height on the group of rational points $E(K)$ with respect to the divisor (O) on E by

$$\widehat{h}_E(P) = \frac{1}{2} \lim_{n \to \infty} \frac{1}{n^2} h(x([n]P)).$$

2.2. Regulators of elliptic curves. Let K be a function field of transcendence degree one over its field of constants k. Let E/K be an elliptic curve over the field K. We assume that E has trace zero. Let m be the Mordell-Weil rank of $E(K)$, which is finite by the Lang-Néron theorem, see [**Con06**] for instance. Let $\widehat{h}_E$ be the Néron-Tate height on E. Let $< .,. >$ be the associated bilinear form, given by

$$< P, Q >= \frac{1}{2}\left(\widehat{h}_E(P+Q) - \widehat{h}_E(P) - \widehat{h}_E(Q)\right)$$

for any $P, Q \in E(K)$.

DEFINITION 2.1. Let $P_1, \ldots, P_r$ be a basis of the lattice $E(K)/E(K)_{\text{tors}}$, where $E(K)$ is the Mordell-Weil group. The regulator of E/K is defined by

$$\text{Reg}(E/K) = \det(< P_i, P_j >_{1 \leq i,j \leq r}).$$

In the case $r = 0$, the regulator is equal to 1.

We gather here three results needed for the sequel.

LEMMA 2.2 (Lemma 3.1 of [**AHP21**]). *Let K be a function field of transcendence degree one over its field of constants k. Let E be an elliptic curve over the field K. We assume that E has trace zero. Let r be the Mordell-Weil rank of $E(K)$. Assume $r \geq 1$. Let $\Lambda = E(K)/E(K)_{\text{tors}}$ and for any $i \in \{1, \ldots, r\}$, let us denote the Minkowski ith-minimum of $(\Lambda, \sqrt{\widehat{h}_E})$ by λ_i. Then we have*

$$\lambda_1 \cdots \lambda_r \leq r^{r/2} (\text{Reg}(E/K))^{1/2}. \tag{4}$$

THEOREM 2.3 (Theorem 6.1 of [**AHP21**]). *Let $K = k(\mathcal{C})$ be a function field of characteristic $p > 0$ and genus g. Let E/K be an elliptic curve of discriminant $\Delta(E/K)$ and assume that the j-map of E has inseparable degree p^s. Let $P \in E(K)$ be a non-torsion point. Then one has*

$$\widehat{h}_E(P) \geq p^{-2s} 10^{-15.5 - 23g} h(E),$$

where $h(E) = \frac{1}{12} \deg \Delta(E/K)$.

LEMMA 2.4. *Let $K = k(\mathcal{C})$ be a function field of characteristic $p > 0$ and genus g. Let E/K be an elliptic curve over K. Let r_{an} denote the analytic rank of E/K and let r denote its algebraic rank over K. Then $r \leq r_{\text{an}}$.*

PROOF. This is a direct consequence of Theorem 5.2 page 436 of [**Tat66**]. □

3. Regulators of elliptic curves over function fields of positive characteristic

Let us start with a useful lemma, which is an explicit version of Proposition 6.9 page 463 in [**Bru92**]. Inequality (6) is weaker than inequality (5), but easier to manipulate. Brumer's work [**Bru92**] provides a bound on the analytic rank. To deduce the control on the algebraic rank we use Lemma 2.4.

LEMMA 3.1. *Let $K = \mathbb{F}_q(\mathcal{C})$ be a function field of characteristic $p \geq 5$ and genus g. Let E be an elliptic curve over K. Let n_E be the degree of the conductor of E and let r denote the rank of $E(K)$. Assume $n_E > 1$. The following inequality holds:*

(5)

$$r \leq \frac{n_E}{2 \log n_E} \log q + \frac{n_E}{(\log n_E)^2} 4\sqrt{q} (\log q)^2 + \frac{7}{2} + \frac{\log q}{\sqrt{q} \log n_E} \Big[(2g - 2)\sqrt{q} + 20g + 17 \Big],$$

and leads to, as $q \geq 5$,

$$r \leq \frac{n_E}{\log n_E} \sqrt{q} (\log q)^2 (5g + 9). \tag{6}$$

Note that we do not assume that n_E is large when compared to q, in contrast with Proposition 6.9 page 463 of [**Bru92**].

PROOF. We follow closely the proof of Proposition 6.9 page 463 in [**Bru92**]. Let us denote by $\mathcal{Z}$ the set of θ such that $1+i\theta/\log q$ is a zero of the L-function of the elliptic curve E, and such that $0 \leq \theta < 2\pi$. For any trigonometric polynomial f with Fourier coefficients denoted $c(n)$, we state the explicit formula (6.7) page 462 in [**Bru92**], for a positive integer parameter Y to be fixed later:

$$r_{\mathrm{an}} f(0) + \sum_{\theta \in \mathcal{Z}} f(\theta) = c(0)(n_E + 4g - 4) + 2\sum_{m=1}^{Y} U_m(E,f), \tag{7}$$

where the $U_m(E,f)$ are defined in (6.6) page 462 in [**Bru92**] and satisfy the following inequality, uniformly in f (there is a term β_K in the original formula, note that we used $\beta_K = (2g+1)(1-q^{-1})^{-1}$, as given in Proposition 6.3 page 461 of [**Bru92**]):

$$\sum_{m=3}^{+\infty} |U_m(E,f)| \leq \frac{2}{\sqrt{q}(1-q^{-1/2})^2} + \frac{(4g+2)(1-q^{-1})^{-1}}{(q-1)(1-q^{-1/2})}, \tag{8}$$

and if we consider (as in (6.5) page 465) the Féjer kernel given by

$$F_Y(\theta) = \frac{(\sin \frac{1}{2} Y\theta)^2}{Y(\sin \frac{1}{2}\theta)^2},$$

for the specific choice $f = F_Y$, we have the inequality

$$|U_2(E,F_Y)| \leq \frac{Y}{2} + \frac{4g+2}{(\sqrt{q}-1)(1-q^{-1})}, \tag{9}$$

and the inequality

$$|U_1(E,F_Y)| \leq \frac{2q^{Y/2}}{\sqrt{q}Y(1-q^{-1/2})^2} + \frac{(2g+1)}{(1-q^{-1})} Y. \tag{10}$$

Following Brumer we fix $f = F_Y$ in equation (7), we get $f(0) = Y$ and $c(0) = 1$, and because the Féjer kernel is **non-negative**,[1] the combination of (7) with (8), (9), (10) leads to

(11)

$$\begin{aligned} r \leq r_{\mathrm{an}} \leq {} & \frac{n_E + 4g - 4}{Y} + \frac{4q^{Y/2}}{Y^2\sqrt{q}(1-q^{-1/2})^2} + \frac{4g+2}{1-q^{-1}} + 1 + \frac{(8g+4)}{Y(\sqrt{q}-1)(1-q^{-1})} \\ & + \frac{4}{Y\sqrt{q}(1-q^{-1/2})^2} + \frac{(8g+4)(1-q^{-1})^{-1}}{Y(q-1)(1-q^{-1/2})}. \end{aligned}$$

[1]The author remembers attending a course in functional analysis of Jean-Michel Morel at ENS Cachan in 2002, where one needed to compare different kernels in Fourier theory. The Féjer kernel will always be remembered as one of the most important, because it is non-negative, this is useful again in this situation!

For any $n_E > 1$, we may now fix $Y = \lceil \frac{2\log n_E}{\log q} \rceil > 0$ and use $\frac{1}{Y} \leq \frac{\log q}{2\log n_E}$ and $q^{Y/2} \leq n_E\, q$ to obtain[2] in (11)

$$\begin{aligned} r \leq & \frac{n_E}{2\log n_E}\log q + \frac{n_E\sqrt{q}(\log q)^2}{(\log n_E)^2(1-q^{-1/2})^2} + \frac{4g+2}{1-q^{-1}} + 1 \\ & + \frac{\log q}{\log n_E}\Big[\frac{4g+2}{(\sqrt{q}-1)(1-q^{-1})} + \frac{2}{\sqrt{q}(1-q^{-1/2})^2} \\ & + \frac{(4g+2)(1-q^{-1})^{-1}}{(q-1)(1-q^{-1/2})} + 2g - 2\Big], \end{aligned}$$

and with $q \geq 5$ we obtain

(12)

$$r \leq \frac{n_E}{2\log n_E}\log q + \frac{n_E}{(\log n_E)^2}4\sqrt{q}(\log q)^2 + \frac{7}{2} + \frac{\log q}{\sqrt{q}\log n_E}\Big[20g + 17 + (2g-2)\sqrt{q}\Big].$$

□

We can now give the proof of Theorem 1.3.

PROOF. If $r = 0$ the result is obvious, we may thus assume that $r \geq 1$. We start by combining Lemma 2.2 and Theorem 2.3 to obtain

$$\text{Reg}(E/K) \geq \frac{1}{r^r}\Big(p^{-2s}10^{-15.5-23g}h(E)\Big)^r. \tag{13}$$

We now want to estimate the denominator by bounding the algebraic rank r from above: one uses Lemma 3.1 (valid when $p \geq 5$ and $n_E > 1$):

$$r \leq \frac{n_E}{\log n_E}\sqrt{q}(\log q)^2(5g+9). \tag{14}$$

Now, as $n_E \leq 12h(E)$ and as $x \mapsto x(\log x)^{-1}$, for $x > e$ is a well defined increasing function, one deduces from (14) that for $n_E > e$

$$r \leq \frac{12h(E)}{\log 12h(E)}\sqrt{q}(\log q)^2(5g+9), \tag{15}$$

which leads to

$$\text{Reg}(E/K) \geq \Big(\frac{\log 12h(E)}{12h(E)\sqrt{q}(\log q)^2(5g+9)}\Big)^r\Big(p^{-2s}10^{-15.5-23g}h(E)\Big)^r \tag{16}$$

and finally

$$\text{Reg}(E/K) \geq \Big(c_0\log 12h(E)\Big)^r, \tag{17}$$

where $c_0 = \Big(p^{2s}12\sqrt{q}(\log q)^2(5g+9)10^{15.5+23g}\Big)^{-1}$. We also need to treat the case $n_E \leq e$: we may use the easy bound $r \leq n_E + 4g - 4$, which gives in particular $r \leq 4g-1$. Inject this in (13) to obtain

$$\text{Reg}(E/K) \geq \Big(\frac{1}{4g-1}p^{-2s}10^{-15.5-23g}h(E)\Big)^r \geq \Big(c_0\log 12h(E)\Big)^r, \tag{18}$$

and the same explicit value of c_0 is valid. This concludes the proof. □

We will now close the discussion with the proof of Corollary 1.4.

[2]Taking $\lfloor\cdot\rfloor$ instead of $\lceil\cdot\rceil$ when choosing Y is a valid option if n_E is assumed big when compared to q.

PROOF. We split into two cases: in Theorem 1.3, either $c_0 \log 12h(E) \leq 1$, in that case p^s bounded implies a bounded height, or $c_0 \log 12h(E) > 1$: in that case, as soon as the rank r is positive and as long as s is bounded from above, a bounded regulator implies a bounded height by inequality (3). In both cases, apply [**MB85**] Théorème 4.6 page 236, which proves that a bounded height implies finiteness, as the constant field is a finite field here. □

Acknowledgments

We thank Pascal Autissier and Marc Hindry for interesting conversations. We thank the referee for useful feedback.

References

[AHP21] Pascal Autissier, Marc Hindry, and Fabien Pazuki, *Regulators of elliptic curves*, Int. Math. Res. Not. IMRN **7** (2021), 4976–4993, DOI 10.1093/imrn/rny285. MR4241121

[Bru92] Armand Brumer, *The average rank of elliptic curves. I*, Invent. Math. **109** (1992), no. 3, 445–472, DOI 10.1007/BF01232033. MR1176198

[Con06] Brian Conrad, *Chow's K/k-image and K/k-trace, and the Lang-Néron theorem*, Enseign. Math. (2) **52** (2006), no. 1-2, 37–108. MR2255529

[Fri89] Eduardo Friedman, *Analytic formulas for the regulator of a number field*, Invent. Math. **98** (1989), no. 3, 599–622, DOI 10.1007/BF01393839. MR1022309

[Gri20] Richard Griffon, *A new family of elliptic curves with unbounded rank*, Mosc. Math. J. **20** (2020), no. 2, 343–374. MR4088798

[HiSi88] Marc Hindry and Joseph H. Silverman, *The canonical height and integral points on elliptic curves*, Invent. Math. **93** (1988), no. 2, 419–450, DOI 10.1007/BF01394340. MR948108

[Mes86] Jean-François Mestre, *Formules explicites et minorations de conducteurs de variétés algébriques* (French), Compositio Math. **58** (1986), no. 2, 209–232. MR844410

[MB85] Laurent Moret-Bailly, *Pinceaux de variétés abéliennes* (French, with English summary), Astérisque **129** (1985), 266. MR797982

[Paz16] Fabien Pazuki, *Northcott property for the regulators of number fields and abelian varieties*, Oberwolfach Rep. **21** (2016), 1122–1125.

[Tat66] John Tate, *On the conjectures of Birch and Swinnerton-Dyer and a geometric analog*, Séminaire Bourbaki, Vol. 9, Soc. Math. France, Paris, (1966), pp. Exp. No. 306, 415–440. MR1610977

[Ulm02] Douglas Ulmer, *Elliptic curves with large rank over function fields*, Ann. of Math. (2) **155** (2002), no. 1, 295–315, DOI 10.2307/3062158. MR1888802

INSTITUTE OF MATHEMATICS, UNIVERSITY OF COPENHAGEN, UNIVERSITETSPARKEN 5, 2100 COPENHAGEN Ø, DENMARK; AND UNIVERSITÉ DE BORDEAUX, IMB, 351, COURS DE LA LIBÉRATION, 33400 TALENCE, FRANCE

Email address: `fpazuki@math.ku.dk`

Contemporary Mathematics
Volume **779**, 2022
https://doi.org/10.1090/conm/779/15675

Introduction to Drinfeld modules

Bjorn Poonen

Abstract. Our goal is to introduce Drinfeld modules and to explain their application to explicit class field theory.

Before introducing Drinfeld modules, let us motivate their study by mentioning some of their applications.

1. Applications

- Explicit class field theory for global function fields (just as torsion of $\mathbb{G}_m$ gives abelian extensions of $\mathbb{Q}$, and torsion of CM elliptic curves gives abelian extensions of imaginary quadratic fields). Here, global function field means $\mathbb{F}_p(T)$ or a finite extension.
- Langlands conjectures for GL_n over global function fields (Drinfeld modular varieties play the role of Shimura varieties).
- Modularity of elliptic curves over global function fields: If E over $\mathbb{F}_p(T)$ has split multiplicative reduction at ∞, then E is dominated by a Drinfeld modular curve.
- Explicit construction of curves over finite fields with many points, as needed in coding theory, namely reductions of Drinfeld modular curves, which have easier-to-write-down equations than the classical modular curves.

Only the first of these will be treated in these notes, though we do also give a very brief introduction to Drinfeld modular curves and varieties. We follow [**Hay92**] as primary reference. For many more details about Drinfeld modules, one can consult the original articles of Drinfeld [**Dri74, Dri77**] or any of the following: [**DH87**], [**GHR92**], [**Gos96**], [**Lau96**], [**Lau97**], [**GvdPRVG97**], [**Ros02**], [**Tha04**].

2. Analytic theory

2.1. Inspiration from characteristic 0. Let Λ be a discrete $\mathbb{Z}$-submodule of $\mathbb{C}$ of rank $r \geq 0$, so there exist $\mathbb{R}$-linearly independent $\omega_1, \ldots, \omega_r$ such that

2020 *Mathematics Subject Classification.* Primary 11G09; Secondary 11G45, 11R37.

Key words and phrases. Drinfeld module, class field theory, $\mathbb{F}_q$-linear polynomial, Tate module, good reduction, stable reduction, Carlitz module, Hilbert class field, ray class field, Drinfeld modular variety.

The writing of this article was supported in part by National Science Foundation grants DMS-841321, DMS-1601946, and DMS-2101040 and Simons Foundation grants #402472 and #550033.

$\Lambda = \mathbb{Z}\omega_1 + \cdots \mathbb{Z}\omega_r$. It turns out that the Lie group $\mathbb{C}/\Lambda$ is isomorphic to $G(\mathbb{C})$ for some algebraic group G over $\mathbb{C}$, as we can check for each value of r:

r	isomorphism of Lie groups	G
0	$\mathbb{C}/\Lambda \xrightarrow{\sim} \mathbb{C}$	the additive group $\mathbb{G}_a$
1	$\mathbb{C}/\Lambda \xrightarrow{\sim} \mathbb{C}^\times$ $z \longmapsto \exp(2\pi i z/\omega_1)$	the multiplicative group $\mathbb{G}_m$
2	$\mathbb{C}/\Lambda \xrightarrow{\sim} E(\mathbb{C})$ $z \longmapsto (\wp(z), \wp'(z))$	an elliptic curve E

(The notation $\wp$ denotes the Weierstrass $\wp$-function associated to the lattice Λ; see [**Sil09**, VI.3], for instance.)

Cases with $r > 2$ do not occur, since $[\mathbb{C} : \mathbb{R}] = 2$.

2.2. Characteristic p analogues. What is a good analogue of the above in characteristic p? Start with a smooth projective geometrically integral curve X over a finite field $\mathbb{F}_q$, and fix a closed point $\infty \in X$. Let $\mathscr{O}(X - \{\infty\})$ denote the coordinate ring of the affine curve $X - \{\infty\}$.

Characteristic 0 ring	Characteristic p analogue	Example
$\mathbb{Z}$	$A := \mathscr{O}(X - \{\infty\})$	$\mathbb{F}_q[T]$
$\mathbb{Q}$	$K := \operatorname{Frac} A$	$\mathbb{F}_q(T)$
$\mathbb{R}$	$K_\infty :=$ completion at ∞	$\mathbb{F}_q((1/T))$
$\mathbb{C}$	$C :=$ completion of $\overline{K}_\infty$	

The completions are taken with respect to the ∞-adic absolute value: For nonzero $a \in A$, define $|a| := \#(A/a) = q^{\deg a}$ (and $|0| := 0$); extend $|\ |$ to K, its completion K_∞, an algebraic closure $\overline{K}_\infty$, and *its* completion C, in turn. The field C is algebraically closed as well as complete with respect to $|\ |$. Some authors use the notation $\mathbb{C}$ or $\mathbb{C}_\infty$ instead of C.

Finite rank $\mathbb{Z}$-submodules of C are just finite-dimensional $\mathbb{F}_p$-subspaces, not so interesting, so instead consider this:

DEFINITION 2.1. An A-lattice in C is a discrete A-submodule Λ of C of finite rank, where

$$\operatorname{rank} \Lambda := \dim_K(K\Lambda) = \dim_{K_\infty}(K_\infty \Lambda).$$

If A is a principal ideal domain, such as $\mathbb{F}_q[T]$, then all such Λ arise as follows:

> Let $\{x_1, \ldots, x_r\}$ be a basis for a finite-dimensional K_∞-subspace in C, and let $\Lambda := Ax_1 + \cdots + Ax_r \subset C$.

Note: In contrast with the characteristic 0 situation, r can be arbitrarily large since $[C : K_\infty]$ is infinite.

THEOREM 2.2. *The quotient C/Λ is analytically isomorphic to C!*

This statement can be interpreted using rigid analysis. More concretely, it means that there exists a power series

$$e(z) = \alpha_0 z + \alpha_1 z^q + \alpha_2 z^{q^2} + \cdots$$

defining a surjective $\mathbb{F}_q$-linear map $C \to C$ with kernel Λ. If we require $\alpha_0 = 1$, then such a power series e is unique.

Sketch of proof. Uniqueness follows from the nonarchimedean Weierstrass preparation theorem, which implies that a convergent power series is determined up to a constant multiple by its zeros: explicitly, if $e(z)$ exists, then

$$e(z) = z \prod_{\substack{\lambda \in \Lambda \\ \lambda \neq 0}} \left(1 - \frac{z}{\lambda}\right). \tag{2.1}$$

(Over $\mathbb{C}$, there would be an ambiguity of multiplication by a function $e^{g(z)}$, but in the nonarchimedean setting, every invertible entire function is constant!)

If we take (2.1) as a definition, there are several things to check:

- The infinite product converges. (Proof: Since Λ is a discrete subgroup of a locally compact group $K_\infty \Lambda$, we have $\lambda \to \infty$.)
- $e(z)$ is surjective. (The nonarchimedean Picard theorem says that a nonconstant entire function omits *no* values.)
- $e(x+y) = e(x) + e(y)$. (Proof: Write Λ as an increasing union of finite-dimensional $\mathbb{F}_p$-subspaces, and $e(x)$ as the limit of the corresponding finite products. If $f(x)$ is a polynomial whose zeros are distinct and form a group G under addition, then $f(x+y) = f(x) + f(y)$, because $f(x+y) - f(x) - f(y)$ vanishes on $G \times G$ but is of degree less than $\#G$ in each variable.)
- $e(cx) = ce(x)$ for each $c \in \mathbb{F}_q$. (Use a proof similar to the preceding, or argue directly.)
- $\ker e = \Lambda$. □

Now C/Λ has a natural A-module structure. Carrying this across the isomorphism $C/\Lambda \to C$ gives an *exotic* A-module structure on C. This is essentially what a Drinfeld module is: the additive group with a new A-module structure.

For each $a \in A$, the multiplication-by-a map $a\colon C/\Lambda \to C/\Lambda$ corresponds under the isomorphism to a map $\phi_a\colon C \to C$ making

$$\begin{array}{ccc} C/\Lambda & \xrightarrow{\;a\;} & C/\Lambda \\ {\scriptstyle e}\downarrow\wr & & \wr\downarrow{\scriptstyle e} \\ C & \xrightarrow{\;\phi_a\;} & C \end{array} \tag{2.2}$$

commute.

Proposition 2.3. *The map ϕ_a is a polynomial!*

Proof. Assume that $a \neq 0$. We have

$$\ker\left(a\colon C/\Lambda \to C/\Lambda\right) = \frac{a^{-1}\Lambda}{\Lambda},$$

which is isomorphic to $\Lambda/a\Lambda = (A/a)^r$, which is finite of order $|a|^r$. So $\ker \phi_a$ should be $e\left(\frac{a^{-1}\Lambda}{\Lambda}\right)$. Define the polynomial

$$\phi_a(z) := az \prod_{t \in \frac{a^{-1}\Lambda}{\Lambda} - \{0\}} \left(1 - \frac{z}{e(t)}\right).$$

Then ϕ_a is the map making (2.2) commute, because the power series $\phi_a(e(z))$ and $e(az)$ have the same zeros and same coefficient of z. □

The proof of Proposition 2.3 shows also that for any nonzero $a \in A$,

$$\deg \phi_a = \# \frac{a^{-1}\Lambda}{\Lambda} = |a|^r.$$

3. Algebraic theory

3.1. $\mathbb{F}_q$-linear polynomials. Let L be a field containing $\mathbb{F}_q$. A polynomial $f(x) \in L[x]$ is called additive if $f(x+y) = f(x) + f(y)$ in $L[x,y]$, and $\mathbb{F}_q$-linear if, in addition, $f(cx) = cf(x)$ in $L[x]$ for all $c \in \mathbb{F}_q$. Think of such polynomials as operators that can be composed: For example, each $a \in L$ defines an operator $x \mapsto ax$ and τ denotes the Frobenius operator $x \mapsto x^p$, so τa is $x \mapsto (ax)^p$ and τ^2 is $x \mapsto x^{p^2}$.

Let $\mathbb{G}_a$ be the additive group scheme over L, viewed as an $\mathbb{F}_q$-vector space scheme over L. Endomorphisms of $\mathbb{G}_a$ *as an $\mathbb{F}_q$-vector space scheme* are $\mathbb{F}_q$-linear by definition:

$$\begin{aligned}
\operatorname{End} \mathbb{G}_a &= \{\mathbb{F}_q\text{-linear polynomials in } L[x]\} \\
&= \left\{ \sum_{i=0}^{n} a_i x^{q^i} : a_i \in L \right\} \\
&= \left\{ \left(\sum_{i=0}^{n} a_i \tau^i \right)(x) : a_i \in L \right\} \\
&=: L\{\tau\};
\end{aligned}$$

this is a ring under addition and composition. More specifically, $L\{\tau\}$ is a twisted polynomial ring, twisted in that the elements $a \in L$ do not necessarily commute with the variable τ: instead, $\tau a = a^q \tau$.

For $f \in L\{\tau\}$, let l.c.(f) denote the leading coefficient a_n of f; by convention, l.c.$(0) = 0$. Also, if $f = \sum_{i=0}^{n} a_i \tau^i$, then the derivative of the $\mathbb{F}_q$-linear polynomial $f(x) \in L[x]$ is the constant $f'(0) = a_0$, which is the "constant term" of f viewed as a twisted polynomial in $L\{\tau\}$.

3.2. Drinfeld modules.

DEFINITION 3.1. An A-field is an A-algebra L that is a field; that is, L is a field equipped with a ring homomorphism $\iota \colon A \to L$. The A-characteristic of L is $\operatorname{char}_A L := \ker \iota$, a prime ideal of A.

We distinguish two cases:

- L is an extension of K and ι is an inclusion; then $\operatorname{char}_A L = 0$. (Example: C.)
- L is an extension of $A/\mathfrak{p}$ for some nonzero prime $\mathfrak{p}$ of A; then $\operatorname{char}_A L = \mathfrak{p}$.

To motivate the following definition, recall that an A-module M is an abelian group M together with a ring homomorphism $A \to \operatorname{End}_{\text{group}} M$.

DEFINITION 3.2. A Drinfeld A-module ϕ over L is the additive group scheme $\mathbb{G}_a$ with a faithful A-module structure for which the induced action on the tangent space at 0 is given by ι. More concretely, ϕ is an injective ring homomorphism

$$\begin{aligned}
A &\longrightarrow \operatorname{End} \mathbb{G}_a = L\{\tau\} \\
a &\longmapsto \phi_a
\end{aligned}$$

such that $\phi'_a(0) = \iota(a)$ for all $a \in A$.

REMARK 3.3. Many authors explicitly disallow ϕ to be the composition $A \xrightarrow{\iota} L \subset L\{\tau\}$, but we allow it when $\operatorname{char}_A L = 0$, since doing so does not seem to break any theorems. Our requirement that ϕ be injective does rule out $A \xrightarrow{\iota} L \subset L\{\tau\}$ when $\operatorname{char}_A L \neq 0$, however; we must rule this out to make Proposition 3.5 below hold.

It turns out that every Drinfeld A-module over C arises from an A-lattice as in Section 2. For a more precise statement, see Theorem 3.11.

3.3. Rank. We could define the rank of a Drinfeld module over C as the rank of the A-lattice it comes from, but it will be nicer to give an *algebraic* definition that makes sense over any A-field.

Let ϕ be a Drinfeld module. For each nonzero $a \in A$, there are nonnegative integers $m(a) \leq M(a)$ such that we may write

$$\phi_a = c_{m(a)}\tau^{m(a)} + \cdots + c_{M(a)}\tau^{M(a)}$$

with exponents in increasing order and $c_{m(a)}, c_{M(a)} \neq 0$. Then $\phi_a(x)$ as a polynomial in x has degree $q^{M(a)}$ and each zero has multiplicity $q^{m(a)}$. In terms of the functions M and m, we will define the rank and height of ϕ, respectively.

For each closed point $\mathfrak{p} \in X$, let $v_\mathfrak{p}$ be the $\mathfrak{p}$-adic valuation on K normalized so that $v_\mathfrak{p}(a)$ is the degree of the $\mathfrak{p}$-component of the divisor (a); thus $v_\mathfrak{p}(K^\times) = (\deg \mathfrak{p})\mathbb{Z}$. Also, define $|a|_\mathfrak{p} := q^{-v_\mathfrak{p}(a)}$. For example, $|\ |_\infty$ is the absolute value $|\ |$ defined earlier.

EXAMPLE 3.4. If $A = \mathbb{F}_q[T]$, then ϕ is determined by ϕ_T, and we define $r = M(T)$. For any nonzero $a \in A$, expanding ϕ_a in terms of ϕ_T shows that $M(a) = (\deg a)r = -rv_\infty(a)$.

A similar result holds for arbitrary A:

PROPOSITION 3.5 (Characterization of rank). *Let ϕ be a Drinfeld module over an A-field L. Then there exists a unique $r \in \mathbb{Q}_{\geq 0}$ such that $M(a) = -rv_\infty(a)$, or equivalently* $\deg \phi_a = |a|^r$, *for all nonzero $a \in A$. (Proposition* 3.13(a) *will imply that r is an* integer.*)*

PROOF. After enlarging L to make L perfect, we may define the ring of twisted Laurent series $L((\tau^{-1}))$ whose elements have the form $\sum_{n\in\mathbb{Z}} \ell_n\tau^n$ with $\ell_n = 0$ for sufficiently large *positive* n; multiplication is defined so that $\tau^n\ell = \ell^{q^n}\tau$. Then $L((\tau^{-1}))$ is a division ring with a valuation $v\colon L((\tau^{-1})) \to \mathbb{Z} \cup \{+\infty\}$ sending τ^n to $-n$ (same proof as for usual Laurent series over a field). Thus $\phi\colon A \to L\{\tau\}$ extends to a homomorphism $\phi\colon K \to L((\tau^{-1}))$, and v pulls back to a nontrivial valuation v_K on K. We have $v_K(a) = -M(a) \leq 0$ for all $a \in A - \{0\}$, so $v_K = rv_\infty$ for some $r \in \mathbb{Q}_{\geq 0}$. Then $M(a) = -rv_\infty(a)$ for all $a \in A - \{0\}$. $\square$

Define the rank of ϕ to be r. (This is *not* analogous to the rank of the group of rational points of an elliptic curve.)

3.4. Analogies. Drinfeld modules are *1-dimensional* objects, no matter what the rank is. Comparing with Section 2.1 suggests the following analogies:

rank 0 Drinfeld module $\longleftrightarrow \mathbb{G}_a$

rank 1 Drinfeld module $\longleftrightarrow \mathbb{G}_m$ or CM elliptic curve

(if E has CM by $\mathscr{O}$, view its lattice as rank 1 $\mathscr{O}$-module)

rank 2 Drinfeld module $\longleftrightarrow$ elliptic curve

rank ≥ 3 Drinfeld module $\longleftrightarrow$? (if only such geometric objects existed...)

There is also a higher-dimensional generalization called a t-module [**And86**].

REMARK 3.6. Gekeler [**Gek83, Gek91**] developed a theory of Drinfeld modules over *finite* fields analogous to the theory of abelian varieties over finite fields developed by Deuring, Tate, Waterhouse, and others.

3.5. Height.

PROPOSITION 3.7. *Let ϕ be a Drinfeld module over an A-field L of nonzero characteristic $\mathfrak{p}$. Then there exists a unique $h \in \mathbb{Q}_{>0}$ such that $m(a) = hv_{\mathfrak{p}}(a)$ for all nonzero $a \in A$. (Proposition 3.13(b) will imply that h is an integer satisfying $0 < h \leq r$.)*

PROOF. Enlarge L to make it perfect and extend ϕ to a homomorphism $K \to L((\tau))$ (twisted Laurent series in τ instead of τ^{-1}) to define a valuation on K. It is positive on $\mathfrak{p}$, hence equal to $hv_{\mathfrak{p}}$ for some $h \in \mathbb{Q}_{>0}$. $\square$

Call h the height of ϕ. It is analogous to the height of the formal group of an elliptic curve over a field of characteristic p.

3.6. Drinfeld modules and lattices. For fixed A and L, Drinfeld A-modules over L form a category, with morphisms as follows:

DEFINITION 3.8. A morphism $f\colon \phi \to \psi$ of Drinfeld modules over L is an element of $\operatorname{End} \mathbb{G}_a$ such that $f \circ \phi_a = \psi_a \circ f$ for all $a \in A$: i.e.,

$$\begin{array}{ccc} \mathbb{G}_a & \xrightarrow{\phi_a} & \mathbb{G}_a \\ {\scriptstyle f}\downarrow & & \downarrow{\scriptstyle f} \\ \mathbb{G}_a & \xrightarrow{\psi_a} & \mathbb{G}_a \end{array} \tag{3.1}$$

commutes.

An isogeny between Drinfeld modules ϕ and ψ is a surjective morphism f with finite kernel, or equivalently (since $\mathbb{G}_a$ is 1-dimensional), a nonzero morphism. If such an f exists, ϕ and ψ are called isogenous.

Over $\mathbb{C}$, there is no nonzero algebraic homomorphism from $\mathbb{G}_m$ to an elliptic curve; analogously:

PROPOSITION 3.9. *Isogenous Drinfeld modules have the same rank.*

PROOF. If $f\colon \phi \to \psi$ is an isogeny between Drinfeld modules of rank r and r', respectively, then (3.1) gives

$$(\deg f)|a|^r = |a|^{r'}(\deg f)$$

for all $a \in A$, so $r = r'$. $\square$

Because of Proposition 3.9, we fix the rank in the following.

DEFINITION 3.10. A morphism of rank r A-lattices Λ, Λ' in C is a number $c \in C$ such that $c\Lambda \subseteq \Lambda'$.

THEOREM 3.11. *For each $r \geq 0$, the analytic construction*

$$\{A\text{-lattices in } C \text{ of rank } r\} \xrightarrow{\sim} \{\text{Drinfeld modules over } C \text{ of rank } r\}$$

of Section 2 is an equivalence of categories.

SKETCH OF PROOF. Given a rank r Drinfeld module ϕ over C, choose a non-constant $a \in A$, and consider a power series

$$e(z) = z + \alpha_1 z^q + \alpha_2 z^{q^2} + \cdots$$

with unknown coefficients α_i. The condition $e(az) = \phi_a(e(z))$ determines the α_i uniquely; solve for each α_i in turn. Check that the resulting power series converges everywhere, and that its kernel is an A-lattice in C giving rise to ϕ. The proof of Proposition 2.3 shows more generally that a morphism of A-lattices corresponds to a polynomial map $C \to C$ defining a morphism of Drinfeld modules, and vice versa. □

In particular, homothety classes of rank r A-lattices in C are in bijection with isomorphism classes of rank r Drinfeld modules over C.

3.7. Torsion points. The additive polynomial ϕ_a plays the role of the multiplication-by-n map on an elliptic curve, or the nth power map on $\mathbb{G}_m$.

For $a \neq 0$, the a-torsion subscheme of a Drinfeld module ϕ is $\phi[a] := \ker \phi_a$, viewed as subgroup scheme of $\mathbb{G}_a$. It is a finite group scheme of order $\deg \phi_a = q^{M(a)} = |a|^r$. Let ${}^{\phi}L$ denote the additive group of L viewed as an A-module via ϕ. Then $\phi[a](L)$ is an A-submodule of ${}^{\phi}L$, but its order may be less than $|a|^r$ if L is not algebraically closed or $\phi[a]$ is not reduced.

More generally, if I is a nonzero ideal of A, let $\phi[I]$ be the scheme-theoretic intersection $\bigcap_{a \in I} \phi[a]$. Equivalently, one can define ϕ_I as the monic generator of the left ideal of $L\{\tau\}$ generated by $\{\phi_a : a \in I\}$, and define $\phi[I] := \ker \phi_I$. To understand the structure of $\phi[I](L)$, we need the following basic lemma about modules over Dedekind rings.

LEMMA 3.12. *Let A be a Dedekind ring. Let D be an A-module.*

(a) *If $\ell_1, \ldots, \ell_n$ are distinct nonzero prime ideals of A, and $e_1, \ldots, e_n \in \mathbb{Z}_{\geq 0}$, then*

$$D[\ell_1^{e_1} \cdots \ell_n^{e_n}] \simeq D[\ell_1^{e_1}] \oplus \cdots \oplus D[\ell_n^{e_n}].$$

(b) *If D is divisible, then for each fixed nonzero prime ℓ of A, the A/ℓ^e-module $D[\ell^e]$ is free of rank independent of e.*

PROOF. Localize to assume that A is a discrete valuation ring. Then (a) is trivial. In proving (b), we write ℓ also for a generator of ℓ. Since $D[\ell]$ is an A/ℓ-vector space, we can choose a free A-module F and an isomorphism $i_1 \colon \ell^{-1}F/F \xrightarrow{\sim} D[\ell]$. We construct isomorphisms $i_e \colon \ell^{-e}F/F \xrightarrow{\sim} D[\ell^e]$ for all $e \geq 1$ by induction: given the isomorphism i_e, use divisibility of D to lift i_e to a homomorphism

$i_{e+1}\colon \ell^{-(e+1)}F/F \to D[\ell^{e+1}]$ fitting in a commutative diagram with exact rows

$$\begin{array}{ccccccccc} 0 & \longrightarrow & \ell^{-1}F/F & \longrightarrow & \ell^{-(e+1)}F/F & \xrightarrow{\ell} & \ell^{-e}F/F & \longrightarrow & 0 \\ & & \downarrow{\scriptstyle i_1} & & \vdots{\scriptstyle i_{e+1}} & & \downarrow{\scriptstyle i_e} & & \\ 0 & \longrightarrow & D[\ell] & \longrightarrow & D[\ell^{e+1}] & \xrightarrow{\ell} & D[\ell^e] & \longrightarrow & 0. \end{array}$$

The diagram shows that i_{e+1} is an isomorphism too. □

PROPOSITION 3.13. *Let ϕ be a rank r Drinfeld module over an algebraically closed A-field L.*

(a) *If I is an ideal of A such that $\operatorname{char}_A L \nmid I$, then the A/I-module $\phi[I](L)$ is free of rank r. The same holds even if L is only separably closed.*
(b) *If $\operatorname{char}_A L = \mathfrak{p} \neq 0$, let h be the height of ϕ; then the $A/\mathfrak{p}^e$-module $\phi[\mathfrak{p}^e](L)$ is free of rank $r-h$.*

PROOF. When L is algebraically closed, $\phi_a\colon L \to L$ is surjective for every nonzero $a \in A$. In other words, the A-module ${}^{\phi}L$ is divisible. By Lemma 3.12, the claims for algebraically closed L follow if for each nonzero prime ℓ of A, there exists $e \geq 1$ such that

$$\#\phi[\ell^e](L) = \begin{cases} \#(A/\ell^e)^r, & \text{if } \ell \neq \operatorname{char}_A L; \\ \#(A/\ell^e)^{r-h}, & \text{if } \ell = \operatorname{char}_A L. \end{cases}$$

The class group of A is finite, so we may choose e so that ℓ^e is principal, say generated by a. If $\ell \neq \operatorname{char}_A L$, then ϕ_a is separable, so $\#\phi[\ell^e](L) = \deg \phi_a = |a|^r = \#(A/a)^r$. If $\ell = \operatorname{char}_A L$, then each zero of ϕ_a has multiplicity $q^{m(a)} = q^{hv_{\mathfrak{p}}(a)} = \#(A/a)^h$, so $\#\phi[\ell^e](L) = \#(A/a)^{r-h}$.

Now suppose that L is only separably closed, with algebraic closure $\overline{L}$. If $\operatorname{char}_A L \nmid I$, the proof above shows that $\phi[I](\overline{L})$ consists of L-points, so the structure of $\phi[I](L)$ is the same. □

COROLLARY 3.14. *If ϕ is a rank r Drinfeld module over any A-field L, and I is a nonzero ideal of A, then $\deg \phi_I = \#\phi[I] = \#(A/I)^r$.*

PROOF. The underlying scheme of $\phi[I]$ is $\operatorname{Spec} L[x]/(\phi_I(x))$, so $\#\phi[I] = \deg \phi_I$. For the second equality, assume without loss of generality that L is algebraically closed. For a group scheme G, let G^0 denote its connected component. Define $m(I) := \min\{m(a) : a \in I - \{0\}\}$. If $a \in A - \{0\}$, then $\phi[a]^0 = \ker \tau^{m(a)}$, so $\phi[I]^0 = \ker \tau^{m(I)}$. Thus $\#\phi[I]^0 = q^{m(I)}$, which is multiplicative in I. On the other hand, Proposition 3.13 shows that $\#\phi[I](L)$ is multiplicative in I. Thus the integers $\#\phi[I] = \#\phi[I]^0 \cdot \#\phi[I](L)$ and $\#(A/I)^r$ are both multiplicative in I. They are equal for any power of I that is principal, so they are equal for I. □

COROLLARY 3.15. *Let ϕ be a rank 1 Drinfeld module over a field L of nonzero A-characteristic $\mathfrak{p}$. Then $\phi_{\mathfrak{p}} = \tau^{\deg \mathfrak{p}}$.*

PROOF. Without loss of generality, L is algebraically closed. Since $0 < h \leq r = 1$, we have $h = r = 1$. By Proposition 3.13(b), $\phi[\mathfrak{p}](L) = 0$. On the other hand, $\phi_{\mathfrak{p}}$ is monic, by the general definition of ϕ_I. The previous two sentences show that $\phi_{\mathfrak{p}}$ is a power of τ. By Corollary 3.14, $\deg \phi_{\mathfrak{p}} = \#(A/\mathfrak{p}) = q^{\deg \mathfrak{p}} = \deg \tau^{\deg \mathfrak{p}}$, so $\phi_{\mathfrak{p}} = \tau^{\deg \mathfrak{p}}$. □

COROLLARY 3.16. *In the context of Corollary* 3.15, *if* $\mathfrak{p} = (\pi)$ *for some* $\pi \in A$, *then* $\phi_\pi = c\tau^{\deg \mathfrak{p}}$ *for some* $c \in L^\times$.

PROOF. By definition, $\phi_\mathfrak{p}$ is the monic generator of the left ideal generated by $\{\phi_a : a \in I\}$, which is the left ideal generated by ϕ_π. □

3.8. Tate modules. Let $\ell \subset A$ be a prime ideal not equal to 0 or $\operatorname{char}_A L$. Define the completions $A_\ell := \varprojlim_n A/\ell^n$ and $K_\ell := \operatorname{Frac} A_\ell$. Let L_s be a separable closure of L. Then the Tate module

$$T_\ell\phi := \operatorname{Hom}(K_\ell/A_\ell, {}^\phi L_\mathrm{s})$$

is a free A_ℓ-module of rank r.

Its applications are analogous to those for elliptic curves:

- The endomorphism ring $\operatorname{End}\phi$ is a projective A-module of rank $\le r^2$. In particular, if $r = 1$, then $\operatorname{End}\phi = A$ and $\operatorname{Aut}\phi = A^\times = \mathbb{F}_q^\times$.
- The Galois action on torsion points yields an ℓ-adic representation

$$\rho_\ell \colon \operatorname{Gal}(L_\mathrm{s}/L) \longrightarrow \operatorname{Aut}_{A_\ell}(T_\ell\phi) \simeq \operatorname{GL}_r(A_\ell).$$

4. Reduction theory

4.1. Drinfeld modules over rings. So far we considered Drinfeld modules over A-fields. One can also define Drinfeld modules over arbitrary A-algebras R or even A-schemes. In such generality, the underlying $\mathbb{F}_q$-vector space scheme need only be *locally* isomorphic to $\mathbb{G}_a$, so it could be the $\mathbb{F}_q$-vector space scheme associated to a nontrivial line bundle on the base.

To avoid this complication, let us assume that $\operatorname{Pic} R = 0$; this holds if the A-algebra R is a principal ideal domain, for instance. Then a Drinfeld A-module over R is given by a ring homomorphism

$$\begin{aligned} A &\longrightarrow \operatorname{End}\mathbb{G}_{a,R} = R\{\tau\} \\ a &\longmapsto \phi_a \end{aligned}$$

such that $\phi'_a(0) = a$ in R for all $a \in A$ *and* $\mathrm{l.c.}(\phi_a) \in R^\times$ for all nonzero $a \in A$. The last requirement, which implies injectivity of ϕ (if R is nonzero), guarantees that for any maximal ideal $\mathfrak{m} \subset R$, reducing all the ϕ_a modulo $\mathfrak{m}$ yields a Drinfeld module over $R/\mathfrak{m}$ *of the same rank.*

4.2. Good and stable reduction. Let us now specialize to the following setting:

R: an A-discrete valuation ring
(a discrete valuation ring with a ring homomorphism $A \to R$)

$\mathfrak{m}$: the maximal ideal of R

$L := \operatorname{Frac} R$, the fraction field

$v \colon L \to \mathbb{Z} \cup \{+\infty\}$, the discrete valuation

$\mathbb{F} := R/\mathfrak{m}$, the residue field

ϕ: a Drinfeld module over L of rank $r \ge 1$.

Then

- ϕ has good reduction if ϕ is isomorphic over L to a Drinfeld module over R, that is, if after replacing ϕ by an isomorphic Drinfeld module over L, all the ϕ_a have coefficients in R, and $\text{l.c.}(\phi_a) \in R^\times$ for all nonzero $a \in A$.
- ϕ has stable reduction if after replacing ϕ by an isomorphic Drinfeld module over L, all the ϕ_a have coefficients in R, and $a \mapsto (\phi_a \bmod \mathfrak{m})$ is a Drinfeld module over $\mathbb{F}$ of positive rank.

EXAMPLE 4.1. Let $A = \mathbb{F}_q[T]$. A rank 2 Drinfeld module over L is determined by

$$\phi_T = T + c_1\tau + c_2\tau^2;$$

here $c_1, c_2 \in L$ and $c_2 \neq 0$. Isomorphic Drinfeld modules are given by

$$u^{-1}\phi_T\, u = T + u^{q-1}c_1\tau + u^{q^2-1}c_2\tau^2$$

for any $u \in L^\times$. The condition for stable reduction is satisfied if and only if $v(u^{q-1}c_1) \geq 0$ and $v(u^{q^2-1}c_2) \geq 0$, with at least one of them being an equality. This condition uniquely specifies $v(u) \in \mathbb{Q}$. An element u of this valuation might not exist in L, but u can be found in a suitable ramified finite extension of L.

THEOREM 4.2 (Potential stability). *Let ϕ be a Drinfeld module over L of rank $r \geq 1$. There exists a finite ramified extension L' of L such that ϕ over L' has stable reduction.*

PROOF. Choose generators $a_1, \ldots, a_m$ of the ring A. As in Example 4.1, find L' and $u \in L'$ of valuation "just right" so that all coefficients of $u^{-1}\phi_{a_i}u$ for all i have nonnegative valuation, and there exist i and $j > 0$ such that the coefficient of τ^j in $u^{-1}\phi_{a_i}u$ has valuation 0. □

COROLLARY 4.3. *Let ϕ be a rank 1 Drinfeld module over L. If there exists $a \in A$ such that $\deg \phi_a > 1$ and $\text{l.c.}(\phi_a) \in R^\times$, then ϕ is a Drinfeld module over R.*

Note: Saying that ϕ *is* a Drinfeld module over R is stronger than saying that ϕ is *isomorphic over L to* a Drinfeld module over R, which would be saying that ϕ has good reduction.

PROOF. By enlarging R and L, we may assume that ϕ has stable reduction, so there exists u such that $(u^{-1}\phi u) \bmod \mathfrak{m}$ is a Drinfeld module of positive rank. This reduction has rank at most the rank of ϕ, so it too has rank 1, so ϕ_a and $(u^{-1}\phi_a u) \bmod \mathfrak{m}$ have the same degree. Thus $v(\text{l.c.}(\phi_a))$ and $v(\text{l.c.}(u^{-1}\phi_a u))$ are 0, so $v(u^{\deg \phi_a - 1}) = 0$, so $v(u) = 0$. Now $u^{-1}\phi u$ is a Drinfeld module of rank 1 over R, so ϕ is too. □

5. Example: The Carlitz module

The Drinfeld module analogue of $\mathbb{G}_m$ is the Carlitz module

$$\begin{aligned} \phi\colon A = \mathbb{F}_q[T] &\longrightarrow K\{\tau\} \\ T &\longmapsto T + \tau \end{aligned}$$

(i.e., $\phi_T(x) = Tx + x^q$). This is a Drinfeld module of rank 1 since

$$\deg \phi_T = q = |T|^1.$$

Define

$$
\begin{aligned}
[n] &:= T^{q^n} - T \\
[n]! &:= [1]\,[2]\,\cdots\,[n] \\
e(z) &:= \sum_{n \geq 0} z^{q^n}/[n]! \\
\pi &:= \prod_{n \geq 1} \left(1 - \frac{[n]}{[n+1]}\right) \qquad \in K_\infty \\
i &:= \sqrt[q-1]{-[1]} \qquad \in C.
\end{aligned}
$$

Carlitz [**Car35**], long before Drinfeld, proved that e induces an isomorphism

$$C/\pi i A \longrightarrow (C \text{ with the Carlitz } A\text{-module action}).$$

This is analogous to $\exp\colon \mathbb{C}/2\pi i\mathbb{Z} \xrightarrow{\sim} \mathbb{C}^\times$.

THEOREM 5.1 ([**Car38**, Theorem 9]). *Fix $a \in A$ with $a \neq 0$. Then $K(\phi[a])$ is an abelian extension of K, and* $\operatorname{Gal}(K(\phi[a])/K) \simeq (A/a)^\times$.

Theorem 5.1 is analogous to $\operatorname{Gal}(\mathbb{Q}(\mu_n)/\mathbb{Q}) \xrightarrow{\sim} (\mathbb{Z}/n\mathbb{Z})^\times$, and can be proved in the same way.

THEOREM 5.2 (Analogue of Kronecker–Weber, implicit in [**Hay74**, §7] and [**Dri74**, §8]). *Every abelian extension of K in which the place ∞ splits completely is contained in $K(\phi[a])$ for some a.*

6. Class field theory

The theory of elliptic curves with complex multiplication leads to an explicit construction of the abelian extensions of an imaginary quadratic number field. In this section, we explain work of Drinfeld [**Dri74**] and Hayes [**Hay79**] that adapts this classical theory to construct the abelian extensions of an arbitrary global function field $K = \operatorname{Frac} A$.

6.1. The class group. When A is not a principal ideal domain, class field theory is more complicated than Theorem 5.2 would suggest. Introduce the following notation:

$$
\begin{aligned}
\mathcal{I} &:= \text{the group of nonzero fractional } A\text{-ideals in } K \\
\mathcal{P} &:= \{(c) : c \in K^\times\}, \quad \text{the group of principal fractional } A\text{-ideals} \\
\operatorname{Pic} A &:= \mathcal{I}/\mathcal{P}, \quad \text{the class group of } A.
\end{aligned}
$$

For a nonzero fractional ideal I, let $[I]$ denote its class in $\operatorname{Pic} A$.

6.2. Rank 1 Drinfeld modules over C.

PROPOSITION 6.1. *We have bijections*

$$
\begin{aligned}
\operatorname{Pic} A &\xrightarrow{\sim} \frac{\{\textit{rank } 1\ A\textit{-lattices in } C\}}{\textit{homothety}} \xrightarrow{\sim} \frac{\{\textit{rank } 1 \textit{ Drinfeld modules over } C\}}{\textit{isomorphism}} \\
[I] &\longmapsto (\textit{homothety class of } I \textit{ in } C)
\end{aligned}
$$

PROOF. The second bijection comes from the $r = 1$ case of Theorem 3.11. Thus we need only consider the first map.

Surjectivity: Any rank 1 A-lattice Λ in C can be scaled so that $K\Lambda = K$. Then Λ is a nonzero fractional ideal I.

Injectivity: I is homothetic to I' in C if and only if there exists $c \in K^\times$ such that $I = cI'$. □

COROLLARY 6.2. *Every rank 1 Drinfeld module over C is isomorphic to one defined over K_∞.*

PROOF. When the lattice Λ is contained in K_∞, the power series e and polynomials ϕ_a constructed in Section 2 will have coefficients in K_∞. □

6.3. The action of ideals on Drinfeld modules. The bijection between $\operatorname{Pic} A$ and the set of isomorphism classes of rank 1 Drinfeld modules over C is analytic, not canonical from the algebraic point of view. But a weaker form of this structure exists algebraically, as will be described in Theorem 6.5.

Fix any A-field L. If I is a nonzero ideal of A and ϕ is a Drinfeld module over any A-field L, we can define a new Drinfeld module $I * \phi$ over L isomorphic to the quotient of $\mathbb{G}_a$ by $\phi[I]$; more precisely, there exists a unique Drinfeld module ψ over L such that $\phi_I \colon \mathbb{G}_a \to \mathbb{G}_a$ is an isogeny $\phi \to \psi$, and we define $I * \phi := \psi$.

Suppose that $I = (a)$ for some nonzero $a \in A$. Then ϕ_I is ϕ_a made monic; that is, if $u := \text{l.c.}(\phi_a)$, then $\phi_I = u^{-1}\phi_a$. Therefore ϕ_I is the composition

$$\phi \xrightarrow{\phi_a} \phi \xrightarrow{u^{-1}} u^{-1}\phi u,$$

so $(a) * \phi = u^{-1}\phi u$, which is isomorphic to ϕ, but not necessarily equal to ϕ. This suggests that we define $(a^{-1}) * \phi = u\phi u^{-1}$. Finally, every $I \in \mathcal{I}$ is $(a^{-1})J$ for some $a \in A - \{0\}$ and integral ideal J, and we define $I * \phi = u(J * \phi)u^{-1}$. The following is now easy to check:

PROPOSITION 6.3. *The operation $*$ defines an action of $\mathcal{I}$ on the set of Drinfeld modules over L. It induces an action of* $\operatorname{Pic} A$ *on the set of isomorphism classes of Drinfeld modules over L.*

EXAMPLE 6.4. Suppose that ϕ is over C, and I is a nonzero integral ideal of A. If we identify ϕ analytically with C/Λ, then $\phi[I] \simeq I^{-1}\Lambda/\Lambda$, so

$$I * (C/\Lambda) \simeq (C/\Lambda)/(I^{-1}\Lambda/\Lambda) \simeq C/I^{-1}\Lambda.$$

Let $\mathscr{Y}(C)$ be the set of isomorphism classes of rank 1 Drinfeld A-modules over C.

THEOREM 6.5. *The set $\mathscr{Y}(C)$ is a principal homogeneous space under the action of* $\operatorname{Pic} A$.

PROOF. This follows from Proposition 6.1 and the calculation in Example 6.4 showing that the corresponding action of I on lattices is by multiplication by I^{-1}. □

6.4. Sgn-normalized Drinfeld modules. We will eventually construct abelian extensions of a global function field K by adjoining the coefficients appearing in rank 1 Drinfeld modules. For this, it will be important to have *actual* Drinfeld modules, and not just *isomorphism classes* of Drinfeld modules. Therefore we will choose a (not quite unique) "normalized" representative of each isomorphism class.

Let $\mathbb{F}_\infty$ be the residue field of $\infty \in X$. Since ∞ is a closed point, $\mathbb{F}_\infty$ is a finite extension of $\mathbb{F}_q$. A choice of uniformizer $\pi \in K_\infty$ defines an isomorphism $K_\infty \simeq \mathbb{F}_\infty((\pi))$, and we define sgn as the composition

$$K_\infty^\times \xrightarrow{\sim} \mathbb{F}_\infty((\pi))^\times \overset{\text{l.c.}}{\twoheadrightarrow} \mathbb{F}_\infty^\times.$$

The function sgn is an analogue of the classical sign function $\operatorname{sgn}\colon \mathbb{R}^\times \to \{\pm 1\}$.

From now on, we fix (A, sgn).

DEFINITION 6.6. A rank 1 Drinfeld module ϕ over L is sgn-normalized if there exists an $\mathbb{F}_q$-algebra homomorphism $\eta\colon \mathbb{F}_\infty \to L$ such that $\text{l.c.}(\phi_a) = \eta(\operatorname{sgn} a)$ for all nonzero $a \in A$.

EXAMPLE 6.7. Suppose that $A = \mathbb{F}_q[T]$ and $\operatorname{sgn}(1/T) = 1$. For a Drinfeld A-module ϕ over L, the following are equivalent:

- ϕ is sgn-normalized;
- $\text{l.c.}(\phi_T) = 1$;
- $\phi_T = T + \tau$ (the Carlitz module).

THEOREM 6.8. *Every rank 1 Drinfeld module ϕ over C is isomorphic to a sgn-normalized Drinfeld module. More precisely, the set of sgn-normalized Drinfeld modules isomorphic to ϕ is a principal homogeneous space under $\mathbb{F}_\infty^\times/\mathbb{F}_q^\times$.*

PROOF. When A is generated over $\mathbb{F}_q$ by one element T, then it suffices to choose u so that $u^{-1}\phi_T u$ is monic. The idea in general is that even if A is not generated by one element, its completion will be (topologically).

First, extend ϕ to a homomorphism $K \to C((\tau^{-1}))$ as in the proof of Proposition 3.5. The induced valuation on K is v_∞, so there exists a unique extension to a continuous homomorphism $K_\infty \to C((\tau^{-1}))$, which we again denote by $a \mapsto \phi_a$. Also, l.c. extends to a map $C((\tau^{-1}))^\times \to C^\times$ (not a homomorphism). Let $\pi \in K_\infty$ be a uniformizer with $\operatorname{sgn}(\pi) = 1$. Replacing ϕ by $u^{-1}\phi u$ multiplies $\text{l.c.}(\phi_\pi)$ by $u^{|\pi|-1}$, so we can choose $u \in C^\times$ to make $\text{l.c.}(\phi_\pi) = 1$.

We claim that the new ϕ is sgn-normalized. Define $\eta\colon \mathbb{F}_\infty \to C$ by $\eta(c) := \text{l.c.}(\phi_c)$. For any $a = c\pi^n \in K_\infty^\times$, with $c \in \mathbb{F}_\infty$ and $n \in \mathbb{Z}$, we have

$$\text{l.c.}(\phi_a) = \text{l.c.}(\phi_c\phi_\pi^n) = \text{l.c.}(\phi_c) = \eta(c) = \eta(\operatorname{sgn} a),$$

as required.

The u was determined up to a $(\#\mathbb{F}_\infty - 1)$th root of unity, but $\operatorname{Aut}\phi = A^\times = \mathbb{F}_q^\times$, so $u^{-1}\phi u$ depends only on the image of u modulo $\mathbb{F}_q^\times$. This explains the principal homogeneous space claim. □

Introduce the following notation:

$\mathscr{Y}^+(L) :=$ the set of sgn-normalized rank 1 Drinfeld A-modules over L

$\mathcal{P}^+ := \{(c) : c \in K^\times \text{ and } \operatorname{sgn} c = 1\} \subseteq \mathcal{P}$

$\operatorname{Pic}^+ A := \mathcal{I}/\mathcal{P}^+$, the narrow class group of A.

LEMMA 6.9. *If $\phi \in \mathscr{Y}^+(L)$, then $\operatorname{Stab}_{\mathcal{I}} \phi = \mathcal{P}^+$.*

PROOF. The following are equivalent for a nonzero integral ideal I not divisible by $\operatorname{char}_A \phi$:

- $I * \phi = \phi$;
- $\phi_I \phi_a = \phi_a \phi_I$ for all $a \in A$;

- $\phi_I \in \operatorname{End}\phi$;
- $\phi_I \in A$;
- $\phi_I = \phi_b$ for some $b \in A$.

In particular, if I is an integral ideal in $\mathcal{P}^+$, then $I = (b)$ for some $b \in A$ with $\operatorname{sgn} b = 1$, so $\phi_I = \phi_b$, so $I \in \operatorname{Stab}_{\mathcal{I}} \phi$. Using weak approximation, one can show that the integral ideals in $\mathcal{P}^+$ generate the group $\mathcal{P}^+$, and that a general ideal I can be multiplied by an ideal in $\mathcal{P}^+$ to make it integral and not divisible by $\operatorname{char}_A \phi$.

Thus it remains to show that when I is an integral ideal not divisible by $\operatorname{char}_A \phi$, the condition $\phi_I = \phi_b$ implies $I \in \mathcal{P}^+$. Suppose that $\phi_I = \phi_b$. Taking kernels yields $\phi[I] = \phi[b]$. Since $\operatorname{char}_A \phi \nmid I$, the group scheme $\phi[I]$ is reduced, so $\operatorname{char}_A \phi \nmid b$. By Proposition 3.13, $I = \operatorname{Ann}_A \phi[I] = \operatorname{Ann}_A \phi[b] = (b)$. Also, $\eta(\operatorname{sgn} b) = \text{l.c.}(\phi_b) = \text{l.c.}(\phi_I) = 1$, so $\operatorname{sgn} b = 1$. Thus $I \in \mathcal{P}^+$. □

THEOREM 6.10. *The action of $\mathcal{I}$ on Drinfeld modules makes $\mathscr{Y}^+(C)$ a principal homogeneous space under* $\operatorname{Pic}^+ A$.

PROOF. Lemma 6.9 implies that $\mathscr{Y}^+(C)$ is a disjoint union of principal homogeneous spaces under $\operatorname{Pic}^+ A$, so it suffices to check that $\mathscr{Y}^+(C)$ and $\# \operatorname{Pic}^+ A$ are finite sets of the same size. Theorems 6.8 and 6.5 imply

$$\#\mathscr{Y}^+(C) = \#\mathscr{Y}(C) \cdot \#(\mathbb{F}_\infty^\times/\mathbb{F}_q^\times) = \#\operatorname{Pic} A \cdot \#(\mathbb{F}_\infty^\times/\mathbb{F}_q^\times).$$

On the other hand, the exact sequence

$$1 \longrightarrow \mathcal{P}/\mathcal{P}^+ \longrightarrow \mathcal{I}/\mathcal{P}^+ \longrightarrow \mathcal{I}/\mathcal{P} \longrightarrow 1$$

and the isomorphism $\mathcal{P}/\mathcal{P}^+ \xrightarrow{\sim} \mathbb{F}_\infty^\times/\mathbb{F}_q^\times$ induced by sgn show that

$$\#\operatorname{Pic}^+ A = \#\operatorname{Pic} A \cdot \#(\mathbb{F}_\infty^\times/\mathbb{F}_q^\times).$$ □

6.5. The narrow Hilbert class field. Choose $\phi \in \mathscr{Y}^+(C)$. Define

$$H^+ := K(\text{all coefficients of } \phi_a \text{ for all } a \in A) \subseteq C.$$

Then ϕ is a Drinfeld module over H^+, and so is $I*\phi$ for any $I \in \mathcal{I}$. By Theorem 6.10, these are all the objects in $\mathscr{Y}^+(C)$, so H^+ is also the extension of K generated by the coefficients of ϕ_a for *all* $\phi \in \mathscr{Y}^+(C)$ and all $a \in A$. In particular, H^+ is independent of the choice of ϕ. It is called the narrow Hilbert class field of (A, sgn).

THEOREM 6.11.

(a) *The field H^+ is a finite abelian extension of K.*
(b) *The extension $H^+ \supseteq K$ is unramified above every finite place ("finite" means not ∞).*
(c) *We have* $\operatorname{Gal}(H^+/K) \simeq \operatorname{Pic}^+ A$.

PROOF.

(a) The group $\operatorname{Aut}(C/K)$ acts on $\mathscr{Y}^+(C)$, so it maps H^+ to itself. Also, H^+ is finitely generated over K. These imply that H^+ is a finite normal extension of K.

By Corollary 6.2, each rank 1 Drinfeld module over C is isomorphic to one over K_∞, and it can be made sgn-normalized over the field F obtained by adjoining to K_∞ the $(\#\mathbb{F}_\infty - 1)$th root of some element. Then $H^+ \subset F$. On the other hand, the extensions $K \subseteq K_\infty \subseteq F$ are separable, so H^+ is separable over K.

The automorphism group of $\mathscr{Y}^+(C)$ as a principal homogeneous space under $\operatorname{Pic}^+ A$ equals $\operatorname{Pic}^+ A$, so we have an injective homomorphism

$$\chi\colon \operatorname{Gal}(H^+/K) \hookrightarrow \operatorname{Aut} \mathscr{Y}^+(C) \simeq \operatorname{Pic}^+ A.$$

Thus $\operatorname{Gal}(H^+/K)$ is a finite abelian group.

(b) Let B^+ be the integral closure of A in H^+. Let $P \subset B^+$ be a nonzero prime ideal, lying above $\mathfrak{p} \subset A$. Let $\mathbb{F}_P = B^+/P$. By Corollary 4.3, each $\phi \in \mathscr{Y}^+(H^+) = \mathscr{Y}^+(C)$ is a Drinfeld module over the localization B^+_P, so there is a reduction map

$$\rho\colon \mathscr{Y}^+(H^+) \to \mathscr{Y}^+(\mathbb{F}_P).$$

By Lemma 6.9, $\operatorname{Pic}^+ A$ acts faithfully on the source and target. Moreover, the map ρ is $(\operatorname{Pic}^+ A)$-equivariant, and $\mathscr{Y}^+(H^+)$ is a principal homogeneous space under $\operatorname{Pic}^+ A$ by Theorem 6.10, so ρ is injective.

If an automorphism $\sigma \in \operatorname{Gal}(H^+/K)$ belongs to the inertia group at P, then σ acts trivially on $\mathscr{Y}^+(\mathbb{F}_P)$, so σ acts trivially on $\mathscr{Y}^+(H^+)$, so $\sigma = 1$. Thus $H^+ \supseteq K$ is unramified at P.

(c) Let $\operatorname{Frob}_\mathfrak{p} := \operatorname{Frob}_P \in \operatorname{Gal}(\mathbb{F}_P/\mathbb{F}_\mathfrak{p}) \hookrightarrow \operatorname{Gal}(H^+/K)$ be the Frobenius automorphism. The key point is the formula

$$\operatorname{Frob}_\mathfrak{p} \phi = \mathfrak{p} * \phi$$

for any $\phi \in \mathscr{Y}^+(\mathbb{F}_P)$; let us now prove this. By definition, if $\psi := \mathfrak{p} * \phi$, then $\psi_a \phi_\mathfrak{p} = \phi_\mathfrak{p} \psi_a$ for all $a \in A$. By Corollary 3.15, $\phi_\mathfrak{p} = \tau^{\deg \mathfrak{p}}$, so $\psi_a \tau^{\deg \mathfrak{p}} = \tau^{\deg \mathfrak{p}} \phi_a$. Compare coefficients; since $\tau^{\deg \mathfrak{p}}$ acts on $\mathbb{F}_P$ as $\operatorname{Frob}_\mathfrak{p}$, we obtain $\psi = \operatorname{Frob}_\mathfrak{p} \phi$.

Since $\mathscr{Y}^+(H^+) \to \mathscr{Y}^+(\mathbb{F}_P)$ is injective and $(\operatorname{Pic}^+ A)$-equivariant, it follows that $\operatorname{Frob}_\mathfrak{p}$ acts on $\mathscr{Y}^+(H^+)$ too as $\phi \mapsto \mathfrak{p} * \phi$. Thus $\chi\colon \operatorname{Gal}(H^+/K) \hookrightarrow \operatorname{Pic}^+ A$ maps $\operatorname{Frob}_\mathfrak{p}$ to the class of $\mathfrak{p}$ in $\operatorname{Pic}^+ A$. Such classes generate $\operatorname{Pic}^+ A$, so χ is surjective. $\square$

6.6. The Hilbert class field. Because of the exact sequence

$$0 \longrightarrow \mathcal{P}/\mathcal{P}^+ \longrightarrow \operatorname{Pic}^+ A \longrightarrow \operatorname{Pic} A \longrightarrow 0,$$

the extension $H^+ \supseteq K$ decomposes into two abelian extensions

$$\begin{array}{c} H^+ \\ \Big| \; {\scriptstyle \mathcal{P}/\mathcal{P}^+} \\ H \\ \Big| \; {\scriptstyle \operatorname{Pic} A} \\ K \end{array}$$

with Galois groups as shown. The map of sets $\mathscr{Y}^+(C) \twoheadrightarrow \mathscr{Y}(C)$ is compatible with the surjection of groups $\operatorname{Pic}^+ A \twoheadrightarrow \operatorname{Pic} A$ acting on the sets. By Corollary 6.2, each element of $\mathscr{Y}(C)$ is represented by a Drinfeld module over K_∞, so the decomposition group $D_\infty \subseteq \operatorname{Gal}(H^+/K)$ acts trivially on $\mathscr{Y}(C)$. Thus $D_\infty \subseteq \mathcal{P}/\mathcal{P}^+$. In other words, ∞ splits completely in $H \supseteq K$.

The Hilbert class field H_A of A is defined as the maximal unramified abelian extension of K in which ∞ splits completely. Thus $H \subseteq H_A$. On the other hand,

$\operatorname{Gal}(H/K) \simeq \operatorname{Pic} A \simeq \operatorname{Gal}(H_A/K)$, the latter isomorphism coming from class field theory. Hence $H = H_A$.

6.7. Ray class fields. In this section, we generalize the constructions to obtain *all* the abelian extensions of K, even the ramified ones. Introduce the following notation:

$$\begin{aligned}
\mathfrak{m} :& \quad \text{a nonzero ideal of } A \\
\mathcal{I}_{\mathfrak{m}} :=& \text{ the subgroup of } \mathcal{I} \text{ generated by primes not dividing } \mathfrak{m} \\
\mathcal{P}_{\mathfrak{m}} :=& \ \{(c) : c \in K \text{ and } c \equiv 1 \bmod \mathfrak{m}\} \\
\mathcal{P}^+_{\mathfrak{m}} :=& \ \{(c) : c \in K \text{ and } \operatorname{sgn} c = 1 \text{ and } c \equiv 1 \bmod \mathfrak{m}\} \\
\operatorname{Pic}_{\mathfrak{m}} A :=& \ \mathcal{I}_{\mathfrak{m}}/\mathcal{P}_{\mathfrak{m}}, \text{ the ray class group modulo } \mathfrak{m} \text{ of } A \\
\operatorname{Pic}^+_{\mathfrak{m}} A :=& \ \mathcal{I}_{\mathfrak{m}}/\mathcal{P}^+_{\mathfrak{m}}, \text{ the narrow ray class group modulo } \mathfrak{m} \text{ of } (A, \operatorname{sgn}) \\
\mathscr{Y}^+_{\mathfrak{m}}(C) :=& \ \{(\phi, \lambda) : \phi \in \mathscr{Y}^+(C) \text{ and } \lambda \text{ generates the } A/\mathfrak{m}\text{-module } \phi[\mathfrak{m}](C)\} \\
H^+_{\mathfrak{m}} :=& \ H^+(\lambda) \text{ for any } (\phi, \lambda) \in \mathscr{Y}^+_{\mathfrak{m}}(C) \\
& \text{(the narrow ray class field modulo } \mathfrak{m} \text{ of } (A, \operatorname{sgn})) \\
H_{\mathfrak{m}} :=& \text{ the subfield of } H^+_{\mathfrak{m}} \text{ fixed by } \mathcal{P}_{\mathfrak{m}}/\mathcal{P}^+_{\mathfrak{m}} \\
& \text{(the ray class field modulo } \mathfrak{m} \text{ of } A).
\end{aligned}$$

Arguments similar to those in previous sections show the following:

THEOREM 6.12.

(a) *There is an action of* $\mathcal{I}_{\mathfrak{m}}$ *on* $\mathscr{Y}^+_{\mathfrak{m}}(C)$ *making* $\mathscr{Y}^+_{\mathfrak{m}}(C)$ *a principal homogeneous space under* $\operatorname{Pic}^+_{\mathfrak{m}} A$.

(b) *The field* $H^+_{\mathfrak{m}}$ *is a finite abelian extension of* K*, unramified outside* $\mathfrak{m}$*, and* $\operatorname{Gal}(H^+_{\mathfrak{m}}/K) \simeq \operatorname{Pic}^+_{\mathfrak{m}} A$.

(c) *The extension* $H_{\mathfrak{m}}$ *is the ray class field modulo* $\mathfrak{m}$ *of* A *as classically defined, with* $\operatorname{Gal}(H_{\mathfrak{m}}/K) \simeq \operatorname{Pic}_{\mathfrak{m}} A$.

6.8. The maximal abelian extension. Theorem 6.12 implies that $\bigcup_{\mathfrak{m}} H_{\mathfrak{m}}$ equals $K^{\mathrm{ab},\infty}$, the maximal abelian extension of K in which ∞ splits completely. Finally, if ∞' is a second closed point of X, then the compositum $K^{\mathrm{ab},\infty}K^{\mathrm{ab},\infty'}$ is the maximal abelian extension of K.

6.9. Example of an explicit Hilbert class field. We follow [**Hay91**, Example 3]; see [**Hay91, DH94**] for other examples similar to this one.

Let $q = 2$. Let X be the elliptic curve over $\mathbb{F}_2$ associated to the equation $y^2 + y = x^3$. Let ∞ be the point at infinity on X. Then $A = \mathbb{F}_2[x, y]/(y^2 + y - x^3)$. Since $\mathbb{F}_\infty^\times = \{1\}$, there is only one possible sgn, and $\mathcal{P}/\mathcal{P}^+ \simeq \mathbb{F}_\infty^\times/\mathbb{F}_q^\times \simeq \{1\}$, so $\operatorname{Pic}^+ A \simeq \operatorname{Pic} A \simeq \operatorname{Pic}^0 X \simeq X(\mathbb{F}_2)$, which is of order 3. Thus $H^+ = H$ and $[H : K] = 3$.

Our goal is to use Drinfeld modules to find an explicit equation defining H as an extension of K. By definition, to give a sgn-normalized rank 1 Drinfeld A-module over a given field extension L of K is to give elements $a, c_1, c_2 \in L$ such that the elements

$$\begin{aligned}
\phi_x &= x + aT + T^2 \\
\phi_y &= y + c_1 T + c_2 T^2 + T^3
\end{aligned}$$

of $L\{\tau\}$ satisfy $\phi_x\phi_y = \phi_y\phi_x$ and $\phi_y^2 + \phi_y = \phi_x^3$. In fact, the second condition is redundant: if ϕ_x commutes with ϕ_y, then ϕ_x commutes with $\phi_y^2 + \phi_y - \phi_x^3$, but in $L\{\tau\}$ if an element with nonzero constant term commutes with an element with zero constant term, the second element is 0, as one sees by equating coefficients.

Thus the only condition is $\phi_x\phi_y = \phi_y\phi_x$, which amounts to the system

$$\begin{aligned} xc_1 + ay^2 &= ay + c_1x^2 \\ xc_2 + ac_1^2 + y^4 &= y + c_1a^2 + c_2x^4 \\ x + ac_2^2 + c_1^4 &= c_1 + c_2a^4 + x^8 \\ a + c_2^4 &= c_2 + a^8 \end{aligned}$$

in the unknowns a, c_1, c_2. The first two equations let us eliminate c_1 and c_2 in turn (remember that x and y are constants in K), so we are left with two polynomials in $K[a]$ that must vanish. Their gcd turns out to be $a^3+(x^2+x)a^2+(x+1)^2a+(x+1)^4$, so H is the extension of K generated by a root of this cubic polynomial.

REMARK 6.13. One could also find an equation for H by working analytically, just as one can use lattices in $\mathbb{C}$ to compute CM j-invariants numerically. In both settings, the result can be made rigorous by invoking integrality properties.

REMARK 6.14. Yet another way to find H would be to use geometric class field theory: Let F be the Frobenius endomorphism of X; then the extension of function fields $H \supseteq K$ arises from the finite étale covering $X \xrightarrow{F-1} X$.

Similar calculations can be done when $\deg\infty > 1$, but they are more complicated.

7. Drinfeld modular varieties

7.1. Classical modular curves. The classical modular curve $Y(1)$ is a coarse moduli space whose points over any algebraically closed field k are in bijection with isomorphism classes of elliptic curves over k. Over $\mathbb{C}$, the analytic description of elliptic curves as $\mathbb{C}/\Lambda$ with $\Lambda = \mathbb{Z}\tau + \mathbb{Z}$ for some $\tau \in \mathbb{C} - \mathbb{R}$ shows that

$$Y(1)(\mathbb{C}) \simeq \Gamma\backslash\Omega$$

where $\Omega := \mathbb{C} - \mathbb{R}$ (the union of the upper and lower half planes in $\mathbb{C}$) and $\Gamma := \mathrm{GL}_2(\mathbb{Z})$. (Equivalently, one could replace Ω with the upper half plane, and Γ by the index-2 subgroup $\mathrm{SL}_2(\mathbb{Z})$, but our formulation will be easier to adapt.)

Similarly, the modular curve $Y_1(N)$ is a coarse moduli space whose k-points over any algebraically closed field k of characteristic not dividing N are in bijection with isomorphism classes of pairs (E, P) where E is an elliptic curve over k, and $P \in E(k)$ is a point of exact order N. One can extend this description to define a functor on $\mathbb{Z}[1/N]$-schemes, and this functor is representable by a smooth relative affine curve over $\mathbb{Z}[1/N]$ once $N \geq 4$. Over $\mathbb{C}$, one has

$$Y_1(N)(\mathbb{C}) \simeq \Gamma_1(N)\backslash\Omega$$

where

$$\Gamma_1(N) := \left\{ \begin{pmatrix} 1 & * \\ 0 & * \end{pmatrix} \in \mathrm{GL}_2(\mathbb{Z}) \right\}.$$

(Since we are working in $\mathrm{GL}_2(\mathbb{Z})$ instead of $\mathrm{SL}_2(\mathbb{Z})$, it is not OK to replace the lower right * with 1.)

7.2. Drinfeld modular curves. Elliptic curves over $\mathbb{C}$ are described analytically by rank 2 lattices, so elliptic curves are analogous to rank 2 Drinfeld modules. Drinfeld modular curves classify rank 2 Drinfeld modules with level structure.

For simplicity, let us assume that $A = \mathbb{F}_q[T]$. Each rank 2 Drinfeld module has the form

$$\begin{aligned} \phi^{(a,b)} \colon A &\longrightarrow L\{\tau\} \\ T &\longmapsto T + a\tau + b\tau^2 \end{aligned}$$

for some $a \in L$ and $b \in L^\times$. The definition of morphism shows that $\phi^{(a,b)} \simeq \phi^{(a',b')}$ if and only if there exists $u \in L^\times$ such that $a' = u^{q-1}a$ and $b' = u^{q^2-1}b$. So $j := a^{q+1}/b$ is invariant under isomorphism, like the j-invariant of an elliptic curve.

The Drinfeld modular curve $Y(1)$ classifying rank 2 Drinfeld modules without level structure is a coarse moduli space isomorphic to $\mathbb{A}^1$ with coordinate j. Analytically,

$$Y(1)(C) \simeq \Gamma \backslash \Omega$$

where $\Omega := C - K_\infty$ (the Drinfeld upper half plane) and $\Gamma := \mathrm{GL}_2(A)$.

Similarly, for each nonzero $n \in A$, the Drinfeld modular curve $Y_1(n)$ classifies rank 2 Drinfeld modules equipped with a torsion point of exact order n. One can make this more precise by specifying a functor on $A[1/n]$-schemes. The functor is representable by a smooth relative curve over $A[1/n]$ when n is nonconstant.

EXAMPLE 7.1. Let us describe $Y_1(T^2)$ explicitly. First consider triples (a, b, z) where $\phi_{T^2}(z) = 0$ and $\phi_T(z) \neq 0$. These are described by the equations $\phi_T(z) = y$ and $\phi_T(y) = 0$ with $y \neq 0$. In other words,

$$\begin{aligned} Tz + az^q + bz^{q^2} &= y \\ T + ay^{q-1} + by^{q^2-1} &= 0. \end{aligned}$$

Eliminating y rewrites this system as the single equation

$$T + a(Tz + az^q + bz^{q^2})^{q-1} + b(Tz + az^q + bz^{q^2})^{q^2-1} = 0.$$

Another triple (a', b', z') gives rise to an isomorphic Drinfeld module with torsion point if and only if there exists an invertible u such that $a' = u^{q-1}a$, $b' = u^{q^2-1}b$, $z' = u^{-1}z$. So $Y_1(T^2)$ is the quotient of the above affine scheme by an action of $\mathbb{G}_m$. The quotient can be obtained simply by setting $z = 1$, to obtain

$$T + a(T + a + b)^{q-1} + b(T + a + b)^{q^2-1} = 0.$$

So $Y_1(T^2)$ is the relative curve defined by this equation in $\mathbb{A}^2_{A[1/T]} = \operatorname{Spec} A[1/T][a, b]$.

For much more on Drinfeld modular curves, see [**Gek86**].

7.3. Drinfeld modular varieties and stacks. More generally, given any $r \geq 1$ and nonzero ideal $\mathfrak{n} \leq A$, Drinfeld [**Dri74**, §5] defined the notion of (full) level $\mathfrak{n}$ structure on a rank r Drinfeld A-module, and he proved that the functor

$$\begin{aligned} \textbf{\textit{A}-schemes} &\longrightarrow \textbf{Sets} \\ S &\longmapsto \{\text{Drinfeld } A\text{-modules over } S \text{ with level } \mathfrak{n} \text{ structure}\}/\text{isomorphism} \end{aligned}$$

is representable by an A-scheme Y, provided that $\mathfrak{n}$ is not too small (Drinfeld assumes that $\mathfrak{n}$ is divisible by at least two distinct primes of A). Applying deformation theory to analogues of formal groups and p-divisible groups, he proved also that $Y \to \operatorname{Spec} A$, after removing the fibers above primes dividing $\mathfrak{n}$, is smooth of relative dimension $r-1$.

Without any restriction on $\mathfrak{n}$, one can define a moduli *stack* $\mathscr{Y}$ and take its coarse space Y. Like classical modular curves and Shimura varieties, these can also be compactified.

EXAMPLE 7.2 ([**Dri74**, §8]). Suppose that $r = 1$ and $\mathfrak{n} = (1)$ (no level structure). Then $\mathscr{Y}$ is of relative dimension 0 over $\operatorname{Spec} A$, and its coarse space Y is a finite A-scheme.

- For $A = \mathbb{F}_q[T]$, there is only one rank 1 Drinfeld module over C up to isomorphism (the Carlitz module). We have $Y = \operatorname{Spec} A$.
- For more general A, define

$$H := \text{the Hilbert class field of } A$$
$$\mathcal{O}_H := \text{the integral closure of } A \text{ in } H.$$

Then $Y = \operatorname{Spec} \mathcal{O}_H$, so we have bijections

$$Y(C) \longleftrightarrow \{A\text{-embeddings } \mathcal{O}_H \to C\} \longleftrightarrow \{K\text{-embeddings } H \to C\}.$$

These are principal homogeneous spaces under $\operatorname{Pic} A \simeq \operatorname{Gal}(H/K)$, in accordance with Theorem 6.5.

Acknowledgments

I thank Francesc Fité and the referees for comments.

References

[And86] Greg W. Anderson, *t-motives*, Duke Math. J. **53** (1986), no. 2, 457–502, DOI 10.1215/S0012-7094-86-05328-7. MR850546 ↑172

[Car35] Leonard Carlitz, *On certain functions connected with polynomials in a Galois field*, Duke Math. J. **1** (1935), no. 2, 137–168, DOI 10.1215/S0012-7094-35-00114-4. MR1545872 ↑177

[Car38] Leonard Carlitz, *A class of polynomials*, Trans. Amer. Math. Soc. **43** (1938), no. 2, 167–182, DOI 10.2307/1990037. MR1501937 ↑177

[DH87] Pierre Deligne and Dale Husemoller, *Survey of Drinfel′d modules*, Current trends in arithmetical algebraic geometry (Arcata, Calif., 1985), Contemp. Math., vol. 67, Amer. Math. Soc., Providence, RI, 1987, pp. 25–91, DOI 10.1090/conm/067/902591. MR902591 ↑167

[Dri74] V. G. Drinfel′d, *Elliptic modules* (Russian), Mat. Sb. (N.S.) **94(136)** (1974), 594–627, 656. MR0384707 ↑167, 177, 184, 185

[Dri77] V. G. Drinfel′d, *Elliptic modules. II* (Russian), Mat. Sb. (N.S.) **102(144)** (1977), no. 2, 182–194, 325. MR0439758 ↑167

[DH94] D. S. Dummit and David Hayes, *Rank-one Drinfel′d modules on elliptic curves*, Math. Comp. **62** (1994), no. 206, 875–883, DOI 10.2307/2153547. With microfiche supplement. MR1218342 ↑182

[Gek83] Ernst-Ulrich Gekeler, *Zur Arithmetik von Drinfel′d-Moduln* (German), Math. Ann. **262** (1983), no. 2, 167–182, DOI 10.1007/BF01455309. MR690193 ↑172

[Gek86] Ernst-Ulrich Gekeler, *Drinfel′d modular curves*, Lecture Notes in Mathematics, vol. 1231, Springer-Verlag, Berlin, 1986, DOI 10.1007/BFb0072692. MR874338 ↑184

[Gek91] Ernst-Ulrich Gekeler, *On finite Drinfel′d modules*, J. Algebra **141** (1991), no. 1, 187–203, DOI 10.1016/0021-8693(91)90211-P. MR1118323 ↑172

[GvdPRVG97] E.-U. Gekeler, M. van der Put, M. Reversat, and J. Van Geel (eds.), *Drinfeld modules, modular schemes and applications*, World Scientific Publishing Co., Inc., River Edge, NJ, 1997. MR1630594 ↑167

[Gos96] David Goss, *Basic structures of function field arithmetic*, Ergebnisse der Mathematik und ihrer Grenzgebiete (3) [Results in Mathematics and Related Areas (3)], vol. 35, Springer-Verlag, Berlin, 1996, DOI 10.1007/978-3-642-61480-4. MR1423131 ↑167

[GHR92] David Goss, David R. Hayes, and Michael I. Rosen (eds.), *The arithmetic of function fields*, Ohio State University Mathematical Research Institute Publications, vol. 2, Walter de Gruyter & Co., Berlin, 1992, DOI 10.1515/9783110886153. MR1196508 ↑167

[Hay74] D. R. Hayes, *Explicit class field theory for rational function fields*, Trans. Amer. Math. Soc. **189** (1974), 77–91, DOI 10.2307/1996848. MR330106 ↑177

[Hay79] David R. Hayes, *Explicit class field theory in global function fields*, Studies in algebra and number theory, Adv. in Math. Suppl. Stud., vol. 6, Academic Press, New York-London, 1979, pp. 173–217. MR535766 ↑177

[Hay91] David R. Hayes, *On the reduction of rank-one Drinfel'd modules*, Math. Comp. **57** (1991), no. 195, 339–349, DOI 10.2307/2938678. MR1079021 ↑182

[Hay92] David R. Hayes, *A brief introduction to Drinfel'd modules*, The arithmetic of function fields (Columbus, OH, 1991), Ohio State Univ. Math. Res. Inst. Publ., vol. 2, de Gruyter, Berlin, 1992, pp. 1–32. MR1196509 ↑167

[Lau96] Gérard Laumon, *Cohomology of Drinfeld modular varieties. Part I*, Cambridge Studies in Advanced Mathematics, vol. 41, Cambridge University Press, Cambridge, 1996. Geometry, counting of points and local harmonic analysis. MR1381898 ↑167

[Lau97] Gérard Laumon, *Cohomology of Drinfeld modular varieties. Part II*, Cambridge Studies in Advanced Mathematics, vol. 56, Cambridge University Press, Cambridge, 1997. Automorphic forms, trace formulas and Langlands correspondence; With an appendix by Jean-Loup Waldspurger, DOI 10.1017/CBO9780511661969. MR1439250 ↑167

[Ros02] Michael Rosen, *Number theory in function fields*, Graduate Texts in Mathematics, vol. 210, Springer-Verlag, New York, 2002, DOI 10.1007/978-1-4757-6046-0. MR1876657 ↑167

[Sil09] Joseph H. Silverman, *The arithmetic of elliptic curves*, 2nd ed., Graduate Texts in Mathematics, vol. 106, Springer, Dordrecht, 2009, DOI 10.1007/978-0-387-09494-6. MR2514094 ↑168

[Tha04] Dinesh S. Thakur, *Function field arithmetic*, World Scientific Publishing Co., Inc., River Edge, NJ, 2004, DOI 10.1142/9789812562388. MR2091265 ↑167

Department of Mathematics, Massachusetts Institute of Technology, Cambridge, Massachusetts 02139-4307

Email address: poonen@math.mit.edu

URL: http://math.mit.edu/~poonen/

Selected Published Titles in This Series

779 **Samuele Anni, Valentijn Karemaker, and Elisa Lorenzo García, Editors,** Arithmetic, Geometry, Cryptography, and Coding Theory 2021, 2022

778 **Carlos Galindo, Alejandro Melle Hernández, Julio José Moyano-Fernández, and Wilson A. Zúñiga-Galindo, Editors,** p-Adic Analysis, Arithmetic and Singularities, 2022

777 **Bang-Yen Chen, Nicholas D. Brubaker, Takashi Sakai, Bogdan D. Suceavă, Makiko Sumi Tanaka, Hiroshi Tamaru, and Mihaela B. Vajiac, Editors,** Differential Geometry and Global Analysis, 2022

776 **Aaron Wootton, S. Allen Broughton, and Jennifer Paulhus, Editors,** Automorphisms of Riemann Surfaces, Subgroups of Mapping Class Groups and Related Topics, 2022

775 **Fernando Galaz-García, Cecilia González-Tokman, and Juan Carlos Pardo Millán, Editors,** Mexican Mathematicians in the World, 2021

774 **Randall J. Swift, Alan Krinik, Jennifer M. Switkes, and Jason H. Park, Editors,** Stochastic Processes and Functional Analysis, 2021

773 **Nicholas R. Baeth, Thiago H. Freitas, Graham J. Leuschke, and Victor H. Jorge Pérez, Editors,** Commutative Algebra, 2021

772 **Anatoly M. Vershik, Victor M. Buchstaber, and Andrey V. Malyutin, Editors,** Topology, Geometry, and Dynamics, 2021

771 **Nicolás Andruskiewitsch, Gongxiang Liu, Susan Montgomery, and Yinhuo Zhang, Editors,** Hopf Algebras, Tensor Categories and Related Topics, 2021

770 **Stéphane Ballet, Gaetan Bisson, and Irene Bouw, Editors,** Arithmetic, Geometry, Cryptography and Coding Theory, 2021

769 **Kiyoshi Igusa, Alex Martsinkovsky, and Gordana Todorov, Editors,** Representations of Algebras, Geometry and Physics, 2021

768 **Dražen Adamović, Andrej Dujella, Antun Milas, and Pavle Pandžić, Editors,** Lie Groups, Number Theory, and Vertex Algebras, 2021

767 **Moshe Jarden and Tony Shaska, Editors,** Abelian Varieties and Number Theory, 2021

766 **Paola Comparin, Eduardo Esteves, Herbert Lange, Sebastián Reyes-Carocca, and Rubí E. Rodríguez, Editors,** Geometry at the Frontier, 2021

765 **Michael Aschbacher,** Quaternion Fusion Packets, 2021

764 **Gabriel Cunningham, Mark Mixer, and Egon Schulte, Editors,** Polytopes and Discrete Geometry, 2021

763 **Tyler J. Jarvis and Nathan Priddis, Editors,** Singularities, Mirror Symmetry, and the Gauged Linear Sigma Model, 2021

762 **Atsushi Ichino and Kartik Prasanna,** Periods of Quaternionic Shimura Varieties. I., 2021

761 **Ibrahim Assem, Christof Geiß, and Sonia Trepode, Editors,** Advances in Representation Theory of Algebras, 2021

760 **Olivier Collin, Stefan Friedl, Cameron Gordon, Stephan Tillmann, and Liam Watson, Editors,** Characters in Low-Dimensional Topology, 2020

759 **Omayra Ortega, Emille Davie Lawrence, and Edray Herber Goins, Editors,** The Golden Anniversary Celebration of the National Association of Mathematicians, 2020

758 **Jan Šťovíček and Jan Trlifaj, Editors,** Representation Theory and Beyond, 2020

757 **Kaïs Ammari and Stéphane Gerbi, Editors,** Identification and Control: Some New Challenges, 2020

756 **Joeri Van der Veken, Alfonso Carriazo, Ivko Dimitrić, Yun Myung Oh, Bogdan D. Suceavă, and Luc Vrancken, Editors,** Geometry of Submanifolds, 2020

For a complete list of titles in this series, visit the
AMS Bookstore at **www.ams.org/bookstore/conmseries/**.